Fundamentals of Electrical Engineering

SECOND EDITION

Fundamentals of Electrical Engineering

SECOND EDITION

Dr. (Mrs.) Bharti Dwivedi
Professor and Head of Electrical Engineering Department
Institute of Engineering & Technology, Lucknow

Dr. Anurag Tripathi
Assistant Professor
Electrical Engineering Department
Institute of Engineering & Technology, Lucknow

WILEY

Fundamentals of Electrical Engineering
SECOND EDITION

Authors: Dr. (Mrs.) Bharti Dwivedi and Dr. Anurag Tripathi

Published by Wiley India Pvt. Ltd.,
Plot No. 382, 383 Shahabad Daulatpur, New Delhi-110042

Printed at: Replika Press Pvt. Ltd.

First Edition: 2012
Second Edition: 2013
Reprint: 2026

ISBN: 978-81-265-4271-0

Other Wiley Editorial Offices:
John Wiley & Sons, Inc. 111 River Street, Hoboken, NJ 07030, USA
Wiley-VCH Verlag GmbH, Boschstr. 12, 69469 Weinheim, Germany
John Wiley & Sons Australia, Ltd (ACN:009 673 081) at Level 4,600 Bourke St, Melbourne Victoria 3000
John Wiley & Sons Singapore Pte. Ltd, 134 Jurong Gateway Road, #04-307H, Singapore 600134
John Wiley & Sons Canada Ltd, 22 Worcester Road, Etobicoke, Ontario, Canada, M9W ILI
John Wiley & Sons, Ltd, New Era House, 8 Oldlands Way, Bognor Regis, West Sussex, PO22 9NQ, UK

www.wileyindia.com

Preface

It is said that the *engineering* is an art of applying science for the betterment of mankind. First-year engineering students come with a fair enough background of physics of electricity. Among the three core engineering branches, *electrical* is the most abstract one, with almost all of its topics like current, voltage and power (to name a few) being basic yet imperceptible to us. This fact leads to the requirement of a book which is not only easy to understand but clarifies those hidden concepts in a lucid manner.

The core purpose of writing this book was to cater to the need of the first course on electrical engineering of major universities in India. Efforts have been made to bring on surface the *unspoken explanations between the lines* to make the concepts clear to the students and the subject teachers. The book contains certain methods of solving electrical circuits, which are not only based on the first principles but are also described in a stepwise manner. The book tries to relate the topics which the students have already studied in class 10+2, with the present syllabus and thus tries to ensure a smooth progressive transition from physics to electrical engineering. This book is primarily targeted towards students of Gautam Buddha Technical University (GBTU), Uttarakhand Technical University (UTU), and Mahamaya Technical University (MTU). With the inclusion of Chapter 6, the book also covers the syllabi of RGPV, Bhopal, Indraprastha University, Delhi, and Aryabhatta Knowledge University, Patna. With this, the authors feel that a comprehensive presentation of Basic Electrical Engineering has been made for most major universities.

Chapter 1 contains some basic definitions in electrical engineering like linearity, *R*, *L*, and *C* as linear elements, bilateral-unilateral elements, active-passive elements and source transformation basics. The chapter also explains the commonly used methods of circuit solving like KVL, KCL, Thevenin theorem, Norton theorem, superposition theorem along with the maximum power transfer theorem. Finally, star-delta transformation is explained. Only independent sources are dealt with as per the syllabus requirement.

Chapter 2 deals with the sinusoidal steady-state analysis. The phasor concept has been lucidly explained, and the basic building blocks of the ac circuit analysis, like impedance and power triangles, the *j* operator, impedance and admittance components, power and power factor, have been presented along with the phenomena of resonance.

Chapter 3 presents the three-phase power and its measurement. It also covers the construction and principle of operation of the various measuring instrument types.

Chapter 4 contains three parts: The *first part* presents the concept and necessity of electrical earthing. The *second part* presents magnetic circuits, hysteresis and eddy current losses along with mutual coupling. The *third part* explains the construction, principle of operation, equivalent circuit, emf equation, calculation of losses and efficiency of a single-phase transformer. The chapter ends with an introduction to the autotransformer.

Chapter 5 starts with the introduction of the basic concepts of electromechanical energy conversion principles. It is followed by the four machine operations. DC motor/generator, three-phase and single-phase induction motors and three-phase synchronous motor/generator are the four machine types presented.

Chapter 6 presents some miscellaneous topics in DC and AC circuit analyses, single-phase transformers and machines.

Each chapter contains a sufficient number of solved examples, based on examination questions, to help the students imbibe the concepts thoroughly. Depiction of concepts through relevant figures at appropriate places makes the flow of the content smoother. The reason-based answers to short answer questions is a unique method adopted to make the concept crystal clear to the students. Every chapter contains sufficient numbers of unsolved problems for practice. Model question papers incorporated at the end of the book will help the students to rehearse for the actual examination.

Acknowledgements

The book was written keeping all other activities in place. Hence the authors wish to thank their family members who spared them from domestic responsibility during the course of writing the book. Authors show their gratitude to their colleagues for giving thoughtful suggestions while the book was taking its final shape. The authors would like to give special thanks to the editorial team at Wiley India, especially Ms. Meenakshi Sehrawat. She has been extremely co-operative and patient throughout the journey.

The authors feel obliged to the persons who have provided their valuable suggestions for improvement/ error removal and the same have been duly incorporated in this revised edition.

The authors are open to receive further suggestions/modifications from students and subject teachers.

Dr. (Mrs.) Bharti Dwivedi
Dr. Anurag Tripathi

Contents

1 DC Circuit Analysis and Network Theorems

LEARNING OBJECTIVES

After a careful study of the chapter, you would be able to understand:

- Elementary concepts of an electric circuit.
- Some basic properties of electric circuits and elements.
- Kirchhoff's voltage and current laws, as applied to the solution of electrical circuits.
- Four basic theorems for the solution of electric circuits.
- Star–delta transformation and its usage in solving various problems.

1.1 Concepts of Circuit and Network

There are five major elements in electrical engineering: *voltage sources*, *current sources*, *resistance*, *inductance* and *capacitance*. A collection of one or more of these five elements is simply called a *network*, whereas a grouping of one or more of the elements that results in a flow of current is known as a *circuit*. Thus every circuit is a network, whereas every network may or may not be a circuit. Figure 1 shows basic diagrams of a network and a circuit.

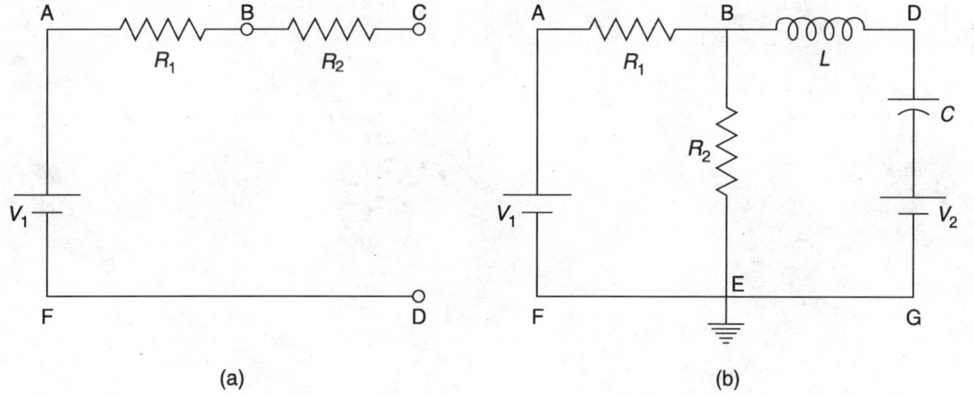

Figure 1 (a) A network; (b) a circuit.

1.2 Active and Passive Elements

Elements that are capable of delivering electrical energy are called *active elements*. Electric sources—voltage source and current source—are thus active elements. The voltage sources V_1 and V_2 in Figures 1(a) and (b) are active elements. Also, the elements that are capable of amplifying or attenuating the voltage or current are treated as active elements (e.g., transistors).

The elements which are capable of receiving and storing or dissipating the electrical energy are called *passive elements*. Resistance, inductance and capacitance come under this category. Resistances R_1, R_2, L and C are passive elements in Figures 1(a) and 1(b).

1.3 Voltage and Current Sources

An ideal voltage source, shown in Figure 2(a), is one which maintains a fixed voltage V_S across the load terminals a and b irrespective of the ohmic value of the load resistance. A practical voltage source is shown in Figure 2(b). It carries an internal resistance of the source, termed as R_S, in series with the voltage V_S.

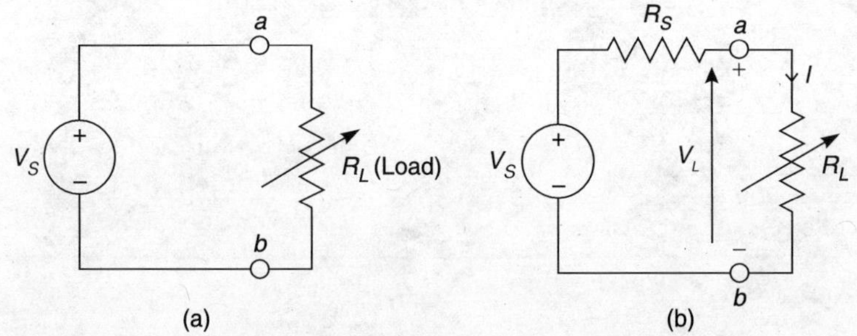

Figure 2 (a) An ideal voltage source; (b) a practical voltage source.

The internal resistance R_S of a practical voltage source as shown in Figure 2(b) is connected in series with it. Due to the fact that "voltage gets distributed in a series connection," some part of the voltage V_S now drops in R_S and the remaining part is available across the load R_L. The current flowing in the above circuit will be

$$I = \frac{V_S}{R_S + R_L}$$

and the voltage drop across R_L will be

$$V_L = IR_L = \frac{V_S R_L}{R_S + R_L} = \frac{V_S}{1 + (R_S/R_L)}$$

Thus, if R_S is zero, the voltage across the load will be V_S and the source will then be said to be an *ideal* one. For $R_S \neq 0$, the voltage $V_L < V_S$ and the source is said to be a *practical* one. Therefore the internal resistance of an ideal voltage source is zero. Similar reasoning is adopted for defining the ideal and practical current sources.

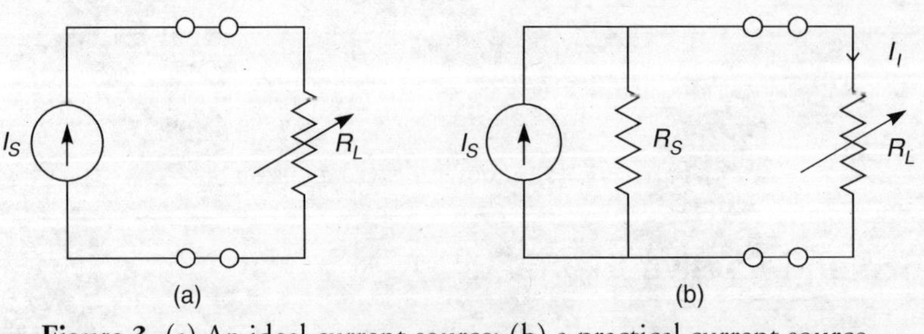

Figure 3 (a) An ideal current source; (b) a practical current source.

As shown in Figure 3(a), an ideal current source is one that delivers a constant current I_S to the load R_L, irrespective of the value of R_L. Figure 3(b) shows a practical current source. Since "current gets distributed in parallel connection," the internal resistance of the practical current source, R_S, is taken in parallel to it, as shown. Thus, according to the *current divide theorem*[1]

$$I_L = \frac{I_S R_S}{R_S + R_L} = \frac{I_S}{1 + (R_L/R_S)}$$

If R_S is infinity, $I_L = I_S$. Thus the internal resistance of an ideal current source is of infinite ohms. Also $R_S < \infty$ for a practical current source, and hence $I_L < I_S$ always.

1.4 Concept of Linearity

A system is said to be *linear* if it follows the principles of *superposition* and *homogeneity*. The principle of superposition states that if the output to a particular input u_1 is y_1 and that due to another input u_2 is y_2, then the output to an input $(u_1 + u_2)$ should be equal to $(y_1 + y_2)$. By homogeneity, it is meant that if the output to an input u_1 is y_1, the output to ku_1 should be ky_1 (where k is scalar).

Thus, if the above two principles are combined, we get the following condition: If for input $u_1 \rightarrow$ output y_1, input $u_2 \rightarrow$ output y_2, then for an input value $(au_1 + bu_2)$, the output is $(ay_1 + by_2)$. In this case the particular system is said to be *linear*. Thus $y = mx$ is a linear system, whereas $y = (mx + c)$ is *nonlinear*. And since $V = IR$ (Ohm's law), the resistance is a linear element.

In an inductive circuit, since the voltage is directly proportional to the time rate of change of current with the constant of proportionality being the inductance, the inductance is also a linear element, that is the plot between di/dt and V is linear. The inductance is the constant of proportionality. That is,

$$V = L\frac{di}{dt}$$

[1]**Current divide theorem:** This theorem tells us about the proportion of current received by each of the resistors (or any other element combination) connected in parallel, as shown in the figure.

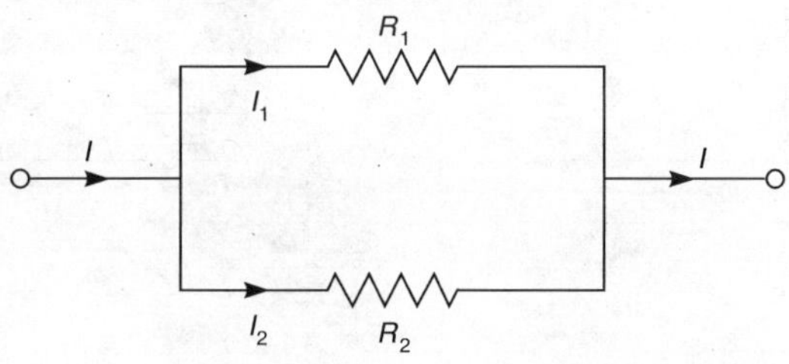

Defining the current divide theorem.

From the figure,

$$I_1R_1 = I_2R_2 \text{ (voltage is same in parallel branches) and } I = I_1 + I_2$$

Also,

$$V = \left(\frac{R_1R_2}{R_1 + R_2}\right)I$$

Solving for I_1 and I_2, we get

$$I_1 = \frac{V}{R_1} = \frac{IR_2}{R_1 + R_2} \quad \text{and} \quad I_2 = \frac{V}{R_2} = \frac{IR_1}{R_1 + R_2}$$

This is the *current divide theorem*.

In a capacitive circuit, charge $Q = Cv$. So

$$\frac{dQ}{dt} = C\frac{dv}{dt}$$

As $dQ/dt = i$, so

$$i = C\frac{dv}{dt}$$

The current is thus directly proportional to the time rate of change of voltage.

1.5 Unilateral and Bilateral Elements

An element through which the magnitude and direction of current does not change on inverting the direction of that element, is called a *bilateral element* (e.g., a resistor, inductor and most of the capacitors). If the magnitude of the current changes on changing the direction of the element, it is said to be *unilateral* (e.g., a diode). Figure 4(a) shows a bilateral element R_1 whereas Figure 4(b) shows an unilateral element D_1.

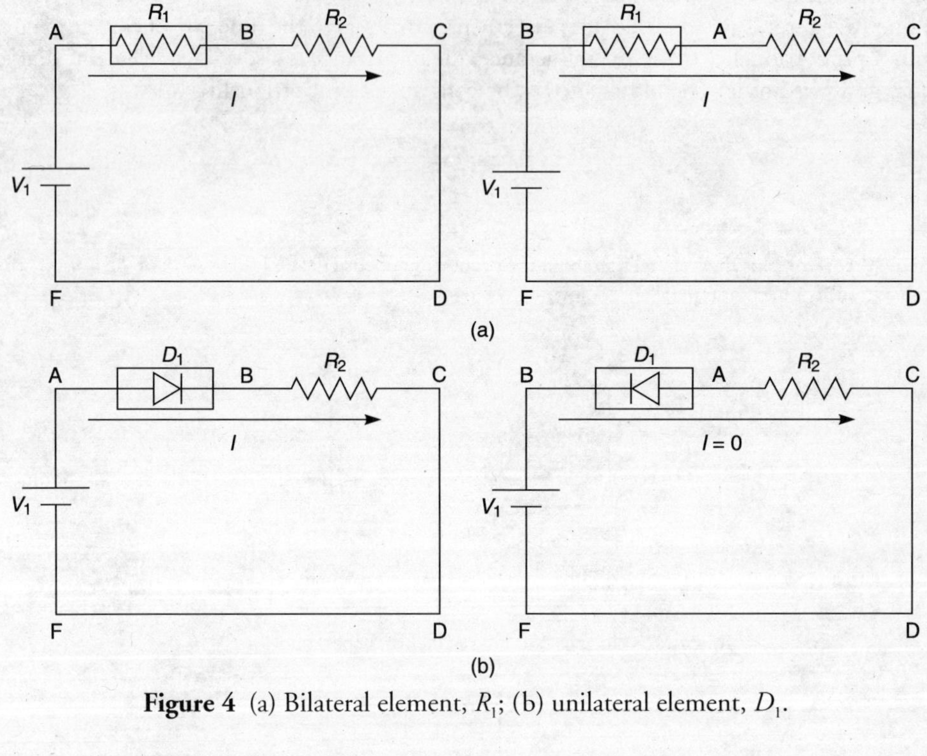

Figure 4 (a) Bilateral element, R_1; (b) unilateral element, D_1.

1.6 Source Transformation

If a voltage source is connected in series with a resistance, as shown in Figure 5, it can be converted into a current source in parallel with the same resistance. The value of the current source is obtained by simple

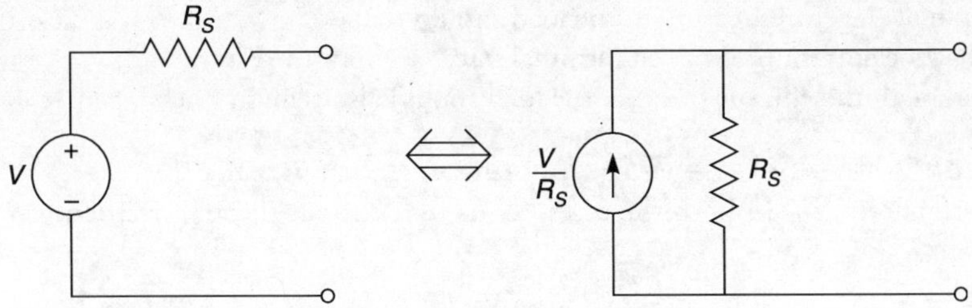

Figure 5 Source transformation.

application of Ohm's law. The reverse is also true. That is, a current source connected in parallel with a resistance can be converted into a voltage source in series with that resistance. To prove this, let us connect a load R_L to the circuit in Figure 5, as shown in Figure 6. Now in Figure 6(a), the current

$$I_L = \frac{V}{R_S + R_L}$$

flows through the load. The voltage across the load would be

$$V_L = I_L R_L = \frac{V R_L}{R_S + R_L}$$

Now converting the voltage source into the current source, as discussed above, we get the circuit as shown in Figure 6(b). Applying the current divide theorem, we get the current through R_L as

$$I_L = \frac{(V/R_S)R_S}{R_S + R_L} = \frac{V}{R_S + R_L}$$

and hence the voltage across R_L is

$$V_L = \frac{V R_L}{R_S + R_L}$$

We see that after the source transformation, the current through and voltage across the load R_L remain unchanged. Hence, we can say that the load R_L could not detect what has happened in the source side. So as long as this continues, source transformation is allowed. Two things have to be kept in mind while applying the source transformation:

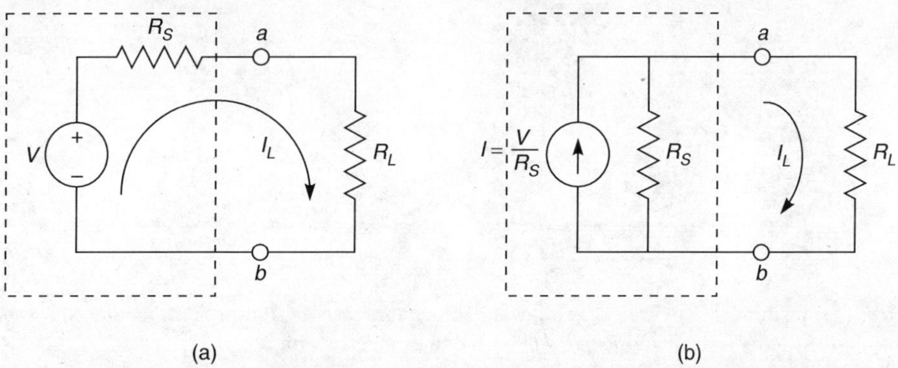

Figure 6 (a) Circuit for source transformation; (b) transformed source.

1. The direction of the current through the load should remain unchanged. It is clear from Figures 6(a) and 6(b) that the current flows from terminal *a* to *b* in both the cases.
2. The magnitude of the current through the load should also remain unchanged.

1.6.1 Condition where Source Transformation is not Applied

Refer to Figures 6(a) and 6(b). In both the cases, let us find current through resistance R_S. In Figure 6(a), the current through R_S is

$$\frac{V}{R_S + R_L}$$

In Figure 6(b), the current through R_S will be

$$\frac{V}{R_S}\,\frac{R_L}{R_S + R_L}$$

Thus, the currents through R_S are different for Figures 6(a) and 6(b). We see that the current through (and hence voltage across) R_S will change as a result of source transformation. Hence it can be concluded that "if the current through such a resistance is desired to be found out, which is also a part of the source transformation process [e.g., R_S in Figures 6(a) and 6(b)], the transformation should *not* be made."[2]

[2]Parallel connection: If there are two elements (say resistors) R_1 and R_2, and if one end of R_1 is joined to one end of R_2 and another end of R_1 is joined to that of R_2, then R_1 and R_2 will be said to be connected in parallel as shown in the figure below.

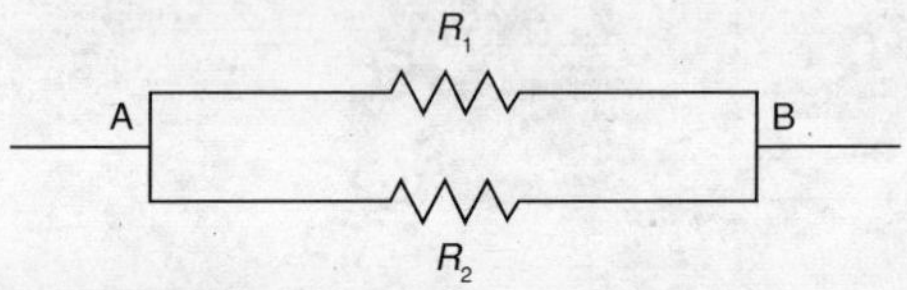

If any active or passive element is connected at points A or B, then also R_1 and R_2 will be said to be always in parallel.

Series connection: If there are two elements (say resistors) R_1 and R_2, and if one end of R_1 is joined to one end of R_2 as shown in the figure below, then R_1 and R_2 can be said to be connected in series, provided there is no active or passive element connected in the connecting point. This is illustrated in the figures below.

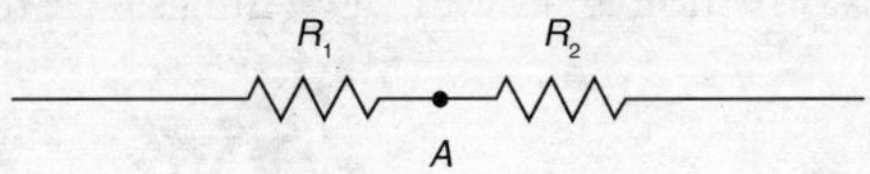

R_1 and R_2 are in series.

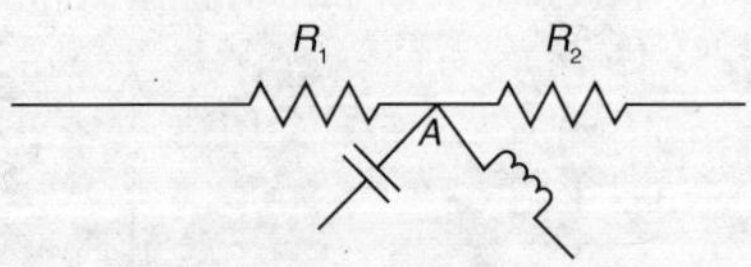

R_1 and R_2 are NOT in series.

But if the current flowing through R_1 is same as that flowing through R_2, then (in spite of the other elements connected at the meeting point A of the two resistors R_1 and R_2), R_1 will be said to be in series with R_2.

So the KEY feature of series connection is that the current flowing through the series elements should be same and that of the parallel elements is that the voltages across the parallel connected elements should be same.

1.7 Kirchhoff's Laws

There are two laws put forward by Kirchhoff.

1. **Kirchhoff's Voltage Law (KVL):** It states that the sum of the voltage gains and voltage drops around a closed loop in a circuit is zero.
2. **Kirchhoff's Current Law (KCL):** It states that the algebraic sum of the currents meeting at an electrical node is zero.

The KVL and KCL form the basis of solving any circuit. Their direct application can be seen in the following text, where the KVL and KCL are applied in the loop method and nodal method of analysis, respectively, for solving the circuits.

1.7.1 Loop Method of Analysis

The *loop method of analysis* pertains to KVL. As shown in Figure 7, ABCDEFA is a *closed loop* that contains two smaller closed loops: ABCFA and FCDEF. These loops cannot be further broken into smaller loops. Hence they are called *meshes*. Thus the difference between a loop and a mesh is the following: Where a loop can be further broken into smaller loops/meshes, a mesh cannot be further divided into smaller entities. The mesh is the smallest subset of a loop. Thus, every mesh is a loop but every loop is not necessarily a mesh.

As per KVL, if we move around a closed loop/mesh (say, ABCFA) and sum up the voltage gains and drops, we get the zero value. We need to follow four rules while writing the KVL equations in this regard.

Rule 1: If we move from negative to positive (or neutral terminal to phase terminal) in a voltage source, we experience a voltage gain.

Rule 2: If we move from positive to negative (or phase to neutral terminal) in a voltage source, we experience a voltage drop.

Rule 3: If we move through a passive element (R, L or C) *in the direction* of the current, we experience a voltage drop.

Rule 4: If we move through a passive element (R, L or C) *against the direction* of the current, we experience a voltage gain.

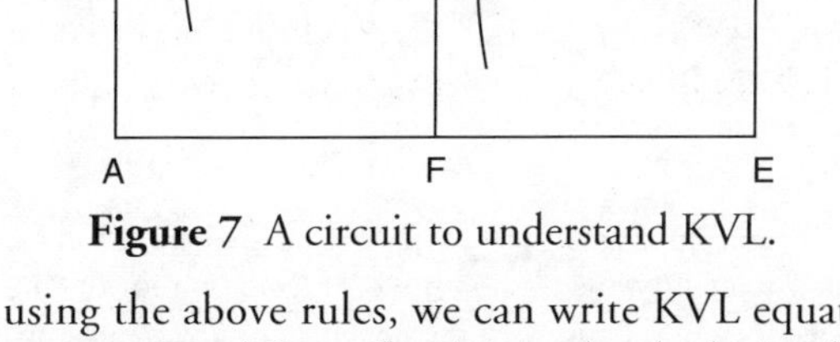

Figure 7 A circuit to understand KVL.

Now referring to Figure 7, using the above rules, we can write KVL equations for the meshes ABCFA and FCDEF. Assume that currents I_1 and I_2 are flowing in the clockwise direction in these meshes. For the mesh ABCFA, we start with A and when we pass through the voltage source we find that we are moving from negative to positive terminal of the source. Hence according to Rule 1, we experience a voltage gain ($+V$) and thus we reach B.

Next, we pass through the resistance R_1 (while moving from B to C) and since we had assumed the direction of current I_1 flowing in the resistance R_1 as clockwise, we find ourselves moving in the direction of current. So, according to Rule 3, we experience a voltage drop $(-I_1R_1)$ and reach C. From C to F, that is in the resistance R_2, two currents I_1 and I_2 are flowing. If we assume that I_1 is greater than I_2, the net current flowing in the resistance R_2 will be $(I_1 - I_2)$ from C to F. And since we are writing the KVL equation, while moving in the clockwise direction (ABCFA), we will be moving in the direction of the current, that is $(I_1 - I_2)$, and therefore will experience a voltage drop $[-(I_1 - I_2)R_2]$ to reach F. Since there is no element connected between F and A, we will next reach A without any voltage gain or drop. Summarizing the discussion, mathematically, we get

$$+V - I_1R_1 - (I_1 - I_2)R_2 = 0 \tag{1.1}$$

where the positive sign is used to represent the voltage gain and the negative sign for voltage drop.

Similarly, for the mesh FCDEF, if again we move in the clockwise direction starting from F, we will have the following KVL equation:

$$-(I_2 - I_1)R_2 - I_2R_3 - I_2R_L = 0 \tag{1.2}$$

Here while moving from F to C, it has been assumed that current I_2 is greater than I_1. So, the net current flowing in the resistance R_2 (from F to C) is $(I_2 - I_1)$. Applying Rule 3, we get that there will be a voltage drop of $-(I_2 - I_1)R_2$ in moving from F to C.[3] Hence any assumption of current directions will lead to a unique KVL equation, provided the four rules mentioned above are correctly followed. Equations (1.1) and (1.2) can be now solved to get the values of I_1 and I_2.

Next, consider a special case of a circuit containing a current source also. Figure 8 shows such a circuit. Since the internal resistance of an ideal current source is infinite, no external current can be assumed to be flowing in the portion BCDEB of the circuit. In other words, only its own current I can flow through the current source and R_3 (since R_3 is in series with the current source).

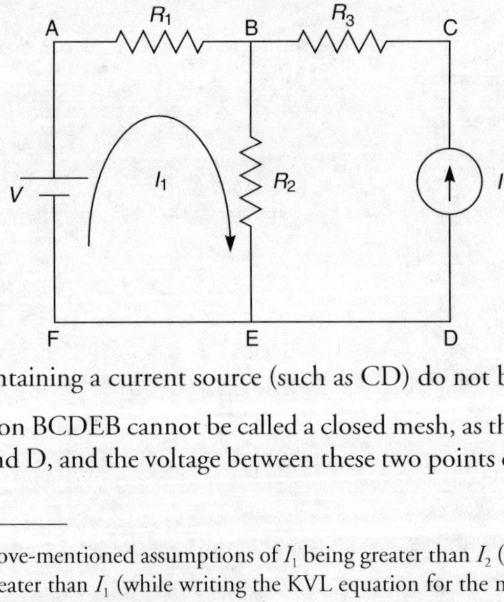

Figure 8 Branches containing a current source (such as CD) do not become a part of a loop.

This also means that the portion BCDEB cannot be called a closed mesh, as there is a presence of a resistance of infinite ohms between C and D, and the voltage between these two points cannot be known (unless given

[3]It needs to be clarified that the above-mentioned assumptions of I_1 being greater than I_2 (while writing the KVL equation for the mesh ABCFA) and I_2 being greater than I_1 (while writing the KVL equation for the mesh FCDEF) are not unique. That is, if in the mesh FCDEF, I_1 is assumed greater than I_2, then the net current flowing in R_2 will be $(I_1 - I_2)$ from top to bottom, that is from C to F. Now, if writing the KVL equation remained clockwise, that is FCDEF, when we moved from F to C, we would be said to be moving against the direction of the current $(I_1 - I_2)$. Therefore, there would be a voltage gain according to Rule 4, and the first term in Eq. (1.2) would be $+(I_1 - I_2)R_2$. A close look at this first term will reveal that it is exactly the same as that of Eq. (1.2).

otherwise). Hence the circuit, in Figure 8, consists of only one mesh, that is ABEFA, where a current I_1 can be arbitrarily assumed to be flowing in the clockwise direction. The KVL applied to the mesh ABEFA gives

$$V - I_1R_1 - (I + I_1)R_2 = 0$$

as the current through R_2 will be I (from CB) plus I_1. Since there is only one mesh, only one KVL equation will be obtained. Also note that there is one unknown, that is I_1. So, it can be found out.

1.7.2 Nodal Method of Analysis

A meeting point of more than two electrical branches is called a *node*. Application of KCL is termed as the *nodal method of analysis*, because here KCL is applied at various nodes.

In Figure 9, as per the definition of the node, only two points B and E qualify to become nodes. Points E, F and D are equipotential points. So, the circuit consists of two nodes: B and E. Any one of the nodes is assumed to be a *datum* (or *reference* or *zero potential* or *ground*) node. The KCL equations can now be written for every other node present in the circuit.

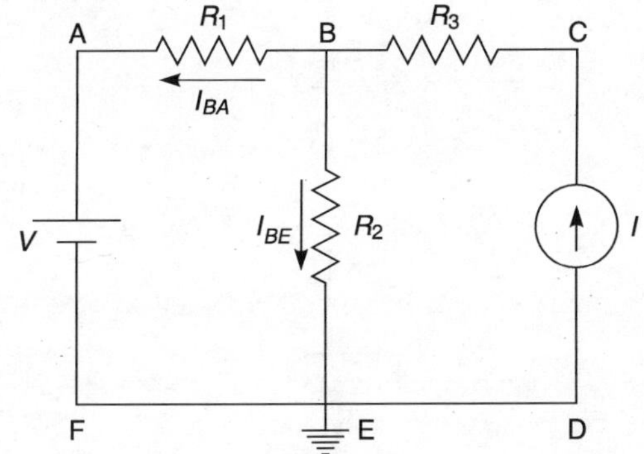

Figure 9 A circuit containing one current and one voltage source for KCL.

Here, if node E is chosen as the datum node, as shown, the KCL equation will have to be written only for the node B. Assume that the node B has got a potential (say, V_B). Since the value of V_B is not known initially, let us also assume that V_B voltage at B is higher than all other node voltages, so that all the currents are going away from the node B. If our assumption (of V_B being the highest voltage) is incorrect, its sign will automatically come out to be negative and hence the answer will remain unchanged. Next, we write the KCL at the node B. It consists of three currents:

1. Current through R_1 will be

$$I_{BA} = \frac{V_B - V}{R_1}$$

because V_B has been assumed greater than V, and all the currents have been assumed to be flowing away from B. Since the polarity of the voltage source V is such that it forces a current towards the node B, V_B and V will oppose each other, and their difference ($V_B - V$) will be the voltage, which will be divided by R_1 to get the current through R_1 as

$$\frac{V_B - V}{R_1}$$

2. The current through R_2 will simply be

$$I_{BE} = \frac{V_B - 0}{R_2}$$

3. In the resistance R_3, the current I (of the current source) can only flow and that also in the direction of C to B. Hence the current going away from the node B, that is from B to C, will be

$$I_{BC} = -I$$

Summing up the three currents, as per KCL, we get

$$I_{BA} + I_{BE} + I_{BC} = 0$$

or

$$\frac{V_B - V}{R_1} + \frac{V_B}{R_2} - I = 0$$

with only V_B as the unknown which can be found out from the above equation. In order to understand the application of nodal method of analysis, consider a circuit example as shown in Figure 10. There are again two nodes B and E, with E being considered as the datum node. The KCL applied at node B yields with the voltage of node B assumed as V_B. Again, V_B can be calculated from the above equation, as the rest of the values are known.

$$\frac{V_B - V_1}{R_1} + \frac{V_B - 0}{R_2} + \frac{V_B - V_2}{R_3} = 0$$

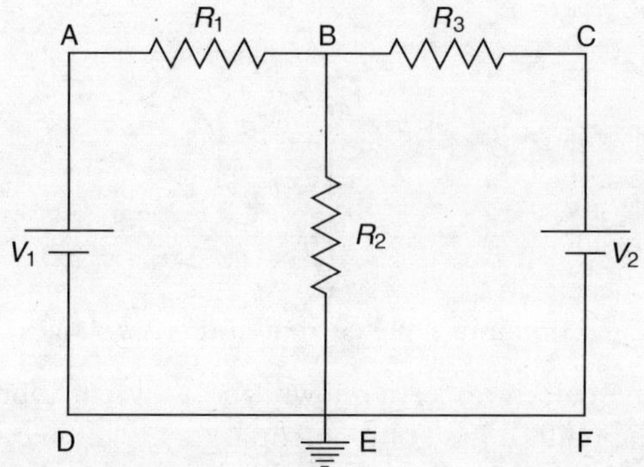

Figure 10 Circuit containing two voltage sources.

Consider Figure 11 as another circuit example for further understanding the KCL application in nodal analysis, where B and E are the two nodes.

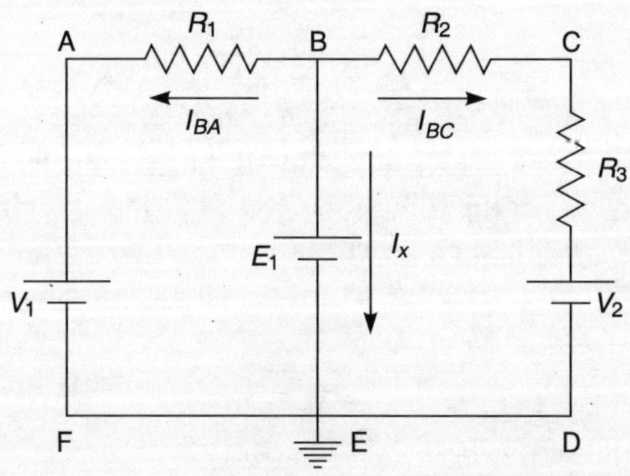

Figure 11 Circuit containing three voltage sources (with one of them having no resistance in series).

Three currents will be coming out from the node B.

1. Current through R_1:

$$I_{BA} = \frac{V_B - V_1}{R_1}$$

2. As R_2 and R_3 are in series:

$$I_{BC} = \frac{V_B - V_2}{R_2 + R_3}$$

3. Nothing can be said about the current in the branch BE, as it does not contain any resistance for the Ohm's law to be applied. Hence an unknown current (say I_x) can be assumed to be flowing from B to E.
4. KCL at B now gives

$$I_{BA} + I_{BC} + I_x = 0$$

or

$$\frac{V_B - V_1}{R_1} + \frac{V_B - V_2}{R_1 + R_2} + I_x = 0$$

It is noted that there is only one unknown in the above equation, because the voltage V_B of the node B is already known as E_1. Hence, only I_x remains as the unknown, which can be found out easily.

Furthermore, let us consider one more case as shown in Figure 12. As can be seen, points H, G, F and I are same (equipotential) points; here they constitute one node. Similarly, points B and C also qualify for nodes, as more than two branches are meeting at each of them. While writing the KCL equation at the node B, we will assume that the voltage V_B is at the highest potential. Hence all the currents go away from the node B. (While writing the KCL equation at the node C, similar assumption will be made for the voltage V_C of node C.) Applying KCL at B, we get the following three currents:

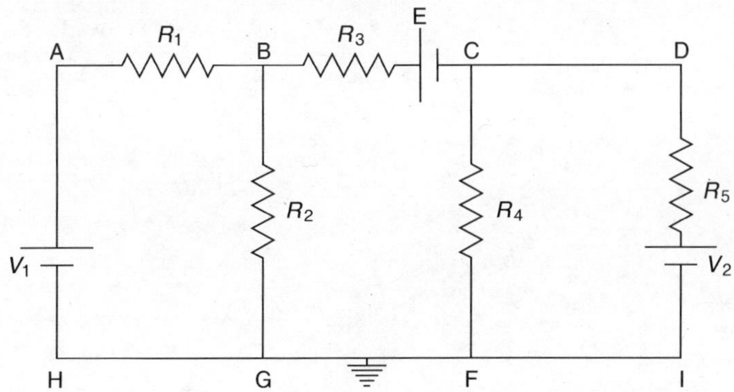

Figure 12 Circuit containing a voltage source between two non-zero nodes.

1. Current through R_1:

$$\frac{V_B - V_1}{R_1}$$

2. Current through R_2:

$$\frac{V_B - 0}{R_2}$$

3. The current is going from B to C. Since the voltage source E pushes the current from C towards B (according to its polarity), therefore it opposes the (away-going) current from the node B. Hence the third current will be

$$\frac{V_B - E - V_C}{R_3}$$

where V_C is the voltage of node C.

So, KCL at the node B sums up to

$$\frac{V_B - V_1}{R_1} + \frac{V_B - 0}{R_2} + \frac{V_B - E - V_C}{R_3} = 0$$

and by similar logic, KCL at the node C is

$$\frac{V_C - V_2}{R_5} + \frac{V_C - 0}{R_4} + \frac{V_C + E - V_B}{R_3} = 0$$

Note the positive sign is in front of *E*, which suggests that the voltage source E supports the away-going current from the node C. So, V_B and V_C are the two unknowns, which can be found by using the above two equations.

1.8 Star–Delta Transformation

One of the most efficient methods of electricity generation is a *three-phase generation*, which means that three sinusoidal voltages are generated simultaneously with their peak values occurring at 120° apart from each other. Such three-phase voltages are fed to the loads that also need to be three-phase loads. Three-phase loads can be connected in only two configurations: (i) star and (ii) delta, as shown in Figure 13.

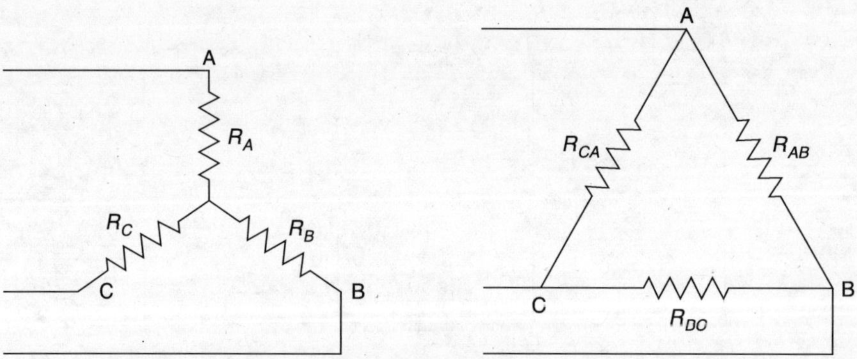

Figure 13 Star and delta connections.

If the resistances in the three phases in the load are the same, it is called a *balanced three-phase load*. More about three-phase circuits will be discussed later in the book. Our present aim is to study about the two configurations, that is star and delta, and the transformation of one to the other.

Figure 14 shows a network consisting of three junction points: A, B and C. Let there be three resistances connected, such that in one case a star circuit results, and in the second case, a delta circuit results. The resistance between any two points (say A and B) has to be kept unchanged in both the star and the delta configurations.

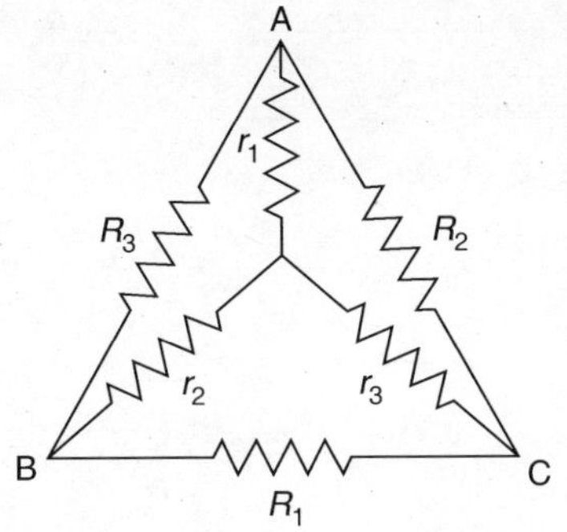

Figure 14 Star–delta transformation.

Let us now find out the equivalent resistances between any two terminals (say, A and B) in the two configurations and equate them so that the transformation holds good.

While remaining in star connection, there is only one path between A and B, and thus the resistance between A and B (in star) will be $(r_1 + r_2)$. On the other hand, in delta transformation, there are two paths between A and B, that is R_3 and $(R_2 + R_1)$. Therefore, these two being in parallel, we can say that the resistance between A and B in delta is R_3 parallel $(R_2 + R_1)$. Now, since the resistance between the points A and B has to be kept identical in either configuration, we can equate these two resistance values. Therefore,

$$r_1 + r_2 = \frac{R_3(R_2 + R_1)}{R_1 + R_2 + R_3} \tag{1.3a}$$

Similarly, for resistance between B and C, we get

$$r_2 + r_3 = R_1 \,||\, (R_2 + R_3) = \frac{R_1(R_2 + R_3)}{R_1 + R_2 + R_3} \tag{1.3b}$$

and for resistance between C and A, we get

$$r_3 + r_1 = R_2 \,||\, (R_3 + R_1) = \frac{R_2(R_3 + R_1)}{R_1 + R_2 + R_3} \tag{1.3c}$$

1.8.1 Delta-to-Star Transformation

In order to understand the delta-to-star transformation we will prove that

$$r_1 = \frac{R_2R_3}{R_1 + R_2 + R_3}, \quad r_2 = \frac{R_3R_1}{R_1 + R_2 + R_3}, \quad r_3 = \frac{R_1R_2}{R_1 + R_2 + R_3}$$

From Eqs. (1.3a)–(1.3c) we have

$$r_1 + r_2 = \frac{R_3(R_2 + R_1)}{R_1 + R_2 + R_3} \Rightarrow r_1 + r_2 = \frac{R_2R_3 + R_3R_1}{R_1 + R_2 + R_3} \tag{1.4}$$

$$r_2 + r_3 = \frac{R_1(R_2 + R_3)}{R_1 + R_2 + R_3} \Rightarrow r_2 + r_3 = \frac{R_1R_2 + R_3R_1}{R_1 + R_2 + R_3} \tag{1.5}$$

$$r_3 + r_1 = \frac{R_2(R_3 + R_1)}{R_1 + R_2 + R_3} \Rightarrow r_3 + r_1 = \frac{R_2R_3 + R_1R_2}{R_1 + R_2 + R_3} \tag{1.6}$$

Subtracting of Eq. (1.5) from Eq. (1.4), we get

$$r_1 - r_3 = \frac{R_2R_3 - R_1R_2}{R_1 + R_2 + R_3} \tag{1.7}$$

Adding Eqs. (1.6) and (1.7), we get

$$2r_1 = \frac{2R_2R_3}{R_1 + R_2 + R_3}$$

or

$$r_1 = \frac{R_2R_3}{R_1 + R_2 + R_3}$$

Similarly, we can find

$$r_2 = \frac{R_3R_1}{R_1 + R_2 + R_3} \quad \text{and} \quad r_3 = \frac{R_1R_2}{R_1 + R_2 + R_3}$$

1.8.2 Star-to-Delta Transformation

From the values of r_1, r_2 and r_3, we can write the following equations:

$$r_1r_2 = \frac{R_2R_3}{R_1 + R_2 + R_3}\,\frac{R_3R_1}{R_1 + R_2 + R_3} = \frac{R_1R_2R_3^2}{(R_1 + R_2 + R_3)^2}$$

Similarly, we can write

$$r_2r_3 = \frac{R_1^2R_2R_3}{(R_1 + R_2 + R_3)^2} \quad \text{and} \quad r_3r_1 = \frac{R_1R_2^2R_3^2}{(R_1 + R_2 + R_3)^2}$$

Adding them, we get

$$\begin{aligned} r_1r_2 + r_2r_3 + r_3r_1 &= \frac{R_1R_2R_3^2}{(R_1 + R_2 + R_3)^2} + \frac{R_1^2R_2R_3}{(R_1 + R_2 + R_3)^2} + \frac{R_1R_2^2R_3^2}{(R_1 + R_2 + R_3)^2} \\ &= \frac{R_1R_2R_3(R_1 + R_2 + R_3)}{(R_1 + R_2 + R_3)^2} = \frac{R_1R_2R_3}{R_1 + R_2 + R_3} \end{aligned} \tag{1.8}$$

But, we know that

$$r_1 = \frac{R_2R_3}{R_1 + R_2 + R_3} \tag{1.9}$$

Therefore, substituting Eq. (1.9) into Eq. (1.8) we obtain

$$r_1r_2 + r_2r_3 + r_3r_1 = R_1r_1 \quad \text{or} \quad R_1 = \frac{r_1r_2 + r_2r_3 + r_3r_1}{r_1}$$

Similarly, we can find

$$R_2 = \frac{r_1r_2 + r_2r_3 + r_3r_1}{r_2} \quad \text{and} \quad R_3 = \frac{r_1r_2 + r_2r_3 + r_3r_1}{r_3}$$

Sections 1.8.1 and 1.8.2 provide the equations for *star–delta transformation.* Using these equations, if a star circuit is given, that is if r_1, r_2 and r_3 are known, then R_1, R_2 and R_3 can be found out and vice versa.

1.9 Superposition Theorem

The superposition theorem states that if a circuit consists of linear elements and has two or more independent sources, the current (or voltage) in any part of the circuit can be found by algebraically adding the contributions of each source acting one at a time.

To find the contribution of a particular source, every other source in the circuit is replaced by its internal resistance. This is done successively for every source, and the algebraic sum of the currents due to each source gives the current due to all the sources acting simultaneously. If ideal sources are present, then the ideal voltage source is replaced by a short circuit (zero-ohm internal resistance) and the ideal current source by an open circuit (infinite-ohm internal resistance).

For example, let us find out the current in the resistance R_2 in the circuit shown in Figure 15 by using the superposition theorem. First, the current in R_2 due to the voltage source alone is found. While doing so, the other source, that is the current source, is replaced by its (infinite ohm) internal resistance, that is an open circuit. Therefore, Figure 16 shows the resultant circuit.

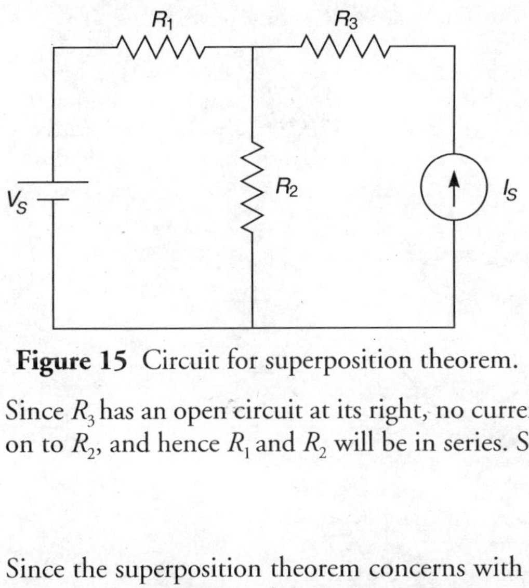

Figure 15 Circuit for superposition theorem.

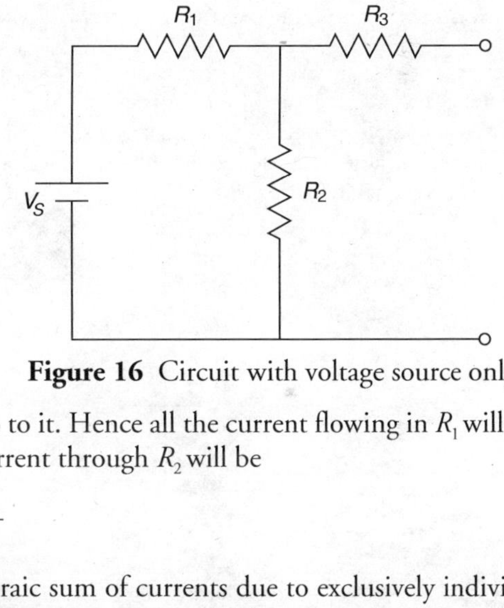

Figure 16 Circuit with voltage source only.

Since R_3 has an open circuit at its right, no current will go to it. Hence all the current flowing in R_1 will pass on to R_2, and hence R_1 and R_2 will be in series. So, the current through R_2 will be

$$\frac{V_S}{R_1 + R_2}$$

Since the superposition theorem concerns with the algebraic sum of currents due to exclusively individual sources, the direction of the individual currents also need to be mentioned. Therefore, the current (with direction) through R_2 due to voltage source V_S only in the circuit is given by

$$I'_{R_2} = \frac{V_S}{R_1 + R_2} \downarrow$$

Now for finding the contribution of the current source, we replace the voltage source by its (zero ohm) internal resistance, that is by a short circuit. Figure 17 is the resultant circuit.

As can be seen, the two ends of R_1 and R_2 get connected bringing them in parallel with each other. The current I_S of the current source will flow through R_3 and get divided in R_1 and R_2. Thus, the current (with direction) through R_2 due to the current source only is given by

$$I''_{R_2} = \frac{I_S R_1}{R_1 + R_2} \downarrow$$

Both the current directions are downward. Therefore, their algebraic sum will be

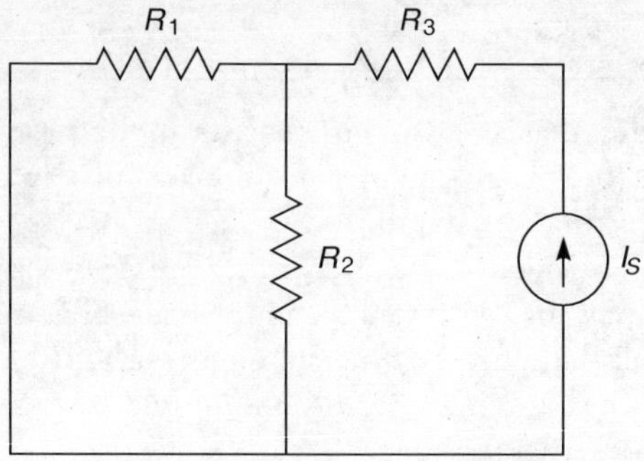

Figure 17 Circuit with current source only.

$$I_{R_2} = I'_{R_2} + I''_{R_2} = \frac{V_S}{R_1 + R_2} + \frac{I_S R_1}{R_1 + R_2} = \frac{V_S + I_S R_1}{R_1 + R_2} \downarrow$$

It is important to note here that the superposition theorem cannot be applied to find power.[4]

---

[4]Since voltage and current are quantities featuring in the Ohm's law, which obeys the linearity principle, the superposition theorem is applicable for only voltage and current calculations. Power is given by I^2R or (V^2/R) or VI, which reflects parabolic relation, which is nonlinear. Hence power calculations cannot be done by the superposition theorem. Let us explain this by an example as shown in figure below.

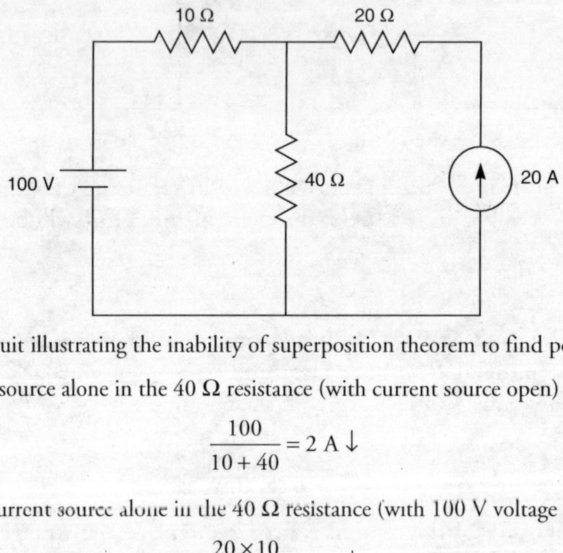

Circuit illustrating the inability of superposition theorem to find power.

The current due to the 100 V source alone in the 40 Ω resistance (with current source open) is

$$\frac{100}{10+40} = 2\text{ A}\downarrow$$

The current due to the 20 A current source alone in the 40 Ω resistance (with 100 V voltage source shorted) is

$$\frac{20\times 10}{10+40} = 4\text{ A}\downarrow$$

Hence the total current due to both sources, acting simultaneously, will be 2 + 4 = 6 A downwards. Thus, the total power in 40 Ω resistance will be $6^2 \times 40 = 1440$ W.

Now let us find out the power in 40 Ω resistance due to individual sources acting alone. The power in 40 Ω resistance due to the voltage source alone, causing a current of 2 A will be $2^2 \times 40 = 160$ W and that due to the current source alone, causing a current of 4 A will be $4^2 \times 40 = 640$ W. The sum of these two powers, 160 + 640 = 800 W, is different from the actual power, 1440 W.

Thus we see that power calculations should not be done by using the superposition theorem.

1.10 Thevenin's Theorem

Thevenin's theorem is a powerful tool for calculating the current (or voltage) in a circuit. The circuit should be linear and time-invariant. This theorem states that if the current (or voltage) is to be found in a particular resistance (say, load resistance R_L) in a circuit, then that resistance is removed, and the voltage (V_{Th}) and the equivalent resistance (R_{Th}) are calculated between the two open-circuited ends thus created. The whole network (with load removed) can then be replaced by a voltage source (having value) V_{Th} and a resistance (having value) R_{Th} in series with it. The load resistance (R_L) is finally connected back and the current through R_L is, therefore, given by

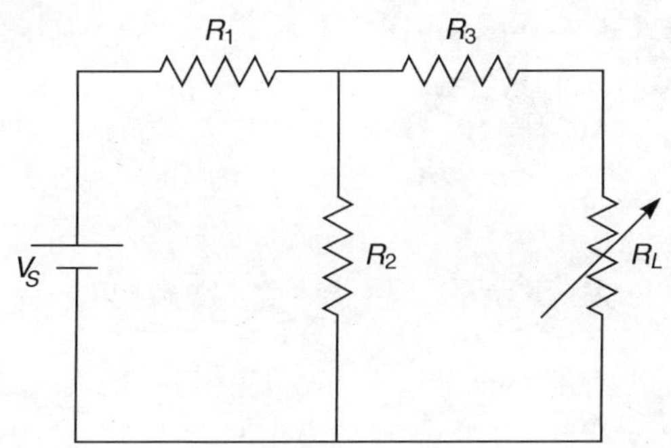

Figure 18 Circuit for explaining Thevenin's theorem.

$$I_L = \frac{V_{\text{Th}}}{R_{\text{Th}} + R_L}$$

Consider an example shown in Figure 18. Here, the load current through R_L is required to be found. The following steps need to be taken.

1. **Removal of R_L:** It creates an open circuit across its two ends. Name the two ends as A and B. The resultant circuit is shown in Figure 19.

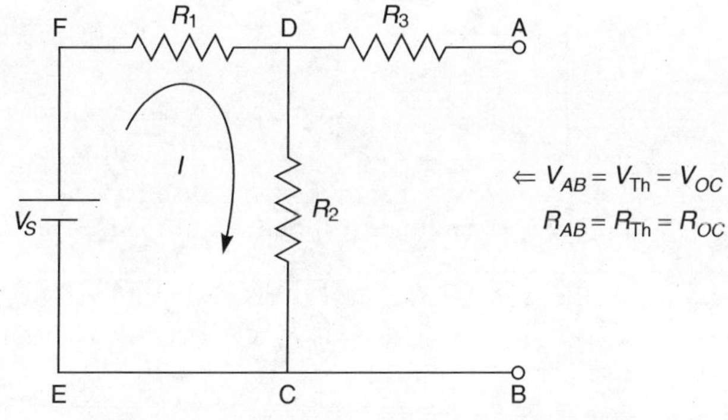

Figure 19 Removal of R_L in the circuit.

2. **To find V_{Th}:** As seen in the circuit, we find the voltage V_{AB} is the voltage between the points A and B, which is also called V_{Th} or V_{OC}. The following steps are taken to find this voltage.

   (a) Find all the currents flowing in the circuit (Figure 19).

   (b) Start from B and reach A, and calculate the voltage gains or drops to get the voltage V_{AB}.

In this example, the only current flowing will be

$$I = \frac{V_S}{R_1 + R_2}$$

No current will go to R_3. Next, we start from B and move towards A. We can do this through two paths BCDA or BCEFDA. Let us choose the smaller path, that is BCDA. We reach C without any voltage gain or drop, as B and C have the same potential. Next, we move from C to D and, in doing so, we find that we are moving against the direction of the current I flowing from D to C in R_2. This will cause a voltage gain according to Rule 4 discussed in Section 1.7.1. Next, we see that points D and A are (electrically) connected through the resistance R_3 and no current is flowing in R_3. Therefore, D and A must be equipotential points. Thus, we reached A. The potential difference between B and A, that is V_{AB}, will be

$$V_{AB} = IR_2 = \frac{V_S R_2}{R_1 + R_2} = V_{OC} = V_{\text{Th}}$$

3. **To find R_{Th}:** Next, we find the resistance R_{AB} between the points A and B. The following steps are taken:
   (a) Replace all the independent sources by their internal resistances. If the sources are ideal, replace the voltage source by a short circuit and the current source by an open circuit. In this example, we have only one voltage source which, if we replace by a short circuit, gives a circuit as shown in Figure 20.
   (b) Start from A and reach B, and calculate the resistance along with. This will be $R_{AB} = R_{\text{Th}} = R_{OC}$.

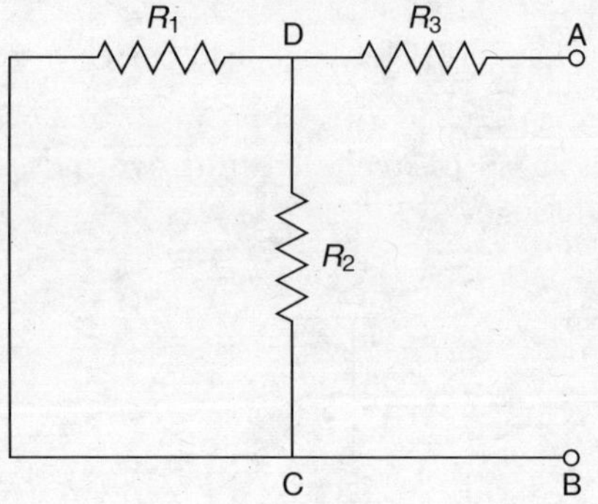

Figure 20 Finding R_{Th} for circuit of Figure 18.

We start from A and move ahead to get R_3. Furthermore, we find at D that there are two parallel paths: R_1 and R_2. Therefore, as shown in Figure 21, R_1 and R_2 come in parallel and their resultant is in series with R_3. So,

$$R_{AB} = R_3 + (R_1 \parallel R_2) = R_3 + \frac{R_1 R_2}{R_1 + R_2} = R_{OC} = R_{\text{Th}}$$

4. **To draw Thevenin's equivalent circuit diagram:** Finally, we get the Thevenin's equivalent of the circuit shown in Figure 19, across the terminals A and B as shown in Figure 22.

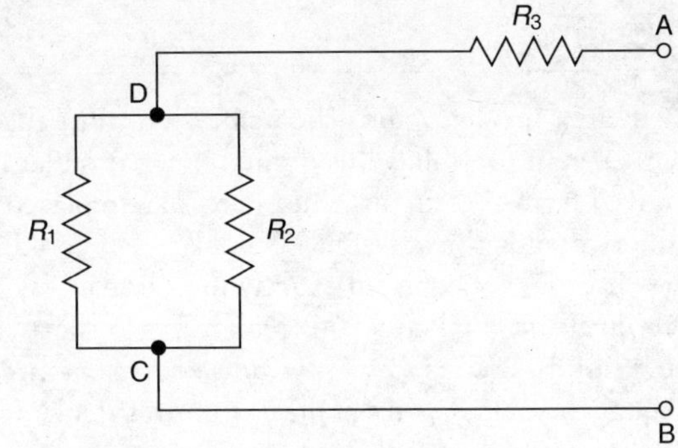

Figure 21 Equivalent circuit of Figure 20.

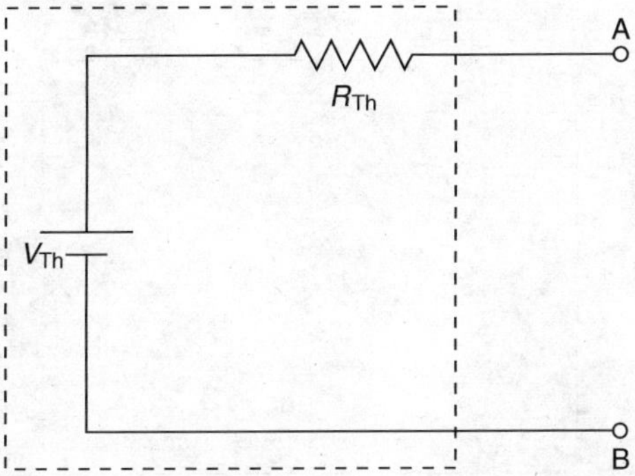

Figure 22 Thevenin's equivalent circuit.

The load resistance, R_L, is then inserted back into the circuit between the points A and B, and the final circuit is shown in Figure 23. The current through R_L will be

$$I_L = \frac{V_{\mathrm{Th}}}{R_{\mathrm{Th}} + R_L}$$

The voltage across R_L is $I_L R_L$. That is,

$$\frac{V_{\mathrm{Th}} R_L}{R_{\mathrm{Th}} + R_L}$$

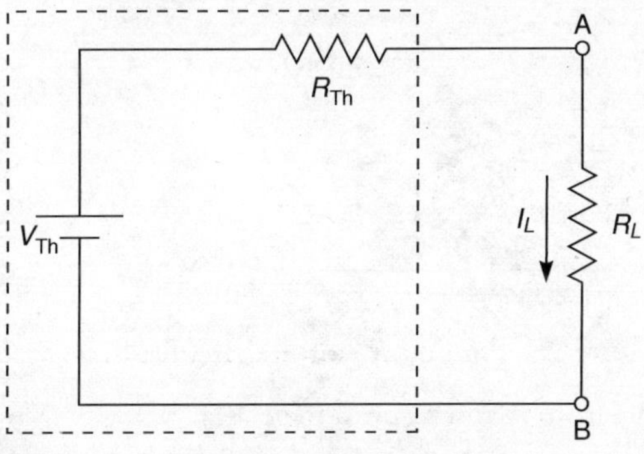

Figure 23 Finding the load current for the circuit of Figure 18.

1.11 Norton's Theorem

Another theorem, which is the dual of the Thevenin's theorem, is Norton's theorem. According to it, if in a linear, time-invariant network, a current (or voltage) is desired in a particular resistance (say load resistance R_L), then that resistance is removed from the circuit. The sources are replaced by their internal resistances. The resistance (R_N) between the two ends thus created is calculated by looking into the network through these open-circuited terminals. Thus, R_N is calculated **exactly** in a similar way as R_{Th}.

Next, the load resistance R_L (in the original circuit) is replaced by a short circuit, and the current (I_N or I_{SC}) through that short-circuited link is calculated. A Norton's equivalent circuit finally can be developed, containing a current source (equal to value I_N) in parallel with a resistance (equal to value R_N). The load resistance (R_L) is connected back in parallel with these two elements. Now, the current through R_L can be easily found by applying the current divide theorem.

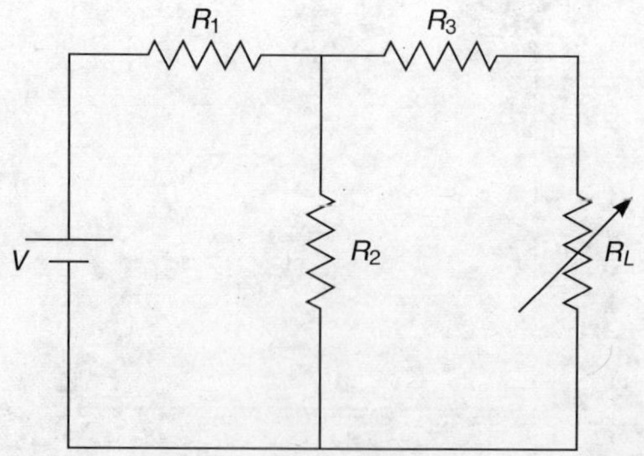

Figure 24 Circuit for explaining the Norton's theorem.

Let us consider the same example (shown in Figure 24) as taken in the case of Thevenin's theorem.

1. **To find R_N:** For finding the resistance R_N, we replace the sources with their internal resistances (i.e. short the voltage source and open the current source, if they are ideal). The present network is reduced to the one shown in Figure 25.

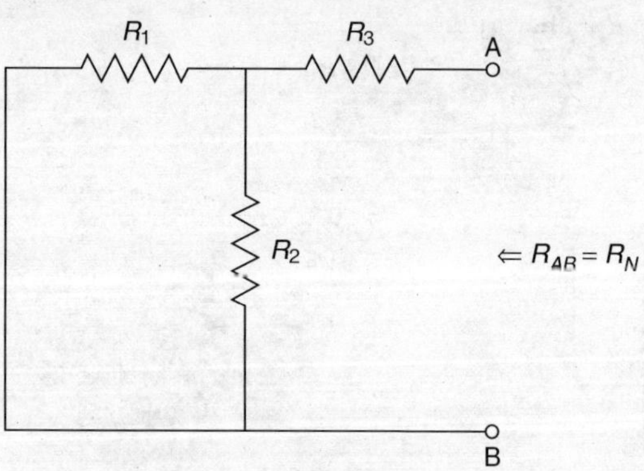

Figure 25 Finding R_N for the circuit of Figure 24.

Now (as for R_{Th}), we start from A and reach B to get R_N:

$$R_N = R_3 + (R_1 \parallel R_2) = R_3 + \frac{R_1 R_2}{R_1 + R_2}$$

2. **To find I_N:** Replace the load R_L (through which the current is desired) by a short circuit and find the current I_N through it. The network for this is shown in Figure 26. There are two nodes, X and Y. Assuming Y as the reference node and writing the KCL equation at the node X, we get

$$\frac{V_X - V_S}{R_1} + \frac{V_X - 0}{R_2} + \frac{V_X - 0}{R_3} = 0$$

Here V_X is the only unknown quantity, which can be found out. After knowing V_X, now I_N can be obtained as

$$I_N = \frac{V_X}{R_3}$$

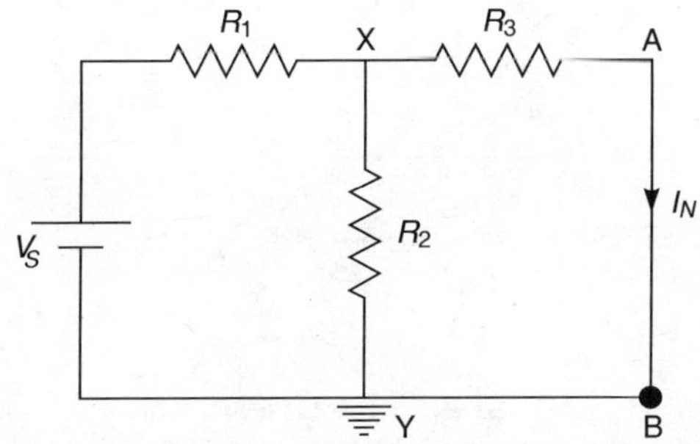

Figure 26 Finding I_N for the circuit of Figure 24.

3. **To draw Norton's equivalent circuit:** Finally the Norton's equivalent circuit across terminals A and B is drawn, as shown in Figure 27. Relationship between Thevenin and Norton Theorem results are mentioned at the end of the page footnote.[5]

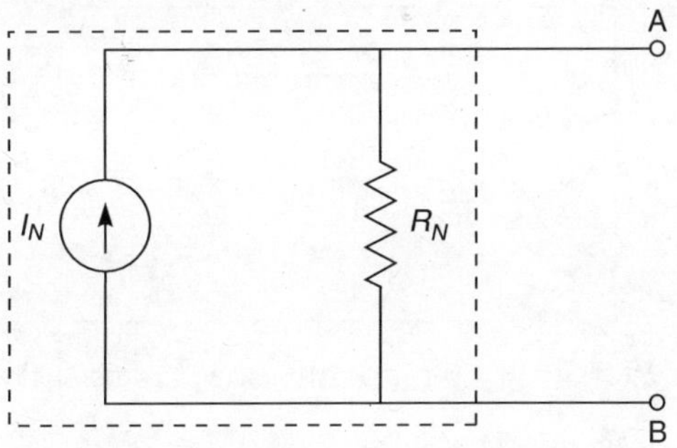

Figure 27 The Norton's equivalent circuit.

[5]Since the circuits considered for explaining the Thevenin and Norton theorems (in Figures 18 and 24, respectively) are same, hence the current through R_L will be same in both cases. Thus

$$\frac{V_{\text{Th}}}{R_{\text{Th}} + R_L} = \frac{I_N R_N}{R_N + R_L}$$

Now since $R_N = R_{\text{Th}}$, hence the denominators of the above equation are same. Hence we get the following very important relation:

$$V_{\text{Th}} = I_N R_N \quad \text{or} \quad I_N = \frac{V_{\text{Th}}}{R_{\text{Th}}}$$

The load R_L is then connected back between the terminals A and B, as shown in Figure 28. The desired current through R_L is then found by applying the current divide theorem as

$$I_L = \frac{I_N R_N}{R_N + R_L}$$

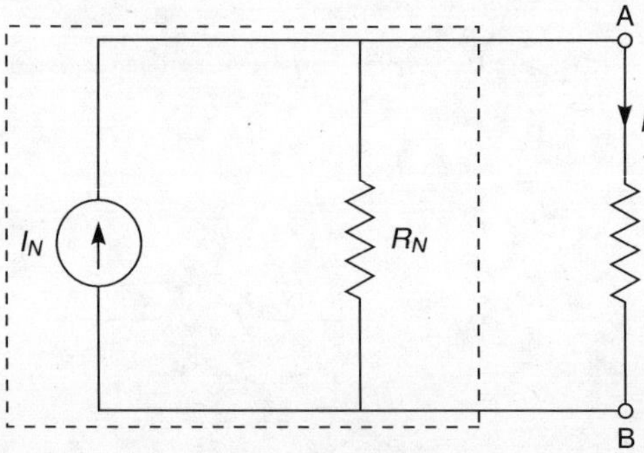

Figure 28 Finding the load current for the circuit of Figure 24.

1.12 Maximum Power Transfer Theorem

This theorem states that in a linear circuit, a source can transfer the maximum power to a load resistance when the load resistance is equal to the internal resistance of the source. In order to understand the theorem, consider a circuit as shown in Figure 29.

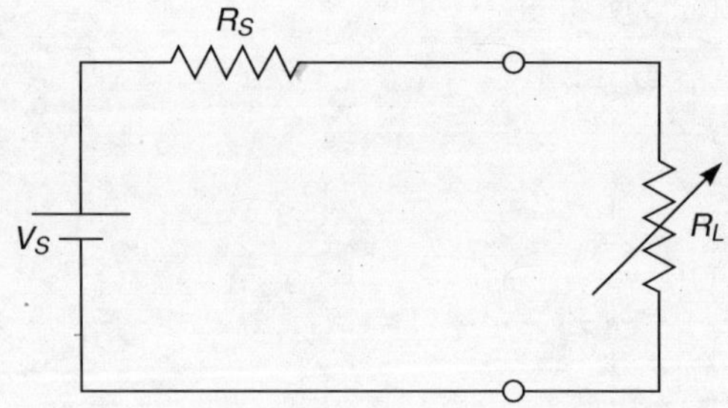

Figure 29 Circuit for maximum power transfer theorem.

In case the circuit is complex, Thevenin's theorem is applied and V_{Th} and R_{Th} are used in place of V_S and R_S, respectively. The resistance R_S is the internal resistance of the voltage source. The current through the load resistance R_L is given by

$$I_L = \frac{V_S}{R_S + R_L}$$

The power dissipated in R_L is given by

$$P_L = I_L^2 R_L = \left(\frac{V_S}{R_S + R_L}\right)^2 R_L = \frac{V_S^2 R_L}{(R_S + R_L)^2}$$

Now with R_L being variable, the power P_L will also vary with R_L. For finding the maximum value of P_L, we differentiate P_L with respect to R_L and equate the answer to zero. This gives us the condition for maximum value of P_L. Next, we substitute this condition back into the expression of P_L to get the value of the maximum power $P_{L(\max)}$.

Differentiating P_L with respect to R_L, we get

$$\frac{dP_L}{dR_L} = \frac{(V_S^2 \times 1)(R_S + R_L)^2 - [(V_S^2 R_L) \times 2(R_S + R_L) \times 1]}{(R_S + R_L)^4}$$

Putting the RHS = 0 and solving, we get

$$R_L = R_S$$

This becomes the condition for maximum power transfer to R_L.

Now, in order to find the amount of maximum power transferred from the source to the load, the above condition is substituted in the expression of P_L. The value of maximum power, thus, becomes

$$P_{L(\max)} = \left(\frac{V_S}{R_L + R_L}\right)^2 R_L = \left(\frac{V_S}{2R_L}\right)^2 R_L = \frac{V_S^2}{4R_L}$$

At the time of maximum power transfer, that is when $R_L = R_S$, the power input is given by

$$V_S \times I_S = V_S \times \frac{V_S}{2R_L}$$

or

$$P_{\text{in}} = \frac{V_S^2}{2R_L}$$

Since the output (maximum) power available at load is

$$\frac{V_S^2}{4R_L}$$

the efficiency of power transfer (at the time of maximum power transfer) will be given by

$$\eta = \frac{V_S^2/(4R_L)}{V_S^2/(2R_L)} = 0.5 \text{ or } 50\%$$

Thus, we see that at the time of maximum power transfer, the power transfer efficiency is only 50%.[6]

The following steps help in solving the problems related to maximum power transfer theorem.

1. **To find the value of R_L so that it draws maximum power from the source:** If the circuit in question is a multi-loop (large) one, the value of R_L for which the maximum power is transferred to it is found by thevenizing the circuit across R_L and finding R_{Th}. The value of R_L for $P_{L(\max)}$ will then be equal to R_{Th}.

---

[6]Does that mean we cannot get efficiencies above 50%? The answer is No, we can indeed do so but at a certain cost. To understand this consider a 100 V voltage source having internal resistance of 1 Ω as shown below.

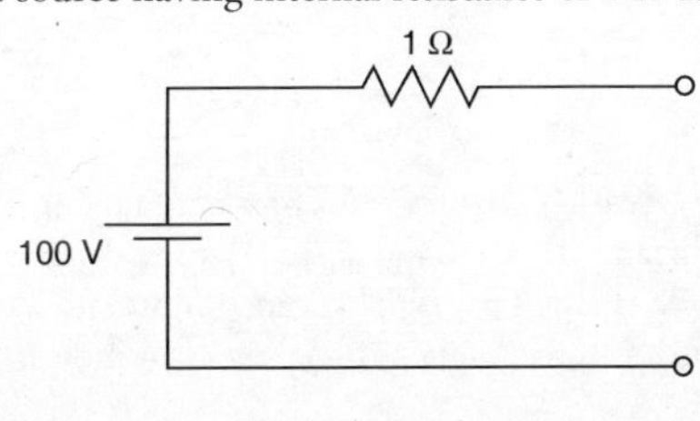

2. **To find the maximum power transferred:** The value of $P_{L(max)}$ can be found in a complex circuit by developing Thevnin's equivalent circuit (i.e. by finding V_{Th} and R_{Th}) and connecting $R_L = R_{Th}$ across it. The power transferred to the load under this condition is the maximum power transferred.

Let us understand this theorem by considering an example shown in Figure 30. Say it is desired to find the value of R_L for which the maximum power will be transferred to it.

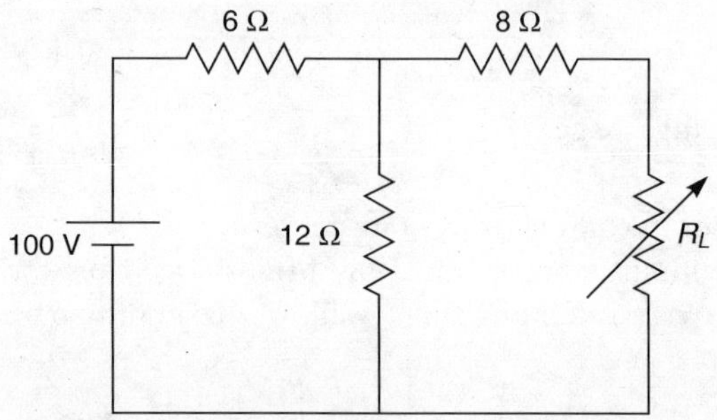

Figure 30 Circuit for explaining the maximum power transfer theorem.

[6 (cont.)] Now if maximum power has to be extracted out of this source, then according to the MPT theorem, a load resistance $R_L = 1\ \Omega$ will be required. In that case the circuit will look like

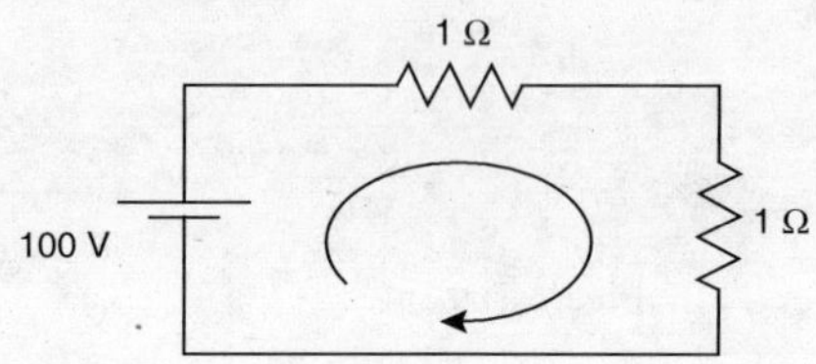

The current flowing will be

$$\frac{100}{1+1} = 50\text{ A}$$

and $$P_{L(max)} = 50^2 \times 1 = 2500\text{ W}$$

Also since the power input will be $100 \times 50 = 5000$ W, hence under the maximum power transfer condition, $\eta = 50\%$.

Next let us replace the load $R_L = 1\ \Omega$ by a resistance of (say) 19 Ω. The circuit now looks like

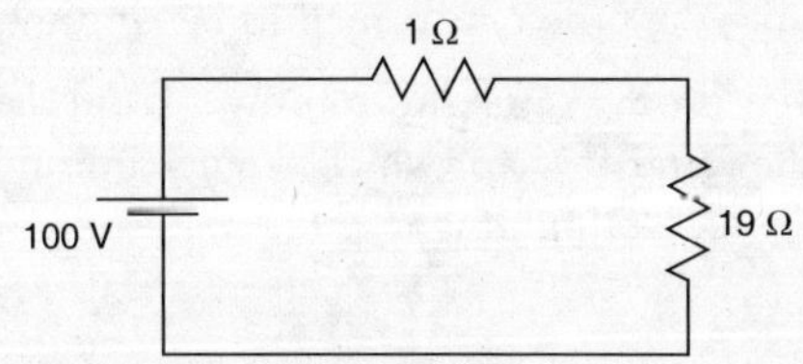

The current flowing will now be 5 A and the power transferred to the load will be $5^2 \times 19 = 475$ W. Under this condition, the input power $= 100 \times 5 = 500$ W. So

$$\eta = \frac{475}{500} = 95\%$$

Thus we conclude that the maximum power which a 100 V voltage source having 1 Ω internal resistance can transfer to the load is 2500 W at 50% efficiency. If the efficiency needs to be increased, the power extracted (and then transferred to the load) needs to be drastically reduced (only 500 W is taken from the source for an efficiency of 95%). Thus, at the cost of under-utilization (also called derating) of the voltage source, higher efficiencies can indeed be achieved.

It can be found by removing R_L and finding the Thevenin equivalent resistance (R_{Th}) across the two open-circuited terminals. This (as already discussed) is done after replacing the independent sources by their internal resistances. The resultant circuit is shown in Figure 31.

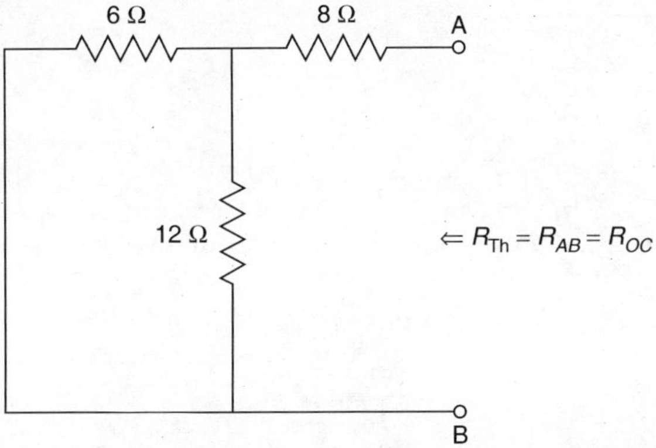

Figure 31 Finding R_N (i.e. R_S) for the circuit of Figure 30.

We have replaced the voltage source (ideal) by a zero-ohm resistance, that is a short circuit. Thus, the resistance between A and B can be found with 8 Ω in series with the parallel combination of 6 Ω and 12 Ω resistances. Therefore

$$R_{\text{Th}} = R_{AB} = 8 + (6||12) = 8 + 4 = 12\ \Omega$$

Now as per the maximum power transfer theorem, the value of R_L, for it to receive the maximum power, will be equal to R_{Th}, that is R_L should be equal to 12 Ω (Figure 32).

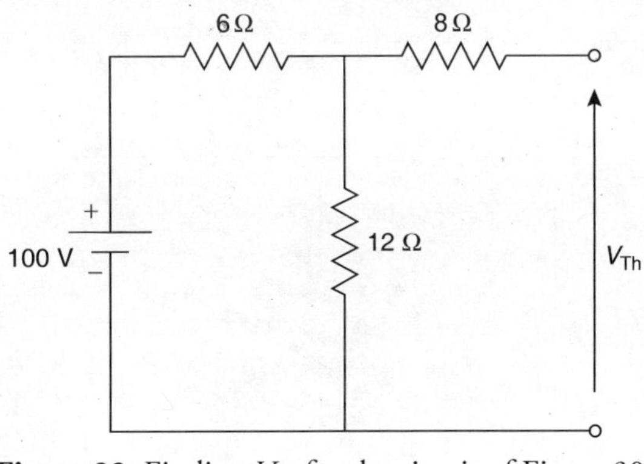

Figure 32 Finding V_{Th} for the circuit of Figure 30.

For finding the value of the maximum power transferred to R_L, we need to find V_{Th} also (see Figure 33). Then in Thevenin equivalent circuit, V_{Th} and two resistances (R_{Th} and R_L) are placed and the power delivered to R_L can be easily found as

$$P_{L(\max)} = \frac{V_{\text{Th}}^2}{4R_L}$$

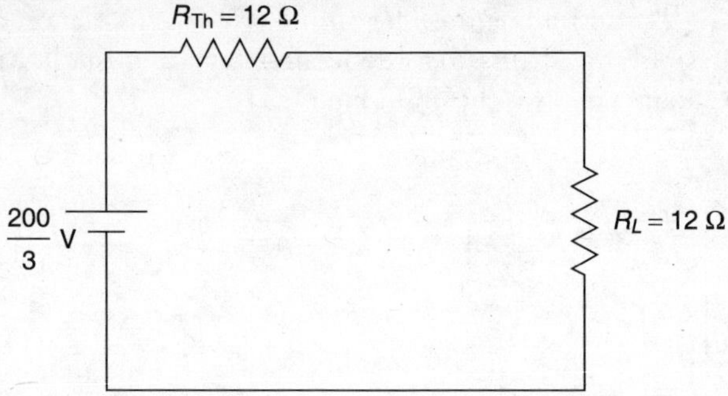

Figure 33 Circuit for finding the maximum power transfer.

Here,

$$V_{Th} = \frac{100}{18} \times 12 = \frac{200}{3}\ \text{V}$$

Hence the maximum power transferred to the load is given by

$$P_{L(\max)} = \frac{(200/3)^2}{4 \times 12} = \frac{200 \times 200 \times 1}{9 \times 4 \times 12} = \frac{2500}{27} = 92.6\ \text{W}$$

Solved Examples

Example 1

Find the current through all the branches of the circuit, as shown in Figure 34, by using KVL.

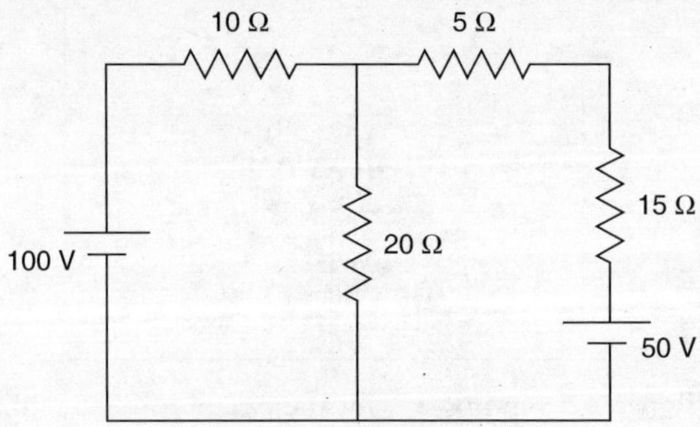

Figure 34 Figure for Example 1.

Solution

The circuit can be redrawn as shown in Figure 35.

As can be seen from the figure, there are three closed loops: ABCFA, FCDEF and ABCDEFA, in the question circuit (see Figure 34). Out of them, the first two are meshes, and we use them to solve

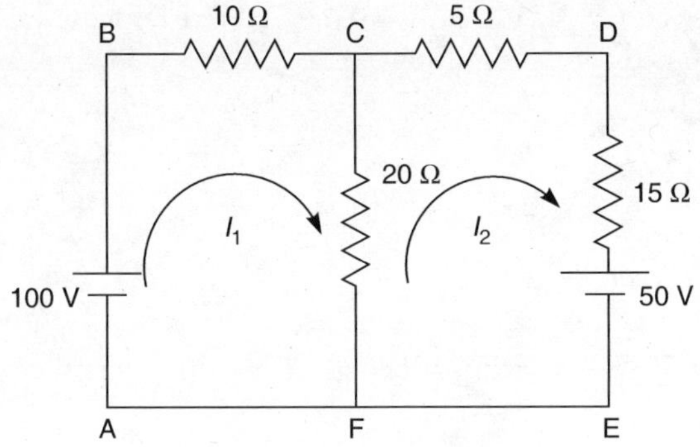

Figure 35 Redrawn Figure 34.

for the branch currents. Let us assume clockwise currents, I_1 and I_2, in these two meshes, as shown in Figure 35.

Writing the KVL equation for the mesh ABCFA, if we start from A and move clockwise (we can start from any point in the loop and move in any direction), we meet the voltage source of 100 V. Since we move from the negative to positive terminal of the voltage source, we experience a voltage gain (as already explained). We move ahead from B and reach the 10 Ω resistor. Here, we find that we are moving in the direction of current I_1, so we experience a voltage drop

$$-I_1R_1 \quad \text{or} \quad -I_1 \times 10$$

Next, we have to go from C to F through the 20 Ω resistor. Now, since we do not know the magnitudes of I_1 and I_2, so we cannot say if the current will flow from C to F (i.e. the current will be $I_1 - I_2$) or it will flow from F to C (i.e. the current will be $I_2 - I_1$). Let us assume the current is flowing from C to F, that is $(I_1 - I_2)$ is the current in the 20 Ω resistor. So, our KVL path is in the direction of the current, and hence there will be a voltage drop $-(I_1 - I_2) \times 20$. Summing up

$$100 - 10I_1 - 20(I_1 - I_2) = 0 \tag{1}$$

Similarly, writing the KVL equation for the mesh FCDEF, we start with FC, that is, 20 Ω resistor. Here, now we can assume that the net current (through FC) is $(I_2 - I_1)$, so that the KVL path will coincide with the direction of the current, and there will be a voltage drop, that is $-(I_2 - I_1) \times 20$.

[Had we stuck to our earlier assumption of taking $(I_1 - I_2)$ as the current in the 20 Ω resistor, we would have found that while moving from F to C for writing the KVL equation, we would be going against the direction of the current, and hence we would have then experienced a voltage gain, that is $(I_1 - I_2) \times 20$. This is precisely same as $-(I_2 - I_1) \times 20$.]

Next, we move from C to D to have a voltage drop, that is, $-I_2 \times 5$. In going from D to E, we have a voltage drop of $-I_2 \times 15$ and another voltage drop of -50 V. Thus summing up, we get

$$-(I_2 - I_1) \times 20 - I_2 \times 5 - I_2 \times 15 - 50 = 0 \tag{2}$$

Arranging Eq. (1), we get

$$30I_1 - 20I_2 = 100 \tag{3}$$

From Eq. (2) we have

$$20I_1 - 40I_2 = 50 \tag{4}$$

Solving Eqs. (3) and (4), we get: $I_1 = 3.75$ A and $I_2 = 0.625$ A. Since both the answers come out to be positive, we say that our initially assumed current directions were correct. So, now we can tell the branch currents:

AB carries $I_1 \Rightarrow 3.75$ A

BC (10 Ω) carries $I_1 \Rightarrow 3.75$ A

CF (20 Ω) carries $(I_1 - I_2) \Rightarrow 3.75 - 0.625 = 3.125$ A

CD (5 Ω) carries $I_2 \Rightarrow 0.625$ A

DE carries $I_2 \Rightarrow 0.625$ A

Note the voltage source of 50 V (by virtue of being smaller than the other source) has to carry current inwards.

Example 2

Find the currents in all the branches of the circuit shown in Figure 36 by using KVL.

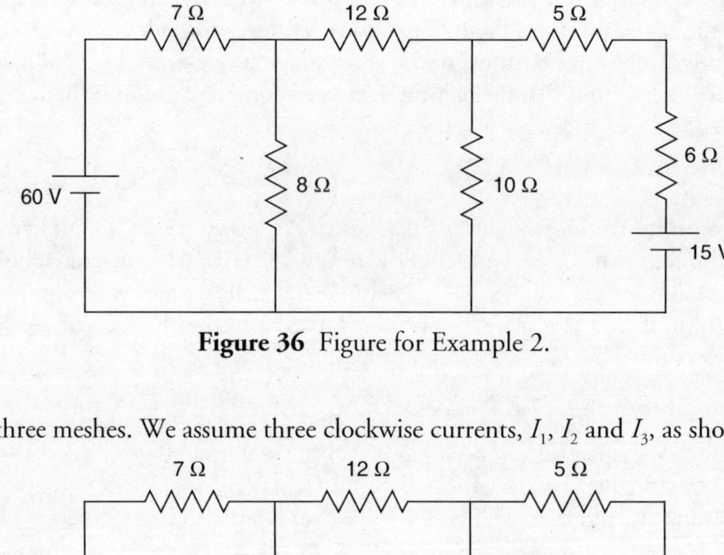

Figure 36 Figure for Example 2.

Solution

Here there are three meshes. We assume three clockwise currents, I_1, I_2 and I_3, as shown in Figure 37.

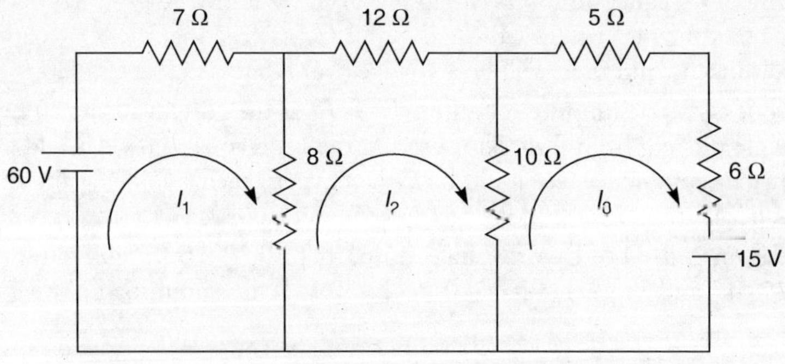

Figure 37 Redrawn Figure 36.

Writing KVL for the left mesh, we get

$$60 - 7I_1 - 8(I_1 - I_2) = 0$$

or

$$15I_1 - 8I_2 = 60 \tag{1}$$

For the central mesh,

$$-8(I_2 - I_1) - 12I_2 - 10(I_2 - I_3) = 0$$

or

$$-8I_1 + 30I_2 - 10I_3 = 0 \quad (2)$$

For the right mesh,

$$-10(I_3 - I_2) - 5I_3 - 6I_3 - 15 = 0$$

or

$$-10I_2 + 21I_3 = -15 \quad (3)$$

Solving the Eqs. (1)–(3), we get

$$I_1 = 4.632 \text{ A}, \quad I_2 = 1.185 \text{ A}, \quad I_3 = -0.15 \text{ A}$$

Therefore, we see that where I_1 and I_2 are correct, I_3 will flow in the anti-clockwise direction. Hence,

Current through the 60 V source = 4.632 A

Current through the 7 Ω resistor = 4.632 A

Current through the 8 Ω resistor = 4.632 – 1.185 = 3.447 A

Current through the 12 Ω resistor = 1.185 A

Current through the 10 Ω resistor = 1.185 – (–0.15) = 1.335 A

Current through the 5 Ω resistor = 0.15 A (from right to left)

Current through the 6 Ω resistor = 0.15 A (from bottom to top)

Current through the 15 V voltage source = 0.15 A (from bottom to top)

Example 3

Solve Example 2 by using KCL.

Solution

We have seen that a node is the meeting point of three or more electrical branches. In Example 2, we thus have three nodes, as shown in Figure 38.

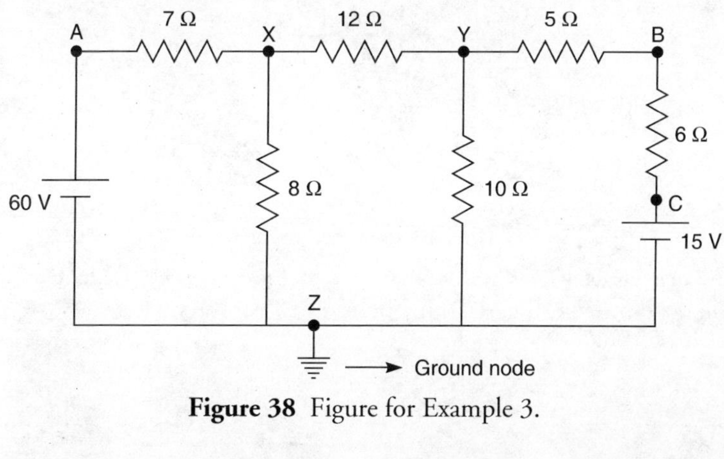

Figure 38 Figure for Example 3.

Figure 38 shows the three nodes: X, Y and Z. X is the meeting point of 7 Ω, 12 Ω and 8 Ω resistors. Hence X is a node. Y is the meeting point of 12 Ω, 5 Ω and 10 Ω. Hence Y is a node. Z is the meeting point of 60 V, 8 Ω, 10 Ω and 15 V. So, Z also qualifies for a node. Points "A" (meeting point of only two electrical branches, 60 V and 7 Ω), "B" (meeting point of 5 Ω and 6 Ω) and "C" (meeting point of only 6 Ω and 15 V) do not qualify for nodes as per the definition.

After recognizing the nodes (X, Y and Z in this case), we designate any one node as the reference, datum, ground or zero-potential node, and write the KCL equations of other nodes with regard to this particular node. Here, node Z is taken as the reference node, as shown by the ground symbol in Figure 38.

Next, we assume voltages at the remaining nodes, say V_X and V_Y. Now, we take one node at a time and write KCL equations, writing KCL at node X, we get

$$\frac{V_X - 60}{7} + \frac{V_X - 0}{8} + \frac{V_X - V_Y}{12} = 0 \tag{1}$$

As already explained, since we do not know the value of V_X, we assume that V_X is big enough to cause all the currents to go away from the node X. Hence, in Eq. (1), the first term in the LHS is the current going from X to "A" to the node Z, the second term is the current going from X to Z (through 8 Ω resistor), and the third term is the current going from X to Y (through the 12 Ω resistor). Rearranging Eq. (1), we get

$$\frac{V_X}{7} + \frac{V_X}{8} + \frac{V_X}{12} - \frac{V_Y}{12} = \frac{60}{7}$$

or

$$0.35V_X - 0.0833V_Y = 8.571$$

or

$$V_X - 0.238V_Y = 24.49 \tag{2}$$

Writing KCL at the node Y, we get

$$\frac{V_Y - V_X}{12} + \frac{V_Y}{10} + \frac{V_Y - 15}{11} = 0$$

or

$$\frac{V_Y}{12} + \frac{V_Y}{10} + \frac{V_Y}{11} - \frac{V_X}{12} = \frac{15}{11}$$

or

$$0.274V_Y - 0.0833V_X = 1.364 \tag{3}$$

or

$$-V_X + 3.29V_Y = 16.37 \tag{4}$$

Solving Eqs. (2) and (4), we get $V_X = 27.69$ V, $V_Y = 13.4$ V. Since V_X is less than 60 V, the current through the 7 Ω resistor will flow from left (i.e. A) to the right (i.e. X) of the value

$$\frac{60 - 27.69}{7} = 4.62 \text{ A}$$

Similarly, the current through 8 Ω resistor is

$$\frac{V_X}{8} = \frac{27.69}{8} = 3.46\text{ A}$$

Since $V_X > V_Y$, so the current through the 12 Ω resistor will flow from X to Y of the value

$$\frac{V_X - V_Y}{12} = \frac{27.69 - 13.4}{12} = 1.19\text{ A}$$

Hence

$$\text{Current through the 10 Ω resistor} = \frac{V_Y}{10} = \frac{13.4}{10} = 1.34\text{ A}$$

And since V_Y is less than 15 V, the currents in the 5 Ω and 6 Ω resistors will be flowing in the anticlockwise directions (from B to Y in 5 Ω and from C to B in the 6 Ω resistors). Hence,

$$\text{Current in 5 Ω and 6 Ω resistors} = \frac{15 - V_Y}{11} = \frac{15 - 13.4}{11} = 0.145\text{ A}$$

Hence the currents can be calculated by solving two equations. This is one less than the number of equations required for KVL case.

NOTE: Slight differences in the current values in Examples 2 and 3 are due to decimal calculation errors.

Example 4

Refer to Figure 39 and compute the voltage required between terminals a and b so that a voltage drop of 45 V occurs across 15 Ω resistor. **(UPTU June 2011)**

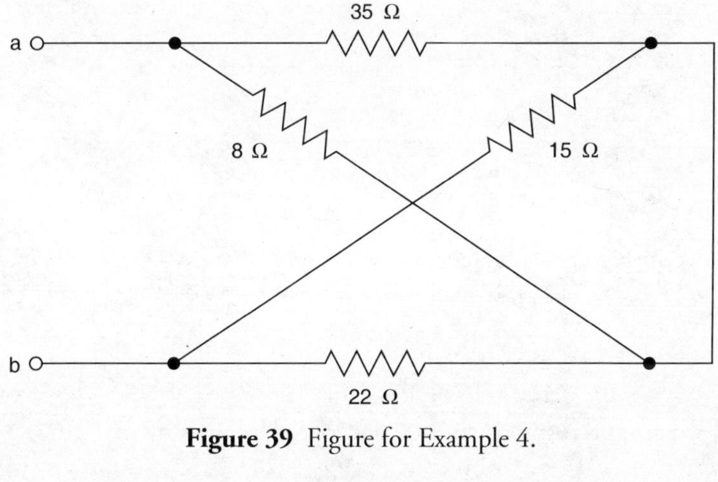

Figure 39 Figure for Example 4.

Solution

The circuit, as shown in Figure 39, can be reduced as shown in Figure 40.

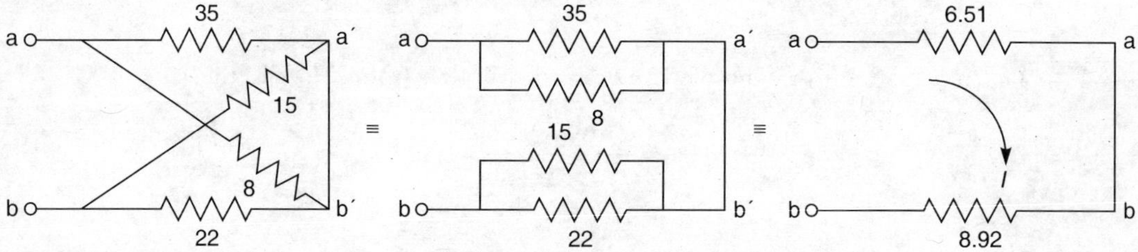

Figure 40 Solution figure for Example 4.

The voltage across 15 Ω resistance is given as 45 V. The same drop would appear across $15||22 = 8.92\ \Omega$ resistor. Hence,

$$I = \frac{45}{8.92} = 5.04\ \text{A}$$

The voltage required between terminals a and b would, therefore, be given by

$$V_{ab} = (6.51 + 8.92) \times 5.04 = 77.76\ \text{V}$$

Example 5

Find the voltage drop across R_1 and R_2 (see Figure 41). The resistance R_3 is not specified.

(UPTU December 2010)

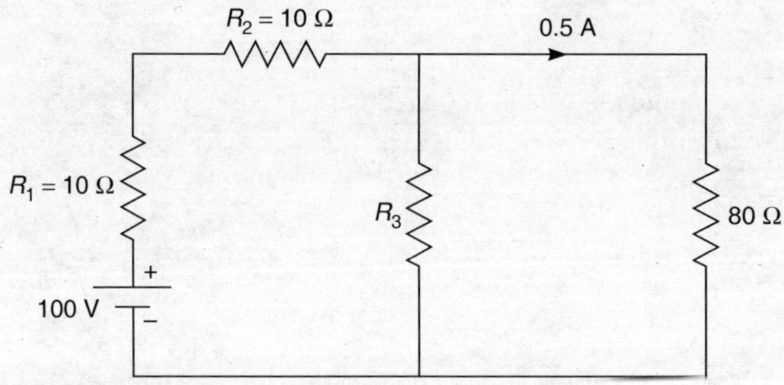

Figure 41 Figure for Example 5.

Solution

The circuit, as shown in Figure 41, be redrawn as shown in Figure 42. Since branch b–e is in parallel with branch c–d, and the drop across branch c–d is $80 \times 0.5 = 40$ V, the voltage of 40 V appears across terminals b and e in the circuit.

Now, in loop a–b–e–f, KVL can be written as

$$100 = 10I + 10I + 40 \quad \text{or} \quad I = 3\ \text{A}$$

Therefore, the desired drops across resistances R_1 and R_2 can be given by

$$V|_{R_1} = IR_1 = 3 \times 10 = 30\ \text{V}, \ V|_{R_2} = IR_2 = 3 \times 10 = 30\ \text{V}$$

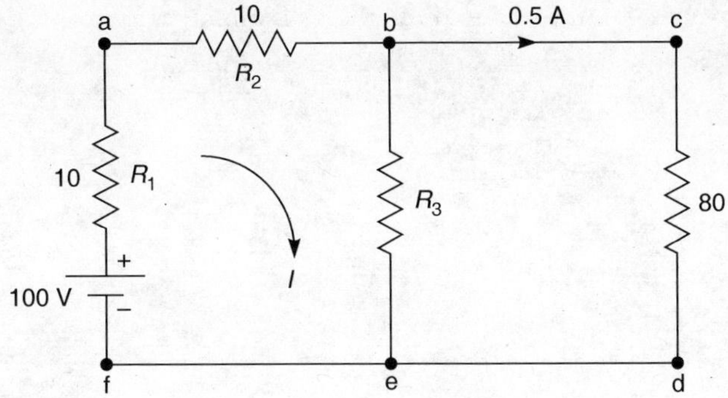

Figure 42 Solution figure for Example 5.

Example 6

In the circuit of Figure 43, find the value of R.

(UPTU June 2009)

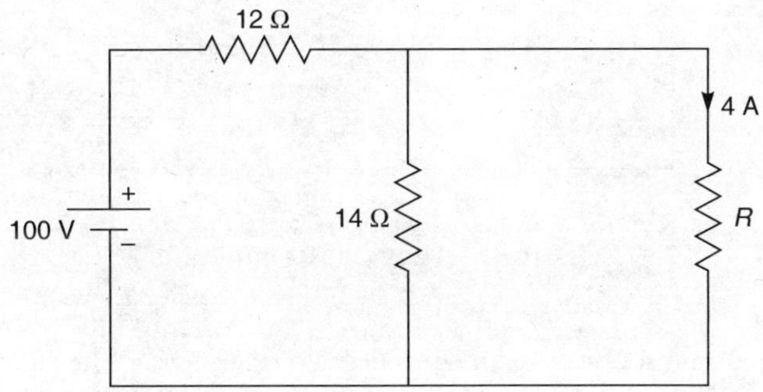

Figure 43 Figure for Example 6.

Solution

The circuit diagram can be redrawn as in Figure 44.

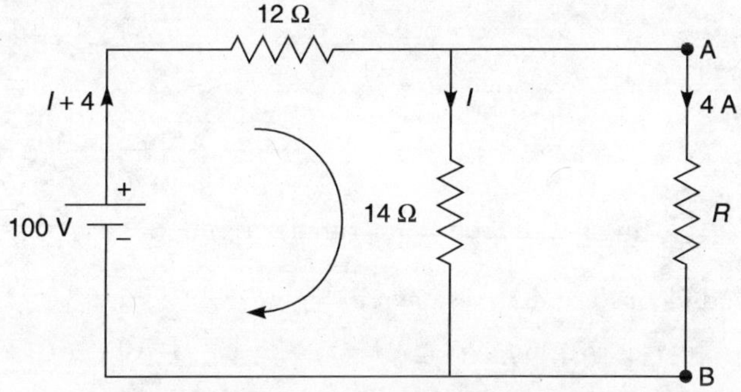

Figure 44 Solution figure for Example 6.

In this diagram, $V_{AB} = 4R = 14I$. In the loop on LHS, KVL can be written as

$$100 = 12\,(I + 4) + 14\,I \quad \text{or} \quad I = \frac{100 - 48}{26} = 2\text{ A}$$

Then $V_{AB} = 14I = 28$ V. Therefore,

$$R = \frac{V_{AB}}{4} = \frac{28}{4} = 7\,\Omega$$

Hence $R = 7\ \Omega$.

Example 7

Find the current in and the voltage across the 2 Ω resistance in Figure 45.

(UPTU June 2008)

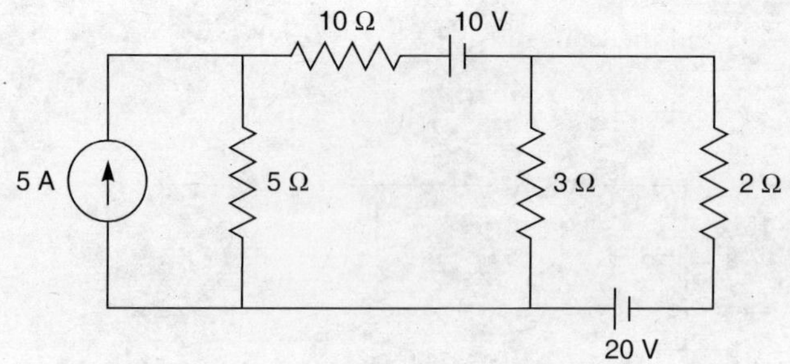

Figure 45 Figure for Example 7.

Solution

Let us convert the 5 A current source into an equivalent voltage source. The circuit becomes as shown in Figure 46.

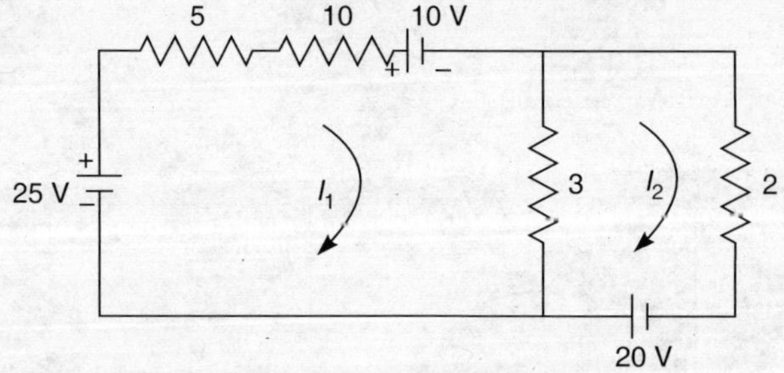

Figure 46 Solution figure for Example 7.

Applying KVL equation in loops 1 and 2, as shown in Figure 46, we get

$$-25 + 15I_1 + 10 + 3(I_1 - I_2) = 0 \text{ and } -20 + 3\,(I_2 - I_1) + 2\,I_2 = 0$$

Solving the two equations above, we get $I_2 = 5$ A. Hence the voltage across 2 Ω resistance is $2 \times 5 = 10$ V.

Example 8

For the circuit shown in Figure 47, the value of R is such that the same amount of power is supplied to the 10 Ω resistance by the current and by the voltage source. Calculate the value of R.

(UPTU December 2008)

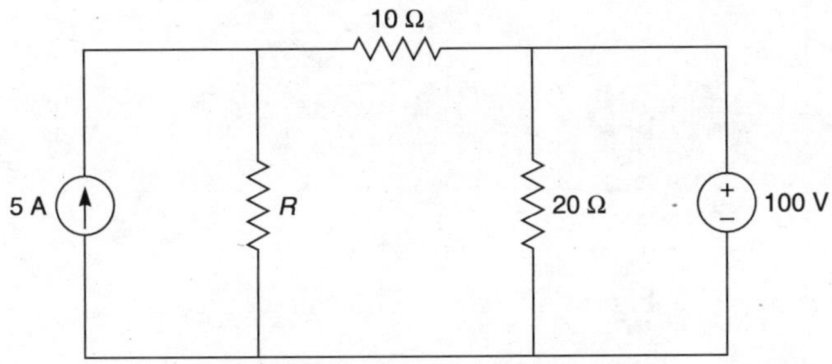

Figure 47 Figure for Example 8.

Solution

It is seen in the given circuit diagram that the powers are supplied by the two sources to the 10 Ω resistance due to currents in opposite directions due to them (acting individually). Hence, in order to receive equal powers by the two sources, it is implied that the currents due to them should be equal. Since they are opposing each other, the net current would be zero through 10 Ω resistance. This is possible only if voltage across R is also 100 V (as it is across 20 Ω resistance). Hence $R = 20$ Ω.

Example 9

Use the node voltage method to calculate the branch currents in the network shown in Figure 48.

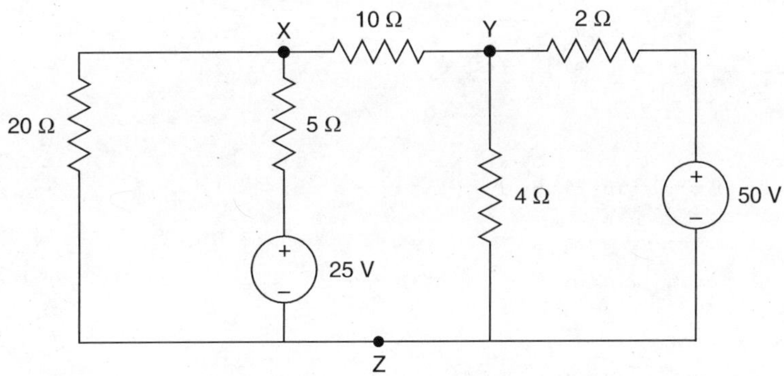

Figure 48 Figure for Example 9.

Solution

We observe that the network contains three nodes: X, Y and Z. Nodes of Figure 48 are shown in Figure 49. Assuming Z to be the reference node, we write the KCL equations at nodes X and Y, taking the voltages as V_X and V_Y. KCL at X gives

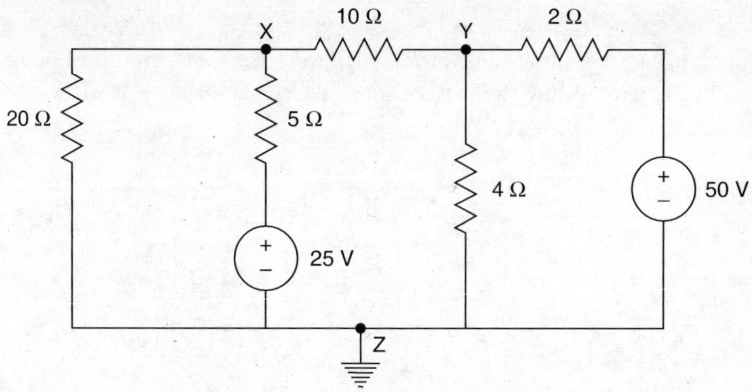

Figure 49 Circuit showing nodes.

$$\frac{V_X}{20}+\frac{V_X-25}{5}+\frac{V_X-V_Y}{10}=0$$

or

$$7V_X-2V_Y=100 \tag{1}$$

KCL at Y gives

$$\frac{V_Y-V_X}{10}+\frac{V_Y}{4}+\frac{V_Y-50}{2}=0$$

or

$$-2V_X+17V_Y=500 \tag{2}$$

Solving Eqs. (1) and (2), we get

$$V_X=\frac{3780}{161}\text{ V},\ V_Y=\frac{740}{23}\text{V}$$

Therefore,

$$V_X=23.478\text{ V},\ V_Y=32.174\text{ V}$$

Hence

$$\text{Branch current in the 20 }\Omega\text{ branch}=\frac{V_X}{20}=1.174\text{ A}$$

$$\text{Branch current in the 5 }\Omega\text{ branch}=\frac{V_X-25}{5}=-0.304\text{ A}$$

$$\text{Branch current in the 10 }\Omega\text{ branch}=\frac{V_X-V_Y}{10}=-0.87\text{ A}$$

$$\text{Branch current in the 4 }\Omega\text{ branch}=\frac{V_Y}{4}=8.04\text{ A}$$

$$\text{Branch current in the 2 }\Omega\text{ branch}=\frac{V_Y-50}{2}=-8.91\text{ A}$$

Thus we see that the currents in 5 Ω, 10 Ω and 2 Ω resistors flow opposite to their assumed directions, that is

(a) Current in the 5 Ω resistor flows upwards.
(b) Current in the 10 Ω resistor flows from the node Y to X.
(c) Current in the 2 Ω resistor flows towards the node Y from the 50 V source.

Example 10

Find the resistance R_{AB} in the circuit shown in Figure 50 by using the star–delta transformation.

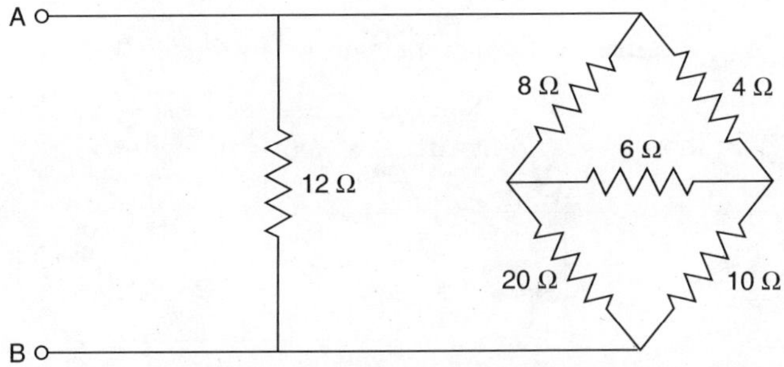

Figure 50 Circuit for Example 10.

Solution

We have various points in the question: we have two deltas, that is, DCE and DEF, and we can convert any one of them to star and move ahead (see Figure 51).

Converting the delta CDE into its equivalent star (as shown by the dotted lines with neutral point N), we get

$$R_{CN} = \frac{8\times 4}{8+4+6} = 1.78\ \Omega,\qquad R_{DN} = \frac{8\times 6}{8+6+4} = 2.67\ \Omega,\qquad R_{EN} = \frac{6\times 4}{8+4+6} = 1.33\ \Omega$$

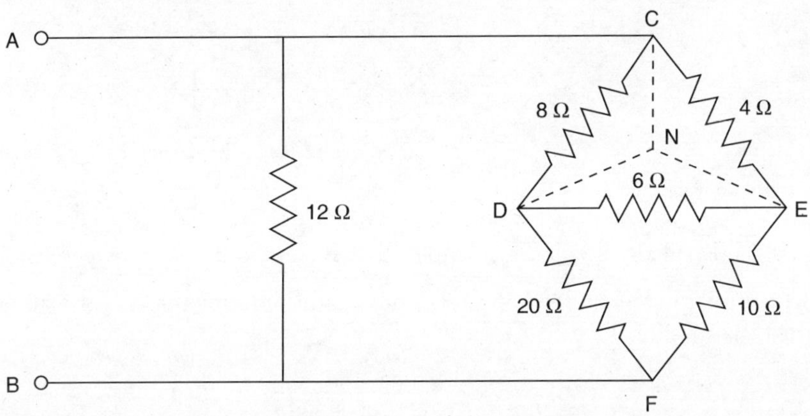

Figure 51 Solution circuit for Example 10.

The resultant circuit is shown in Figure 52.

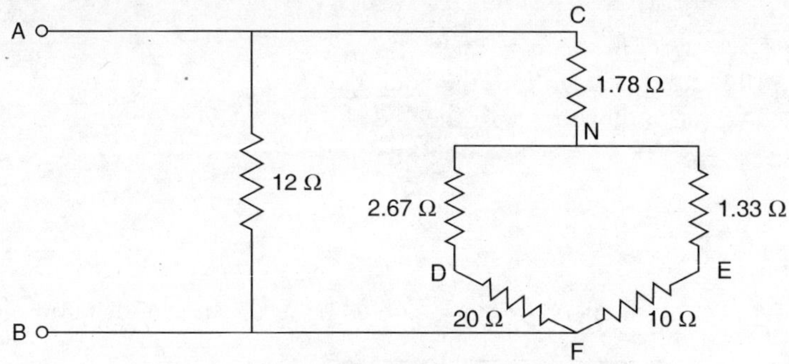

Figure 52 Resultant circuit for Example 10.

As can be seen, the 2.67 Ω resistor is now in series with the 20 Ω resistor. Similarly, the 1.33 Ω resistor is in series with the 10 Ω resistor. So, we get the circuit, as shown in Figure 53.

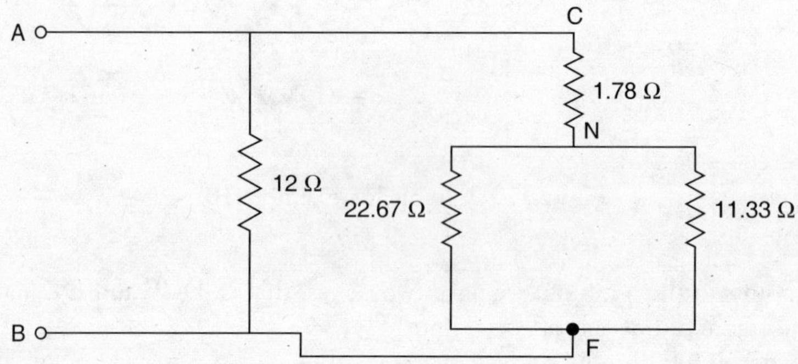

Figure 53 Resultant circuit for Example 10 after combining the series resistors.

Now, 22.67 Ω and 11.33 Ω resistors are in parallel and the resultant resistor is 7.55 Ω, as shown in Figure 54.

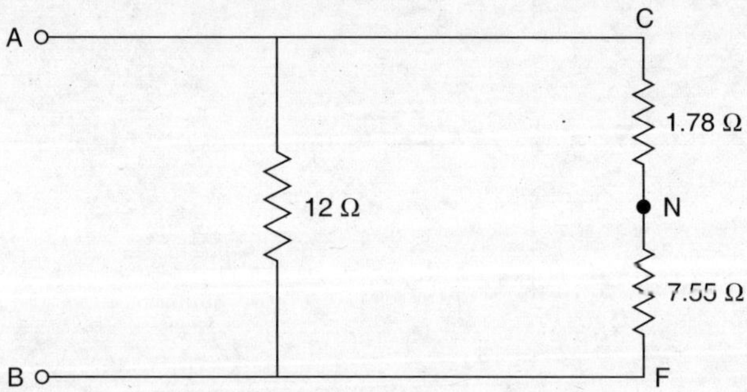

Figure 54 Resultant circuit for Example 10 after combining the parallel resistors.

Also the 1.78 Ω resistor is in series with 7.55 Ω resistor, and the resulting resistors of $1.78 + 7.55 = 9.33\ \Omega$ are parallel with 12 Ω resistor. We get

$$R_{AB} = \frac{12 \times 9.33}{12 + 9.33} = 5.25\ \Omega$$

Example 11

Using Δ-Y transformation, find the resistance between the terminals a and b, as shown in Figure 55, and the total power drawn from the supply in the circuit. **(UPTU December 2006)**

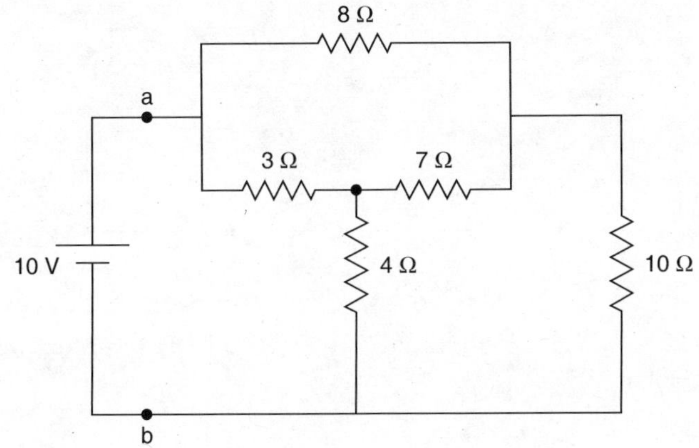

Figure 55 Figure for Example 11.

Solution

We name various nodes in the circuit, as shown in Figure 56. Now we convert the delta formed by the three resistors ($R_{AC} = 8\ \Omega$, $R_{CB} = 7\ \Omega$ and $R_{BA} = 3\ \Omega$) into an equivalent star, as shown by the dotted lines (with neutral as N) in Figure 56. Now, we have

$$R_{AN} = \frac{8\times 3}{8+3+7} = \frac{24}{18} = 1.33\,\Omega, \qquad R_{BN} = \frac{3\times 7}{3+7+8} = \frac{21}{18} = 1.17\,\Omega, \qquad R_{CN} = \frac{8\times 7}{8+7+3} = \frac{56}{18} = 3.11\,\Omega$$

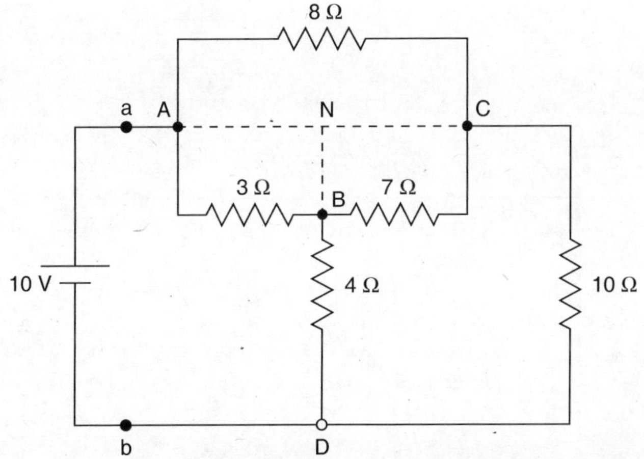

Figure 56 Naming nodes in Figure 55.

Thus, we get the circuit as shown in Figure 57.

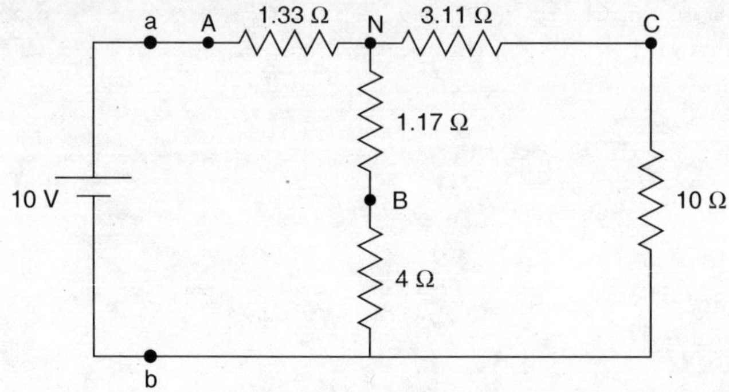

Figure 57 Resultant circuit.

Clearly 1.17 Ω and 4 Ω resistors are in series. Also, 3.11 Ω and 10 Ω resistors are in series. So, we have the circuit as shown in Figure 58.

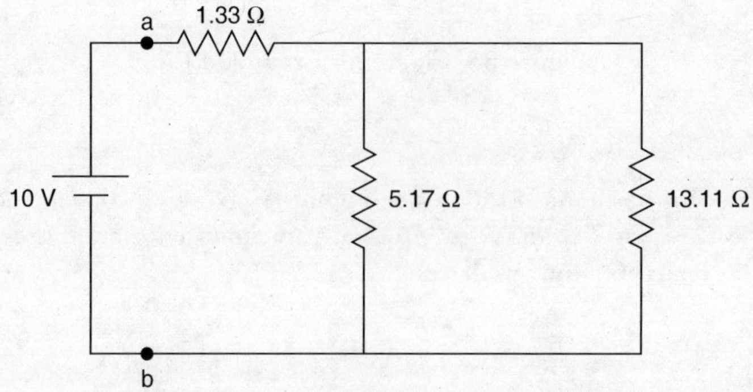

Figure 58 Combining the series resistors.

Now 5.17 Ω and 13.11 Ω resistors are in parallel (see Figure 59). So,

$$\frac{5.17 \times 13.11}{5.17 + 13.11} = 3.71\ \Omega$$

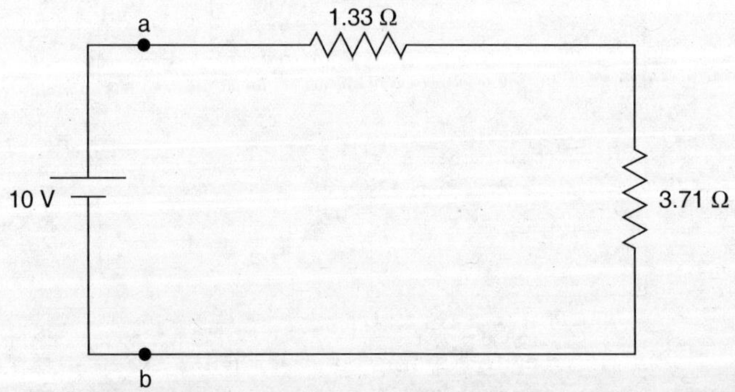

Figure 59 Combining the parallel resistors.

Now, 1.33 Ω and 3.71 Ω are in series (see Figure 60). So, $R_{ab} = 1.33 + 3.71 = 5.04\ \Omega$.

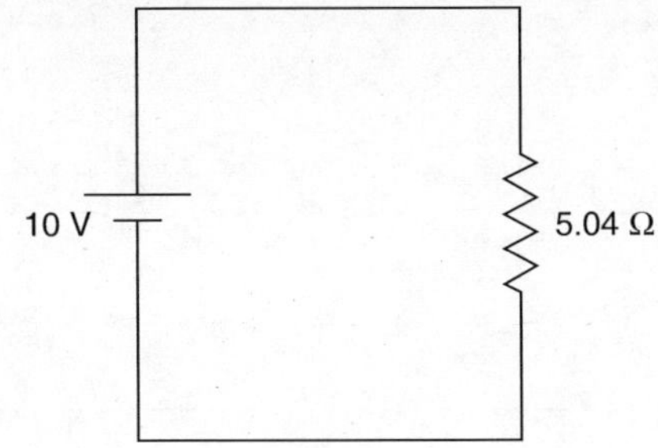

Figure 60 Combining the series resistors.

Power drawn by the 5.04 Ω resistor will be

$$\frac{10^2}{5.04} = 19.84\ \text{W}$$

Example 12

In the given circuit shown in Figure 61, find the resistance between the points B and C.

(UPTU June 2010)

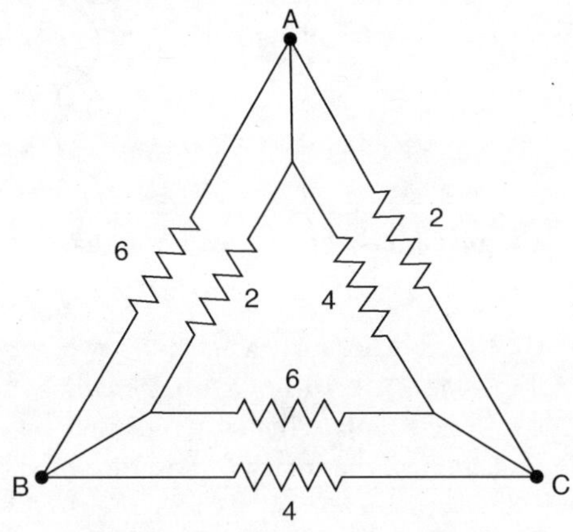

Figure 61 Figure for Example 12.

Solution

In the given circuit, the resistances are reduced as (see Figure 62)

$$R_{AB} = \frac{6\times 2}{6+2} = \frac{12}{8} = 1.5\ \Omega, \qquad R_{AC} = \frac{4\times 2}{4+2} = \frac{8}{6} = 1.33\ \Omega, \qquad R_{BC} = \frac{6\times 4}{6+4} = \frac{24}{10} = 2.4\ \Omega$$

The resistance between terminals B and C will, therefore, be

$$2.4 \,||\, (1.5 + 1.33) = 1.298\ \Omega$$

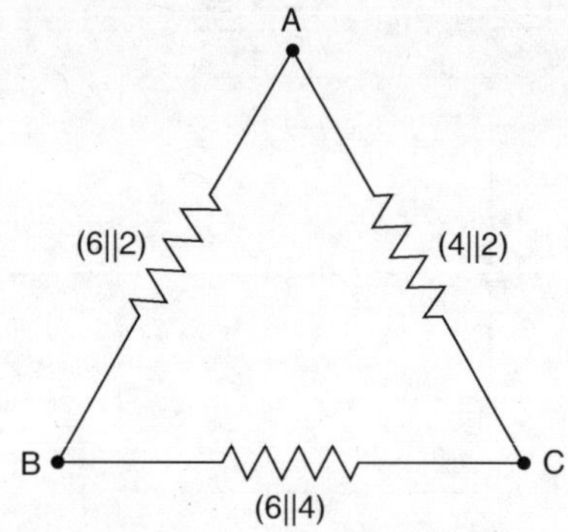

Figure 62 Solution figure for Example 12.

Therefore, the desired value of resistance between terminals B and C would be 1.298 Ω.

Example 13

Find the current through the 6 Ω resistor in the circuit, as shown in Figure 63, by using the superposition theorem.

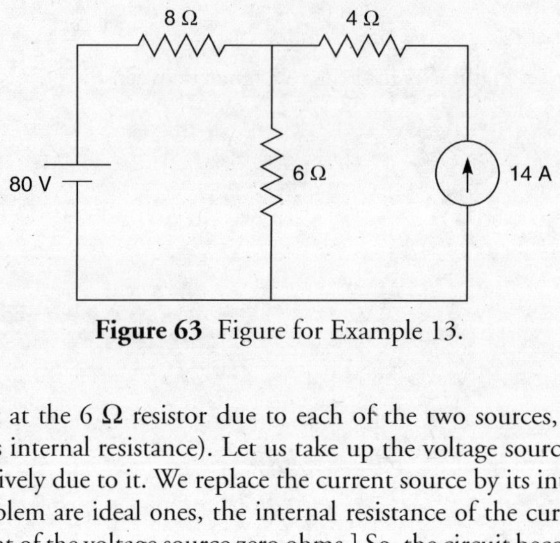

Figure 63 Figure for Example 13.

Solution

We will find the current at the 6 Ω resistor due to each of the two sources, taking one at a time (and replacing the other by its internal resistance). Let us take up the voltage source first and find the current in the 6 Ω resistor exclusively due to it. We replace the current source by its internal resistance. [Since the sources given in the problem are ideal ones, the internal resistance of the current source will be infinity ohms (in parallel) and that of the voltage source zero ohms.] So, the circuit becomes as shown in Figure 64.

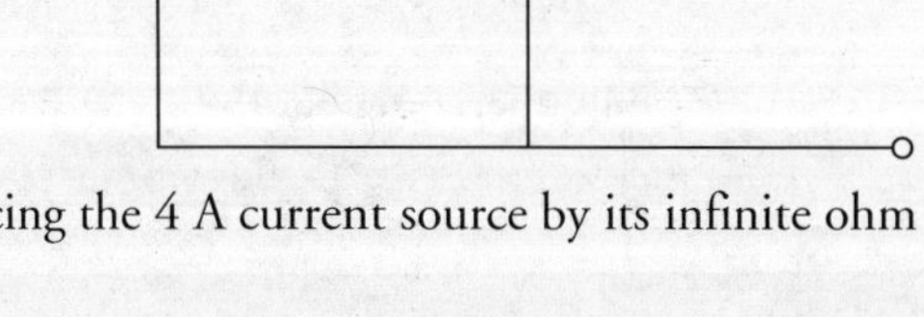

Figure 64 Replacing the 4 A current source by its infinite ohm internal resistance.

Since all the current delivered by the 80 V source pass through 8 Ω resistor, and no part of it can go to the 4 Ω resistor (since it is open-circuited later), the whole current goes through the 6 Ω resistor. Thus 8 Ω and 6 Ω resistors carrying the same currents are said to be in series. The current through the 6 Ω resistor will thus be

$$\frac{80}{8+6} = \frac{80}{14} = 5.714\ \text{A} \quad \text{(from top to bottom)}$$

Now, we find the current in the 6 Ω resistor due to the current source only. For this, we first replace the ideal voltage source of 80 V by its internal resistance of zero ohms (in series). The resultant circuit is shown in Figure 65.

Here, the 14 A current will flow through the 4 Ω resistor, and later get divided into two parallel paths of resistors 8 Ω and 6 Ω. The current through the 6 Ω resistors will be

$$\frac{14 \times 8}{6+8} = 8\ \text{A} \quad \text{(from top to bottom)}$$

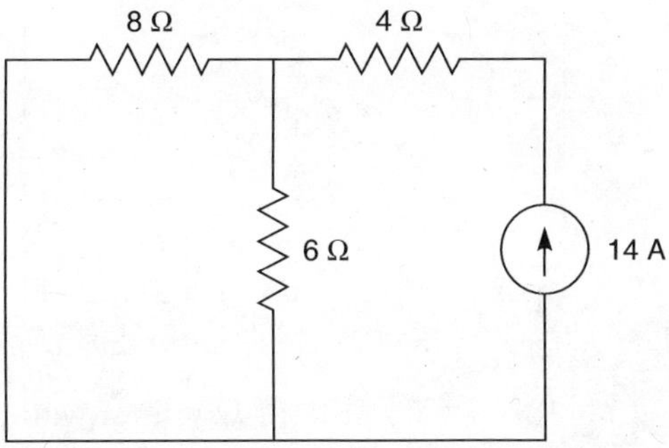

Figure 65 Replacing the 80 V voltage source by its zero ohm internal resistance.

The current in the 6 Ω resistors due to both voltage and current sources acting together (see Figure 66) can now be found by making an algebraic sum of the individual currents. Here, because of both the sources, the current in the 6 Ω resistor will be 5.714 + 8 = 13.714 A (from top to bottom).

Example 14

Find the voltage across the 12 Ω resistor in the circuit, as shown in Figure 66, by using the superposition theorem.

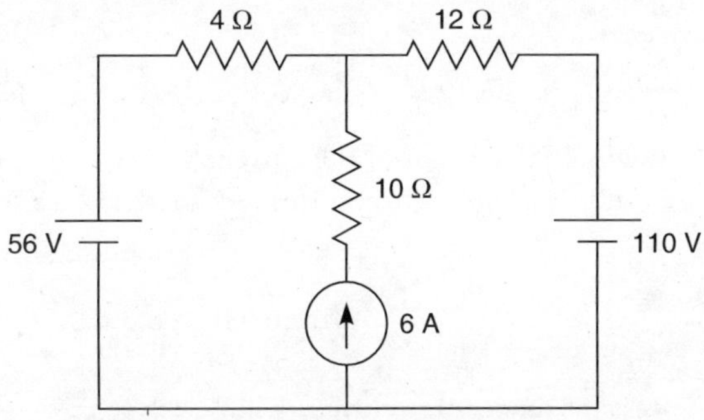

Figure 66 Figure for Example 14.

Solution

The circuit has three sources. Hence, we will take one source at a time and replace the other two sources by their internal resistances. Let us take the 56 V source first. Replacing the 6 A current source by an infinite-ohm resistance (i.e. an open circuit) and the 110 V source by a zero ohm resistance (i.e. a short circuit), we get the resultant circuit as shown in Figure 67.

No current will go to the 10 Ω resistor (due to the presence of open circuit ahead), so all the current flowing through the 4 Ω resistor will go to the 12 Ω resistor too. The current will be

$$\frac{56}{4+12} = \frac{56}{16} = 3.5\,\text{A}$$

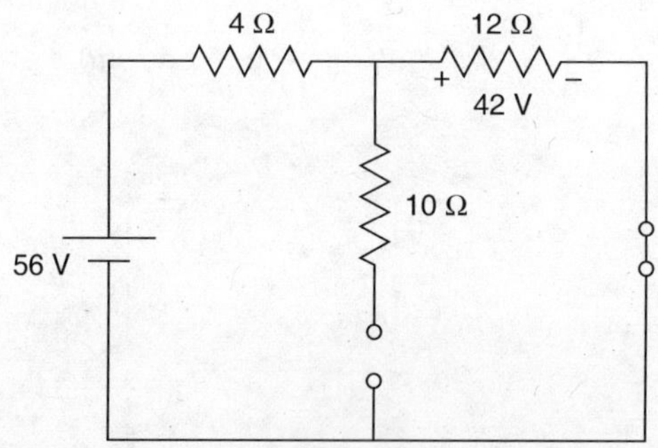

Figure 67 Finding voltage (in the 12 Ω resistance) due to only 56 V source.

This current causes a voltage 3.5 × 12 = 42 V in the 12 Ω resistor with left-end positive and right-end negative, as shown in Figure 67. Next, we take up the 6 A current source. Replacing the two voltage sources by short circuits, we get the circuit as shown in Figure 68.

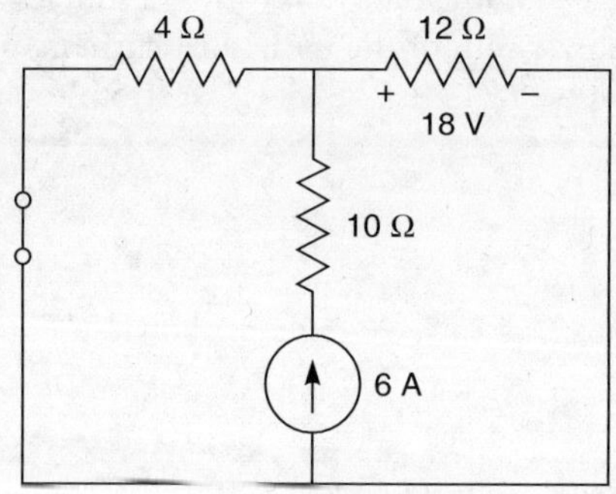

Figure 68 Finding voltage (in the 12 Ω resistance) due to only the 6 A current source.

The 10 Ω resistor will carry the full 6 A current of the current source. Later, the current will get divided in two parallel resistors of 4 Ω and 12 Ω. The current through the 12 Ω resistor (by applying the current divide theorem) will be

$$\frac{6\times 4}{4+12} = \frac{24}{16} = 1.5\,\text{A (from left to right)}$$

The voltage across the 12 Ω resistor due to this current will be 1.5 × 12 = 18 V. It has left-end positive and right-end negative, as shown in Figure 68.

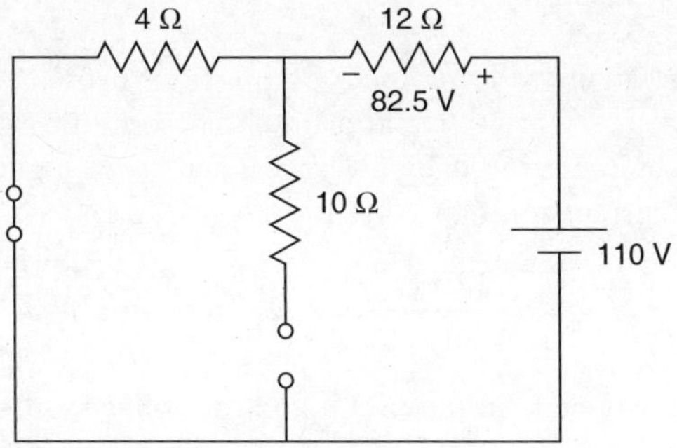

Figure 69 Finding voltage (in the 12 Ω resistance) due to only the 110 V voltage source.

Finally, we take up the 110 V voltage source. Replacing the 56 V voltage source by a short circuit and 6 A current source by an open circuit, we get the resultant circuit, as shown in Figure 69.

Again no current will go to the 10 Ω resistor, so 4 Ω and 12 Ω will be in series, and the current through them will be given by

$$\frac{110}{12+4} = \frac{110}{16} = 6.875 \text{ A (from right to left)}$$

Correspondingly, this time, the voltage across the 12 Ω resistor will be = 6.875 × 12 = 82.5 V with the right-end positive and the left-end negative, as shown in Figure 69.

Thus, the algebraic sum of the three voltages will be −42 − 18 + 82.5 = 22.5 V with the right end of the 12 Ω resistor at a positive (or higher-potential) polarity and the left end at a negative (or lower-potential) polarity. Thus currents and voltages can be found with the help of the superposition theorem.

Example 15

By means of superposition theorem, find the current that flows through R_2 in the circuit of Figure 70.

(UPTU December 2010)

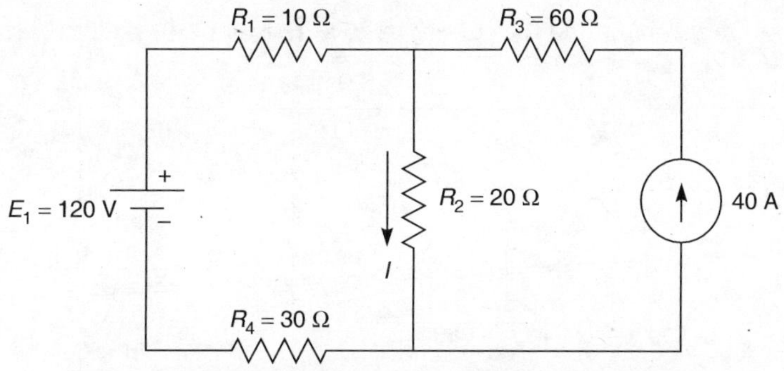

Figure 70 Figure for Example 15.

Solution

In order to find I in the given circuit by using the superposition theorem, let us suppress the sources one by one, leaving behind only one source active at a time. Say, only the voltage source is active and the current source is suppressed by replacing it by an open circuit (see Figure 71). Since no current flows through 60 Ω resistance, the current through 20 Ω resistance due to voltage source would be

$$I_1 = \frac{120}{10+20+30} = 2\ \Omega \text{ (from a to b)}$$

Now let only the current source operate in the circuit and the voltage source be suppressed by replacing it by a short circuit [see Figure 71(b)]. By current division rule, we get

$$I_2 = \frac{10+30}{(10+30)+20} 40 = \frac{40}{60} \times 40 = 26.67 \text{ A (from a to b)}$$

Hence, by applying the superposition theorem, we can say that when both the sources are present in the circuit, the current through R_2 would be

$$I = I_1 + I_2 = 2 + 26.67 = 28.67 \text{ A (from a to b)}$$

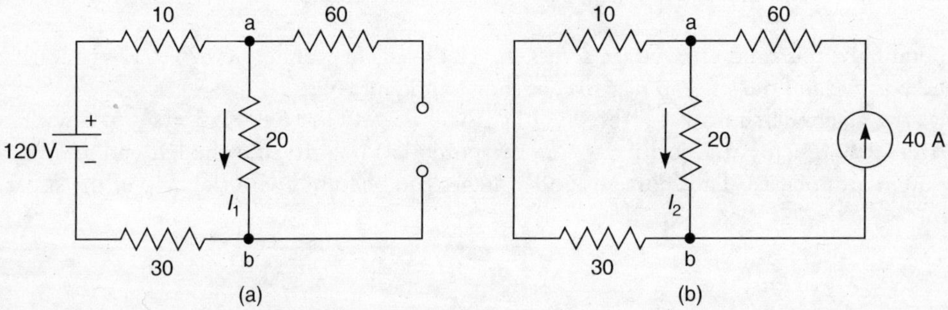

Figure 71 Solution figure for Example 15.

Example 16

Using the superposition theorem, find the current in 20 Ω resistor of the circuit, as shown in Figure 72.

(UPTU June 2010)

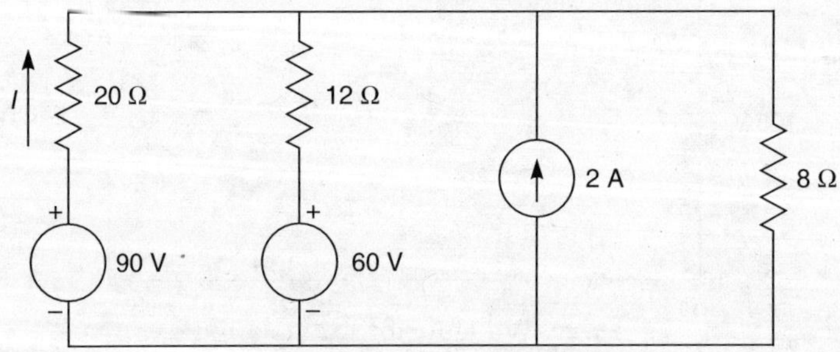

Figure 72 Figure for Example 16.

Solution

For finding I through 20 Ω resistance by the superposition theorem, let us apply one source at a time.

1. Applying 90 V source only (see Figure 73), the current is obtained as

$$I_1 = \frac{V}{20+(12\,||\,8)} = \frac{90}{20+4.8}$$

Therefore, $I_1 = 3.62$ A (from a to b).

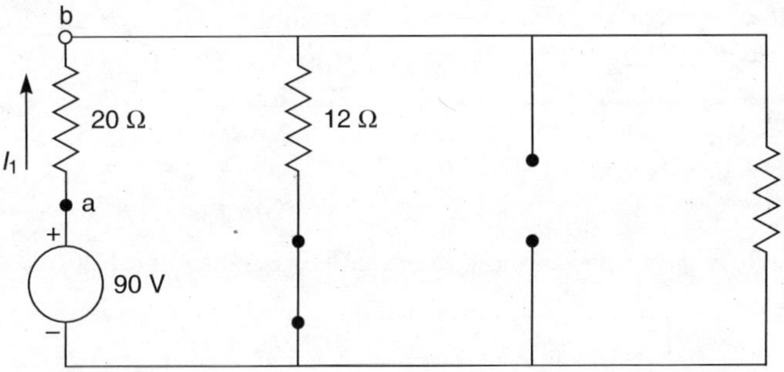

Figure 73 Solution figure (when 90 V source is applied).

2. Applying 60 V source only (Figure 74): The current supplied by the source can be found as

$$I = \frac{60}{12+(20\,||\,8)} = \frac{60}{12+5.71} = 3.39 \text{ A}$$

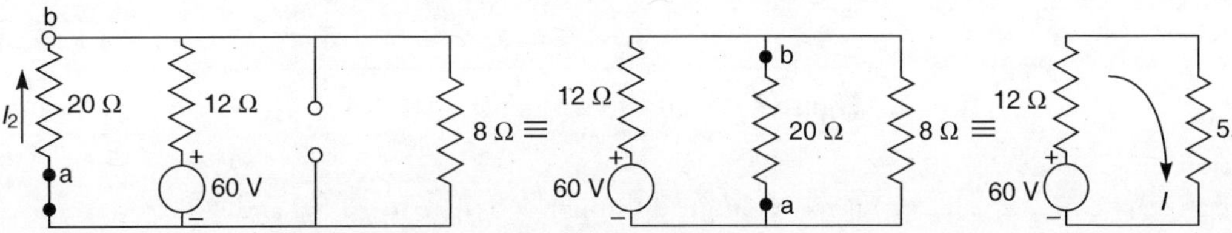

Figure 74 Solution figure (when 60 V source is applied).

Now by applying the current divide theorem, we get

$$I_2 = \frac{8\times 3.39}{20+8} = 0.968 \text{ A (from b to a)}$$

Thus $I_2 = -\,0.968$ A (from a to b).

3. Applying 2 A source only: From the circuit diagram, as in Figure 75, it is clear that the current divide theorem gives

$$I_3 = -\frac{2\times 4.8}{20+4.8} = -0.387 \text{ A (from a to b)}$$

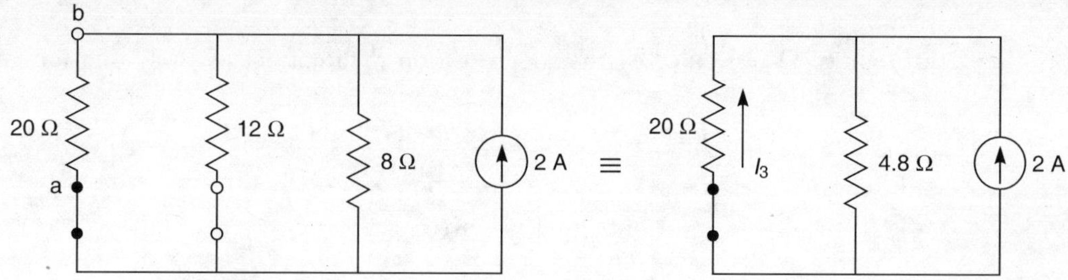

Figure 75 Solution figure (when 2 A source is applied).

Thus the net current flowing through 20 Ω resistance with all the sources in the circuit is given by

$$I = I_1 + I_2 + I_3 \quad \text{or} \quad I = 3.62 - 0.968 - 0.387 = 2.265 \text{ A}$$

Example 17

Find the current through R_L when it takes on values equal to 10, 50 and 200 Ω by using Thevenin's theorem (see Figure 76). **(UPTU June 2011)**

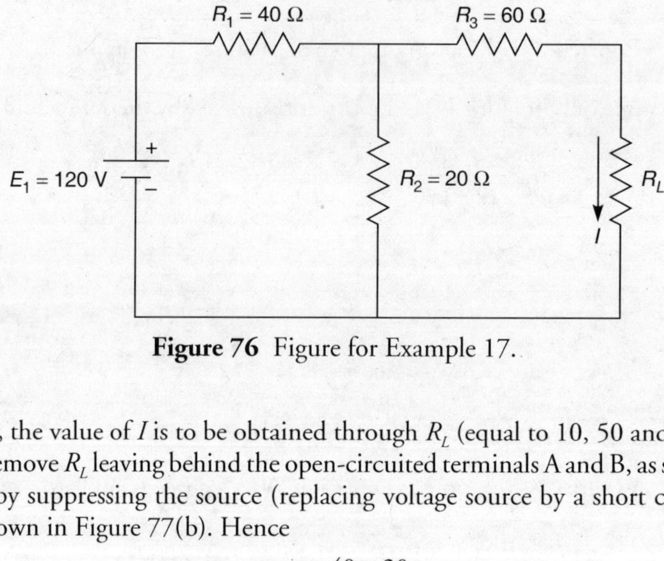

Figure 76 Figure for Example 17.

Solution

In the given circuit, the value of I is to be obtained through R_L (equal to 10, 50 and 200 Ω, respectively). The first step is to remove R_L leaving behind the open-circuited terminals A and B, as shown in Figure 77(a). Next, we find R_{Th} by suppressing the source (replacing voltage source by a short circuit). So, the circuit now becomes as shown in Figure 77(b). Hence

$$R_{\text{Th}} = (40 \| 20) + 60 = \frac{40 \times 20}{40 + 20} + 60 = 73.33 \ \Omega$$

For finding out V_{Th}, from Figure 77(a), we get

$$V_{\text{Th}} = V_{AB} = \frac{120}{(40 + 20)} \times 20 = 40 \text{ V}$$

Thevenin's equivalent circuit, thus, can be drawn as shown in Figure 77(c):

$$I\Big|_{R_L=10} = \frac{40}{73.33 + 10} = 0.480 \text{ A}, \quad I\Big|_{R_L=50} = \frac{40}{73.33 + 50} = 0.324 \text{ A}, \quad I\Big|_{R_L=200} = \frac{40}{73.33 + 200} = 0.146 \text{ A}$$

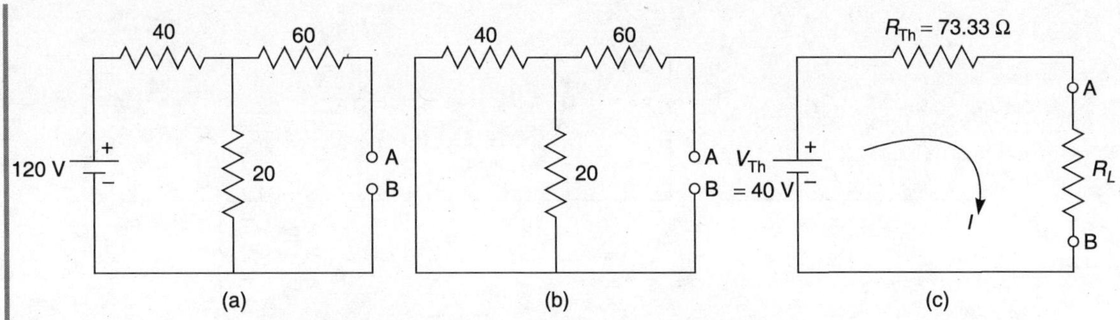

Figure 77 Solution figures (for Example 17).

Example 18

Determine the current in 4 Ω resistance by using Thevenin's theorem in the circuit shown in Figure 78. **(UPTU December 2009)**

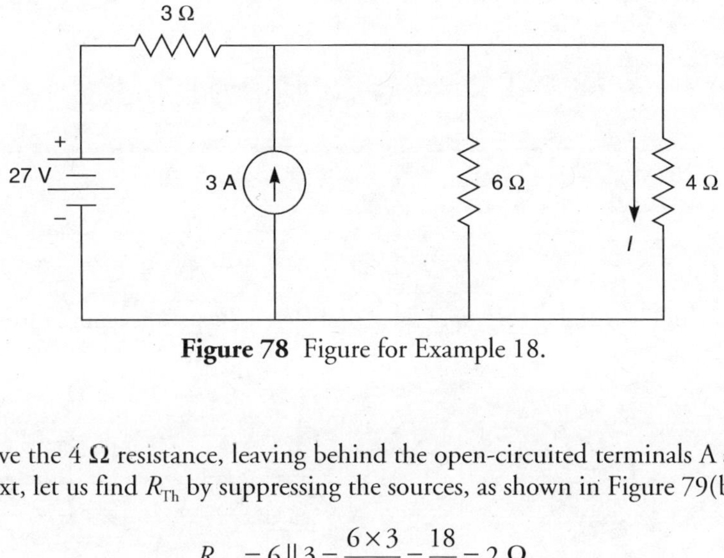

Figure 78 Figure for Example 18.

Solution

Let us first remove the 4 Ω resistance, leaving behind the open-circuited terminals A and B, as shown in Figure 79(a). Next, let us find R_{Th} by suppressing the sources, as shown in Figure 79(b). Then

$$R_{Th} = 6 \| 3 = \frac{6 \times 3}{6+3} = \frac{18}{9} = 2\ \Omega$$

The value of V_{Th} can be found by using superposition theorem:

$$I_1 = \text{Current in the } 6\ \Omega \text{ due to only the voltage source} = \frac{27}{9} = 3\ \text{A} \downarrow \qquad \text{[see Figure 79(c)]}$$

$$I_2 = \text{Current in the } 6\ \Omega \text{ due to only the current source} = \frac{3}{3+6} \times 3 = 1\ \text{A} \downarrow \qquad \text{[see Figure 79(d)]}$$

Thus the current through 6 Ω resistance is $I_1 + I_2 = 3 + 1 = 4$ A. Therefore, $V_{Th} = 6 \times 4 = 24$ V.

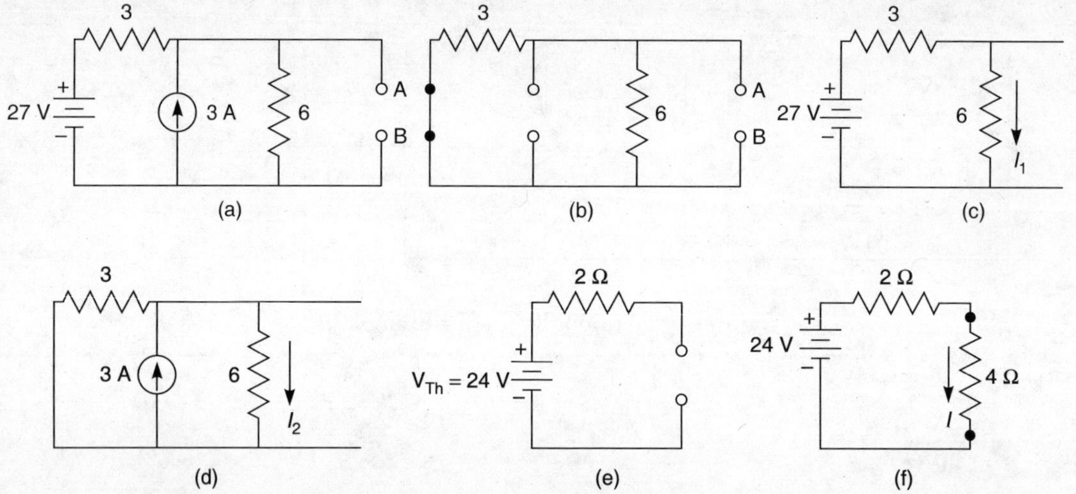

Figure 79 Solution figure for Example 18.

Thevenin's equivalent circuit can be drawn as shown in Figure 79(e). The current through 4 Ω resistance, therefore, will be

$$I = \frac{24}{2+4} = 4 \text{ A} \quad \text{[see Figure 79(f)]}$$

Example 19

Replace the network of Figure 80 to the left of the terminals a–b by its Thevenin equivalent circuit. Hence determine *I*.

(UPTU June 2008)

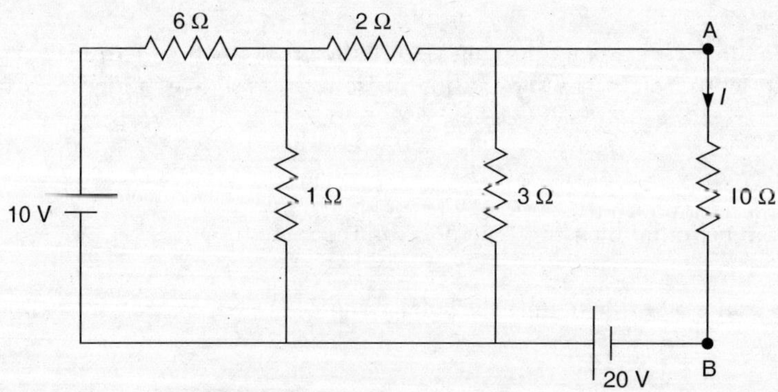

Figure 80 Figure for Example 19.

Solution

First of all, let us remove the 10 Ω resistance and find R_{Th} by suppressing the sources [see Figure 81(a)]. Then

$$R_{\text{Th}} = [(6 \parallel 1) + 2] \parallel 3 = \frac{60}{41}\ \Omega\ = 1.46\ \Omega$$

In order to find the open-circuit voltage between terminals A and B due to the two sources in the circuit, the circuit is redrawn, as shown in Figure 81(b). The current through 3 Ω resistance due to 10 V source acting alone can be found to be equal to 10/41 A. The source of 20 V will not contribute any current, as the circuit is open at A, B, but while finding V_{Th} (i.e., while moving from B toward A), it will contribute a gain of +20 V. Hence

$$V_{\text{Th}} = 3 \times I + 20 = 3 \times \frac{10}{41} + 20 = 20.73\ \text{V}$$

Thevenin's equivalent circuit is drawn as shown in Figure 81(c). The desired current I is given by

$$I_{\text{L}} = \frac{20.73}{1.46 + 10} = 1.81\,\text{A}$$

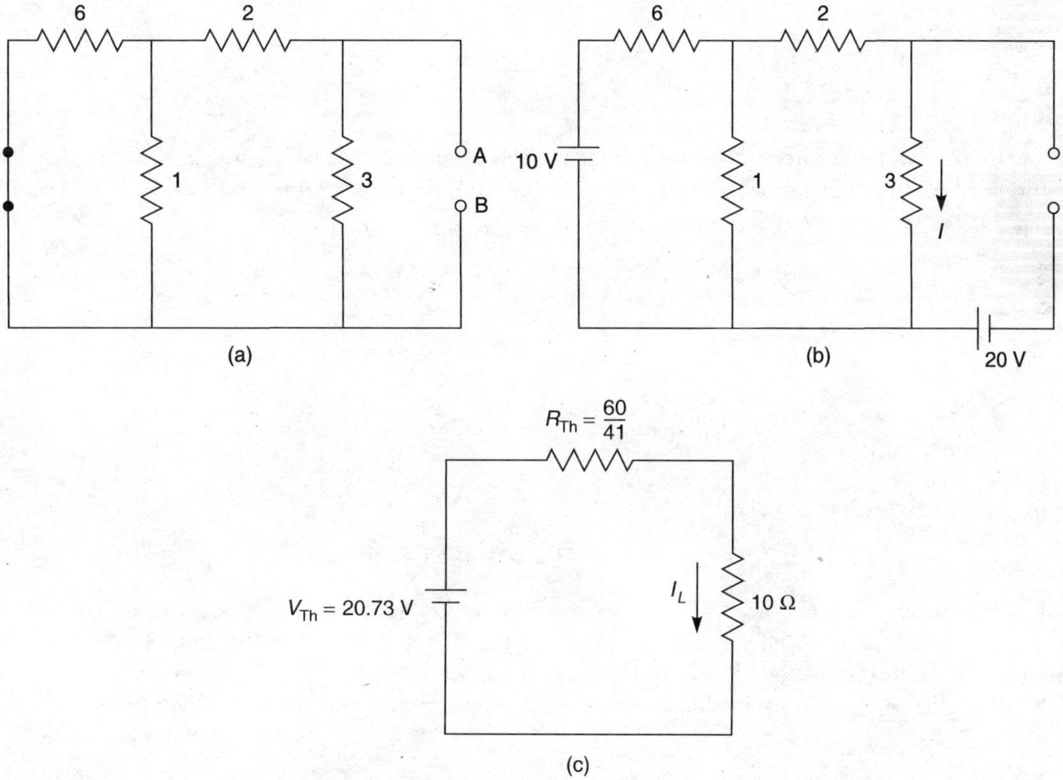

Figure 81 Solution figure for Example 19.

Example 20

The resistances of the various arms of a Wheatstone bridge are shown in Figure 82. The battery has an emf of 2 V and a negligible internal resistance. Using Thevenin's theorem, determine the value and direction of the current in the galvanometer circuit BD. **(UPTU 2001–02)**

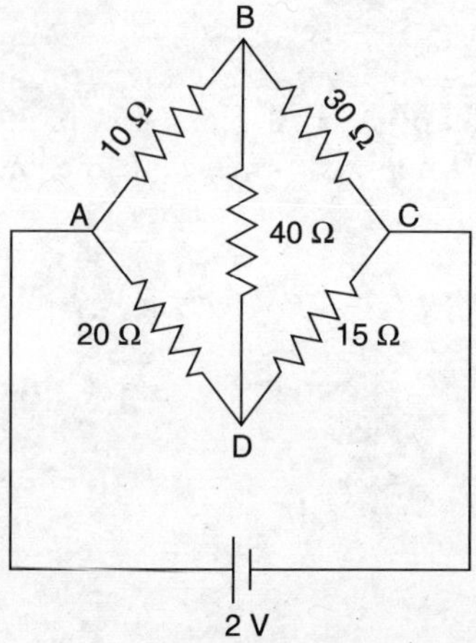

Figure 82 Figure for Example 20.

Solution

To obtain the value and direction of the current in galvanometer circuit BD. Using Thevenin's theorem, first we remove 40 Ω resistor, which is connected across BD [see Figure 83(a)]. Now calculate Thevenin's voltage, V_{Th}. The current I_1 will be

$$I_1 = \frac{2}{40} = 0.05\text{ A}$$

Similarly,

$$I_2 = \frac{2}{35} = 0.057\text{ A}$$

Voltage across BD will be equal to

$$I_2 \times 20 - I_1 \times 10 = 0.057 \times 20 - 0.05 \times 10 = 0.64\text{ V}$$

Starting from D, we move towards B. So in DA, we have voltage gain and in AB we have the voltage drop. Thus, $V_B - V_D = 0.64$ V, and $V_B - V_D$ is equal to the terminal voltage across BD. Hence it should be Thevenin's voltage. This implies $V_{Th} = 0.64$ V.

Now calculating for R_{Th}, the voltage source acts as short circuit and the simplified circuit is shown in Figure 83(b). This gives

$$R_{Th} = (10 \,||\, 30) + (20 \,||\, 15) = \frac{300}{40} + \frac{300}{35} = 16.07\,\Omega$$

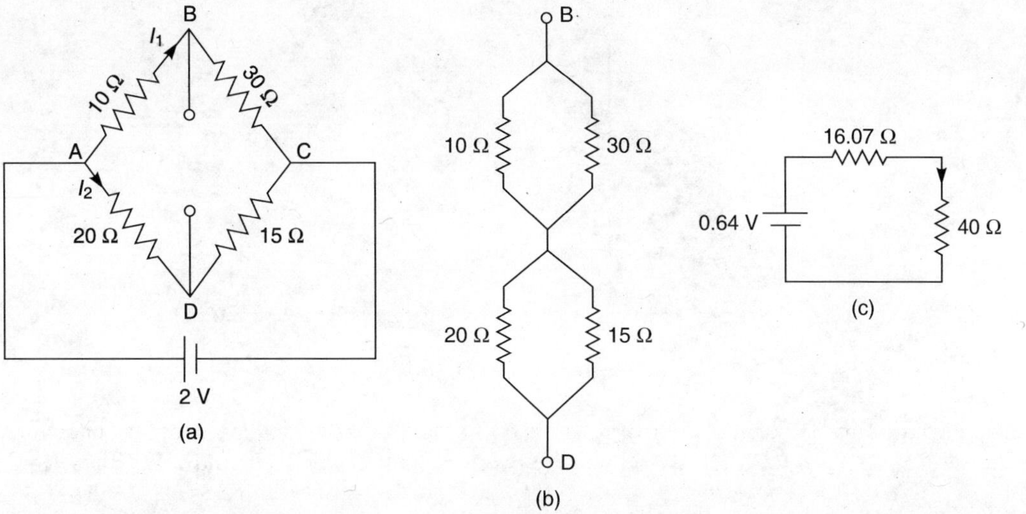

Figure 83 (a) Circuit after removal of resistor; (b) simplified circuit; (c) Thevenin's equivalent circuit.

Thevenin's equivalent circuit is shown in Figure 83(c). The current in 40 Ω resistance is

$$I = \frac{0.64}{40 + 16.07} = 0.0114\text{ A}$$

Example 21

Find the value of R_L that receives maximum power from the source of 100 V, as shown in Figure 84. Also find the value of the power.

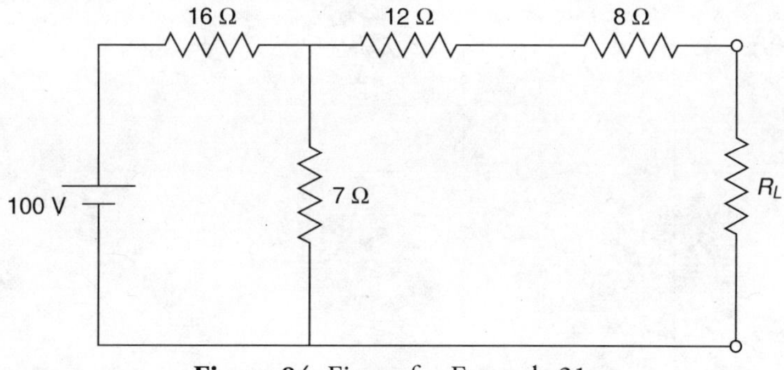

Figure 84 Figure for Example 21.

Solution

As already discussed, the value of R_L, which will receive maximum (real) power, is that which is equal to R_S. R_S, for such multi-loop questions, is not clearly visible. So, it is calculated by finding the value of R_{Th} across R_L. The desired R_L (for maximum power transfer) will then said to be equal to R_{Th}. For finding R_{Th}, we: (i) remove R_L, (ii) replace the ideal voltage source of 100 V, by a short circuit. Hence the circuit results, as shown in Figure 85.

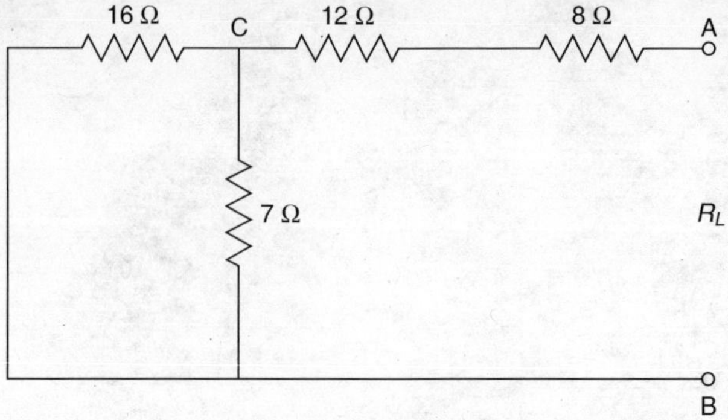

Figure 85 Finding R_{Th}.

Starting from point A, we meet the two resistors of 8 Ω and 12 Ω, which are in series. Therefore, total resistance $R_{AC} = 12 + 8 = 20\ \Omega$. At C, there are two parallel paths of 16 Ω and 7 Ω, whose equivalent will be

$$\frac{16 \times 7}{16 + 7} = 4.87\ \Omega$$

which will be in series with the 20 Ω resistor. So, the total resistance between A and B = 20 + 4.87 = 24.87 Ω, which is the R_{Th}.

Now, for the maximum power transfer to R_L, the value of R_L should be equal to this R_{Th} value, that is $R_L = 24.87\ \Omega$. In order to find the value of the power (maximum power, in fact), which is transferred to this $R_L = 24.87$, we need to find Thevenin's equivalent circuit, so that the complete circuit in question can be converted into a single-loop circuit. As we have already found R_{Th}, now V_{Th} needs to be calculated.

The circuit (for finding V_{Th}) will be as shown in Figure 86.

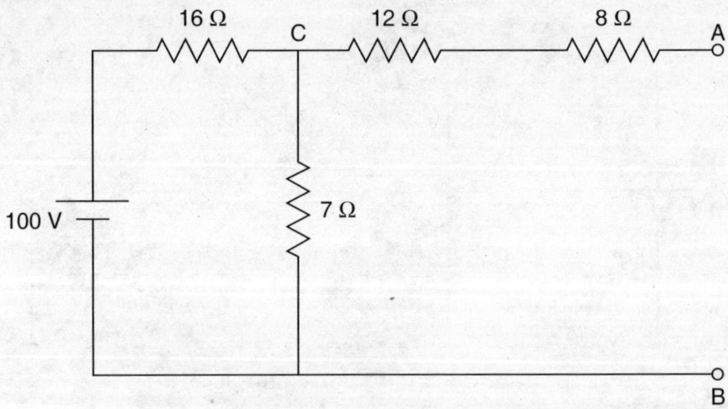

Figure 86 Finding V_{Th}.

No current will flow through 12 Ω and 8 Ω resistors (as there is an open circuit ahead of them). So, the total current delivered by the 100 V source that flows through the 16 Ω resistor will flow through the 7 Ω resistor, and thus 16 Ω and 7 Ω will be in series. The value of the current will be

$$\frac{100}{16+7} = 4.35 \text{ A}$$

Now starting from the point B (to reach A), we first go upwards through the 7 Ω resistor. As we are moving against the direction of current, we will experience the voltage gain of the value $4.35 \times 7 = 30.45$ V. Since there is no current between the points C and A, and they are connected by a 20 Ω resistance, this potential of the point C (of 30.45 V with regard to B) will also be the potential of the point A (with regard to B), and thus $V_{AB} = 30.45 \text{ V} = V_{\text{Th}}$.

Next, we draw the Thevenin's equivalent circuit as in Figure 87. Hence the current through $R_L = 0.612$ A, and the power transferred to it would be

$$(0.612)^2 \times 24.87 = 9.32 \text{ W}$$

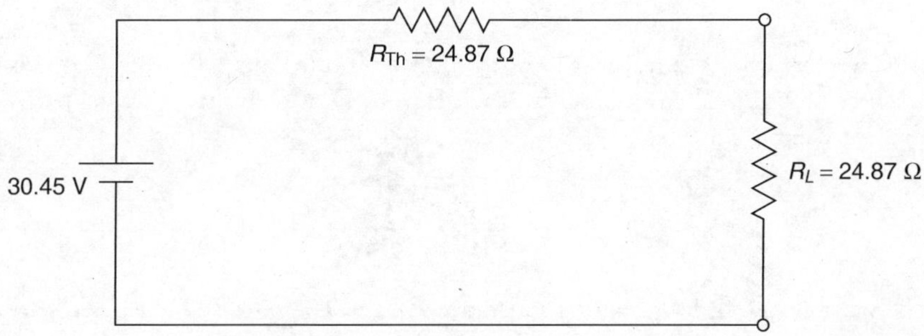

Figure 87 Thevenin's equivalent circuit.

Example 22

In the network shown in Figure 88, find the value of: (a) R_L for the maximum power dissipation and (b) the maximum power.

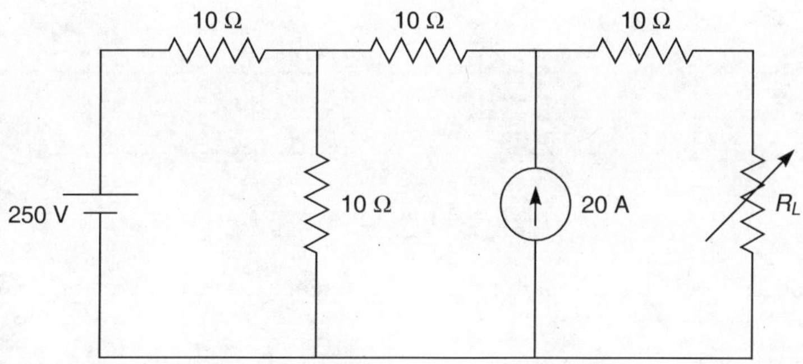

Figure 88 Network for Example 22.

Solution

As already explained in Example 21, we will find R_{Th} (across R_L) in this question, and then equate it to R_L to ensure the maximum power transfer to R_L. We replace the 250 V source by a short circuit (zero ohm resistance) and 20 A source by an open circuit (infinite ohm resistance). The resultant circuit is shown in Figure 89.

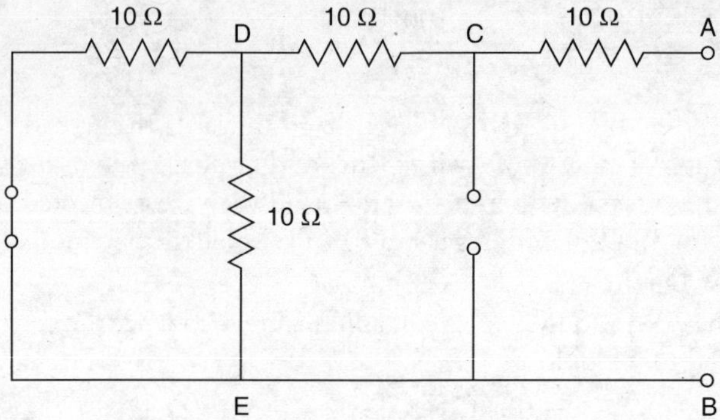

Figure 89 Finding R_{Th}.

Starting from A, we get two 10 Ω resistors in series (AC + CD) (as the path from C downwards is open) for a total resistance of 20 Ω. Then in order to reach E from D, we have two parallel paths of 10 Ω each. So, their equivalent will be

$$\frac{10 \times 10}{10 + 10} = 5\ \Omega$$

Thus we reach E and hence B. So, the total resistance $R_{AB} = 20 + 5 = 25\ \Omega = R_{Th}$. So, R_L will have to be 25 Ω for the maximum power transfer.

To find V_{Th}: We find the currents flowing in the circuit (with R_L removed), as shown in Figure 90.

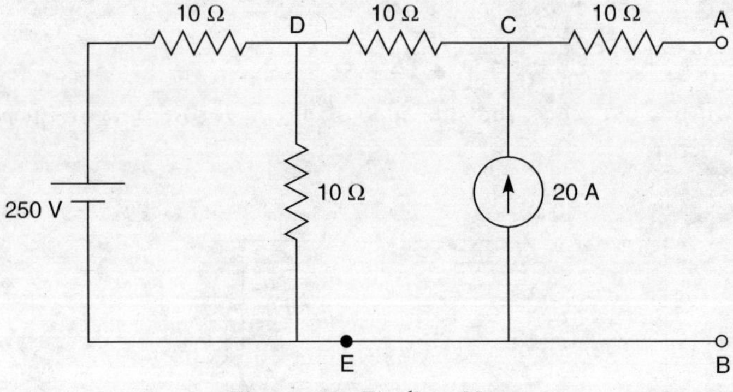

Figure 90 Finding V_{Th}.

There are three nodes: C, D and E. Taking E as the reference node and writing KCL equations for V_D and V_C, we get

$$\frac{V_D - 250}{10} + \frac{V_D}{10} + \frac{V_D - V_C}{10} = 0$$

Hence

$$3V_D - V_C = 250 \qquad (1)$$

and

$$\frac{V_C - V_D}{10} = 20 \quad \text{or} \quad V_C - V_D = 200 \tag{2}$$

Since no current flows through the right-most 10 Ω resistor in Figure 90, V_C can be said to be same as $V_{AB} = V_{\text{Th}}$. Solving Eqs. (1) and (2) for V_C, we get $V_C = 425\text{ V} = V_{\text{Th}}$. So, finally Thevenin's equivalent circuit is shown in Figure 91. The current through R_L is

$$\frac{425}{25+25} = 8.5\text{A}$$

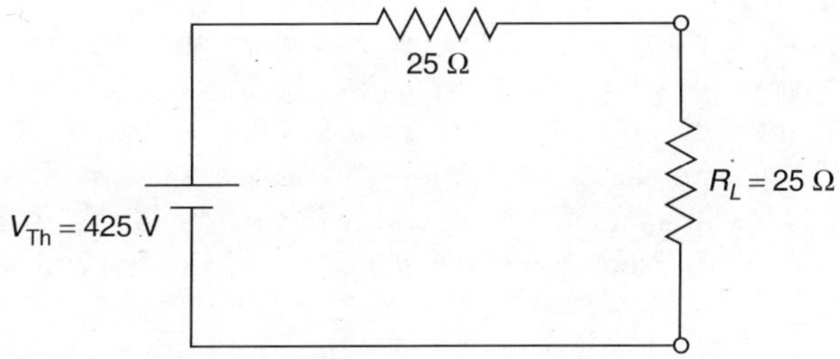

Figure 91 Thevenin's equivalent circuit.

and the power transferred to R_L will be

$$P_{L(\max)} = (8.5)^2 \times 25 = 1806.25\text{ W}$$

Example 23

Consider the circuit shown in Figure 92. Determine the value of

1. R so that the load of 20 Ω should draw the maximum power.
2. Maximum power drawn by load.

(UPTU 2000–01)

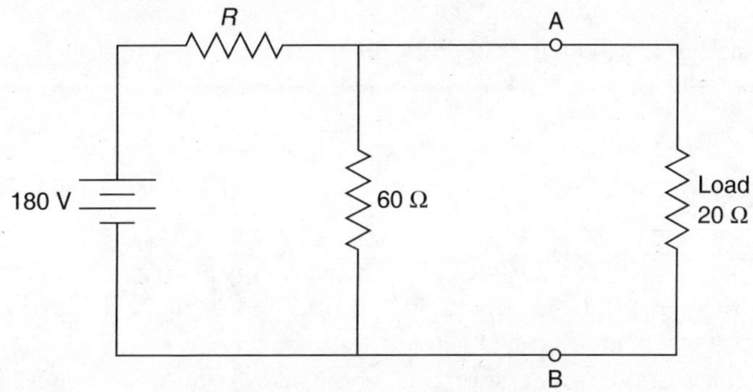

Figure 92 Figure for Example 23.

Solution

See Figure 93.

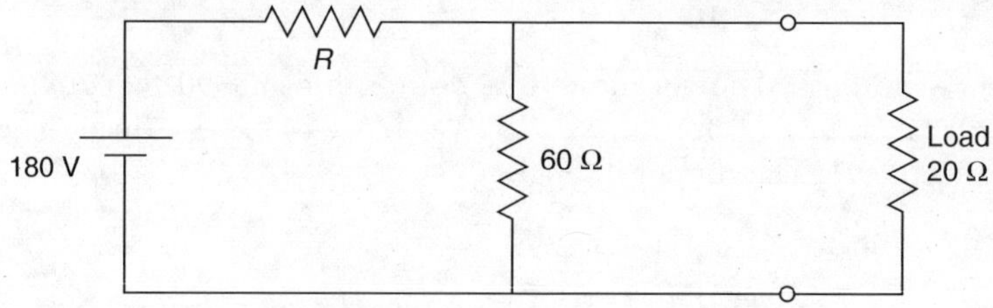

Figure 93 Solution for Example 23.

Here the load resistance has been fixed at 20 Ω. Thus, the internal resistance of the source is variable because 'R' has to be found out for which the fixed load of 20 Ω receives maximum real power. Under such conditions, when the internal resistance of the source is allowed to vary, we must ensure that the source should approach ideal nature, that is for the voltage source in the circuit in Figure 93, the internal resistance should be zero. This will happen only when R_{Th} (across the load $R_L = 20\ \Omega$) is zero. This will only be possible when $R = 0$. Thus, when $R = 0$ the voltage source is ideal and the circuit looks like the one shown in Figure 94.

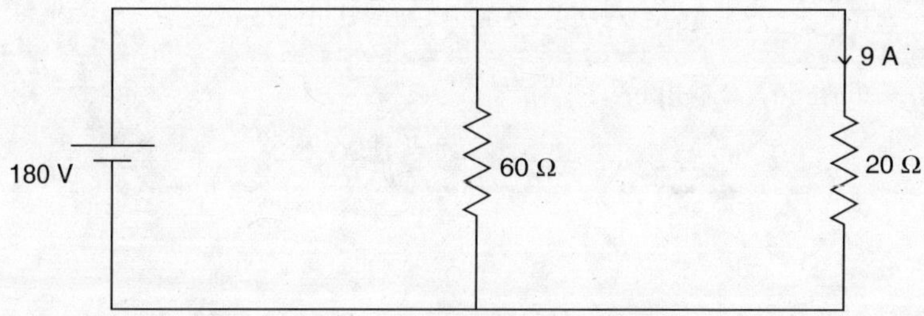

Figure 94

The current through 20 Ω load will be

$$\frac{180}{20} = 9\text{ A}$$

and the power dissipated in it will be $9^2 \times 20 = 1620$ W.

Summary

In this chapter we have discussed DC circuit analysis and network theorems. The components which impart energy to the system are called active elements (like the voltage and current sources), whereas the components which can dissipate or store energy are termed as passive elements (like the resistance for energy dissipation and inductance and capacitance for energy storage). There are two types of construction-wise sources—voltage source and current source. An ideal voltage source establishes a constant voltage across the load (irrespective of the ohmic value of the load), whereas an ideal current source establishes a constant current through the load (again irrespective of the ohmic value of the load). The internal resistance of a voltage source is shown connected in series with it (and its value is zero for an ideal voltage source) and that of a current source is shown connected in parallel with it (and its value is infinity for an ideal current source).

Linearity means capability of a network or an element to follow the principles of superposition and homogeneity. The elements in which the direction of the current changes on connecting them in the inverted manner are called unilateral elements (e.g. a diode or a thyristor), whereas those elements in which the direction of current does not change on connecting them in the inverted manner are called bilateral elements (e.g. a resistor or an inductor).

Making use of the simple Ohm's law, a voltage source in series with a resistance can be converted (or transformed) into a current source. The value of the current delivered by such a current source is given by the voltage (of the voltage source) divided by the resistance (which is connected in series with the voltage source). The resistance gets connected in parallel with the current source thus obtained after the transformation. Reverse process can be applied for converting a current source (having a resistance in parallel) to an equivalent voltage source with the resistance coming in series with it.

In this chapter, the Kirchhoff's voltage and current laws were elaborated with an aim to solving various types of circuit problems. For dealing with three-phase circuits, three-phase loads are required. The three-phase loads can only be connected in two manners. These are called star and delta. The star–delta conversion was also explained in the chapter.

This chapter has explained the four basic theorems, helpful in simplifying and solving the electric circuits. Thevenin's theorem is capable of replacing any complex network (howsoever big) across a load by a single-voltage source, V_{Th}, in series with a resistance, R_{Th}. The Norton theorem is similar to (and dual of) the Thevenin theorem. Here, instead of the voltage source, V_{Th}, the whole network is replaced by a current source, I_N, in parallel with the resistance similar to R_{Th} (called R_N here). The superposition theorem helps find the currents and voltages in the various parts of a network, taking each source at a time (and finding its individual contribution in magnitude and direction). While considering individual source, all other sources are replaced by their internal resistances. The individual contributions are then algebraically summed up to get the voltage or current in a resistor when all the sources act simultaneously. The maximum power transfer theorem tells us about the conditions in a circuit, which lead to the transfer of maximum possible power to the load. The efficiency at the time of maximum power transfer is only 50%.

Short Answer Questions

1. Which electrical quantity is conserved when KCL is satisfied? Explain with reason.
2. A resistive element of 10 Ω carries the current of 5 A. How much current will be carried by this element if another resistance of 15 Ω is connected in (i) series with it and (ii) in parallel with it?
3. Find the current I through the 6 Ω resistance in Figure P1.1.

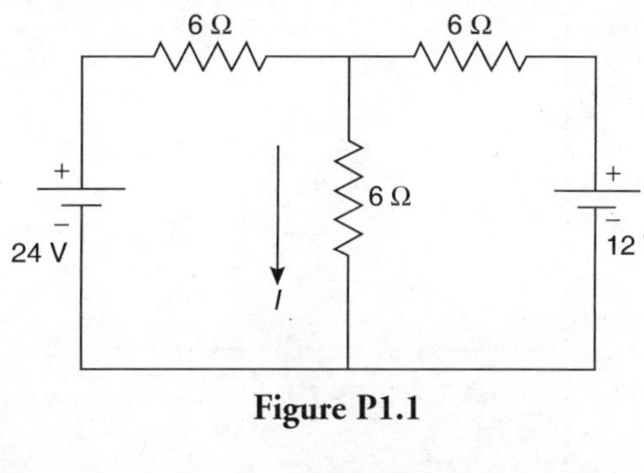

Figure P1.1

4. Can an inductance be called a linear, bilateral and passive element? Explain the reason of your answer.
5. Find the value of current 'I' in the circuit shown in Figure P1.2.

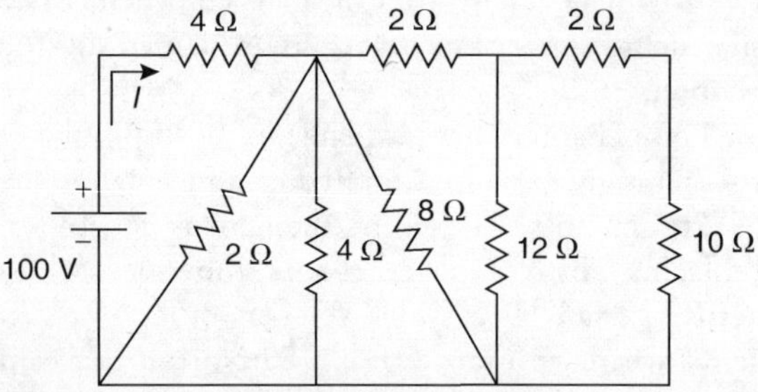

Figure P1.2

6. Find the current 'I' through 4 Ω resistance in the circuit shown in Figure P1.3.

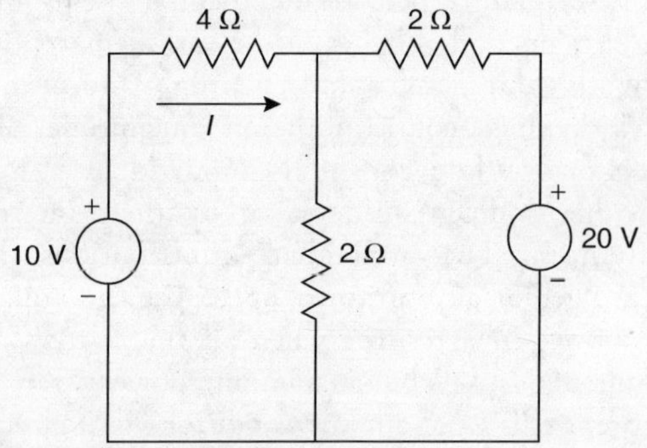

Figure P1.3

7. Find the maximum power that can be delivered to the load resistance in the circuit shown in Figure P1.4.

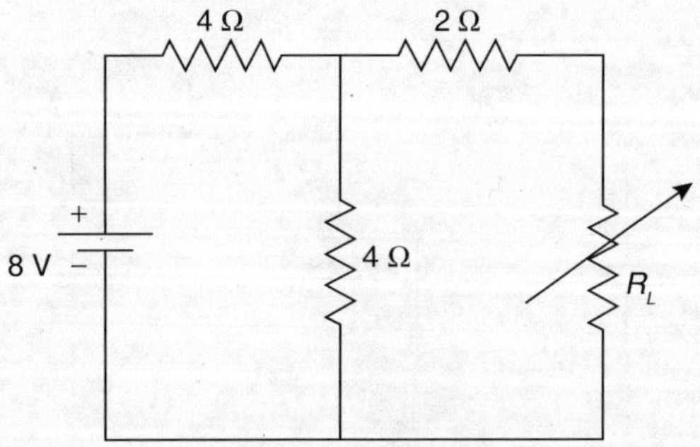

Figure P1.4

8. The value of a source resistance, R_S, can be varied between 30 Ω and 70 Ω in the circuit shown in Figure P1.5. At what value of R_S, will the maximum power be received by the load R_L?

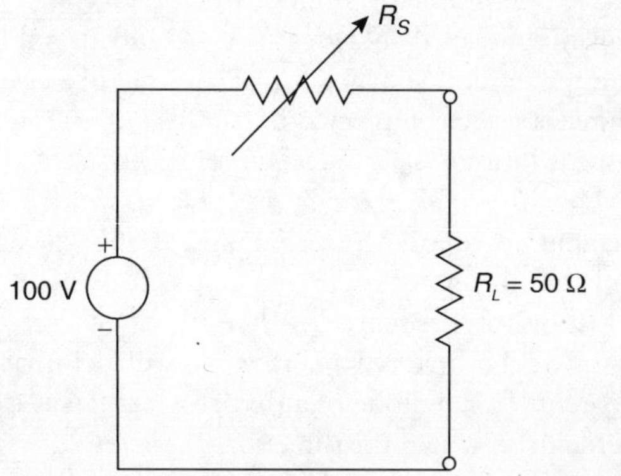

Figure P1.5

9. Find the value of current 'I' in the circuit shown in Figure P1.6.

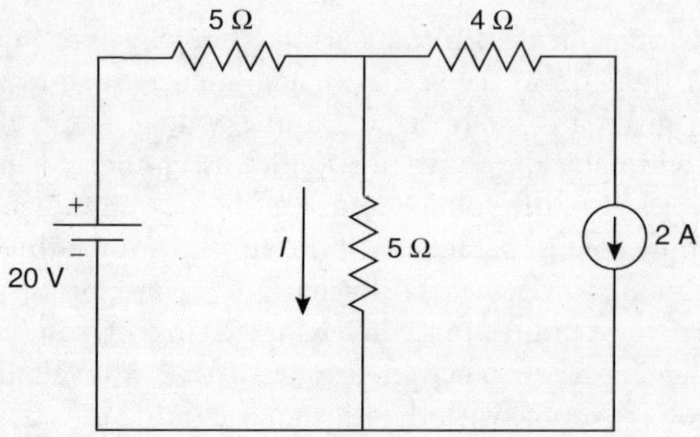

Figure P1.6

10. Draw the star equivalent for the Delta connection shown in Figure P1.7.

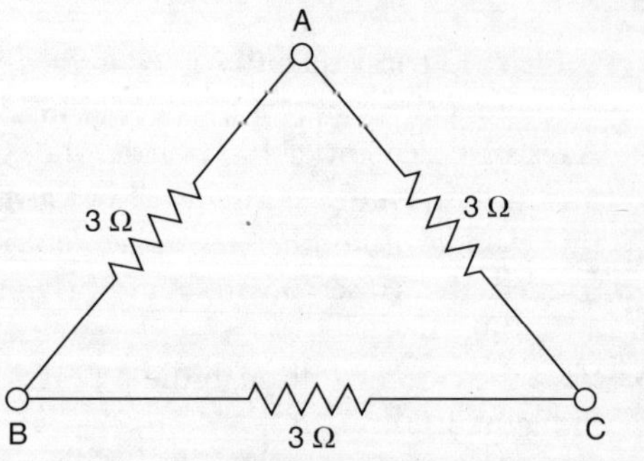

Figure P1.7

Review Questions

1. What do you mean by linearity of a network? Give some examples.
2. What is the difference between bilateral and unilateral elements? Give some examples of each.
3. What are active and passive elements? Discuss with some examples of each.
4. What is source transformation? Why is it necessary? Explain.
5. What are the limitations of source transformation? Explain with a suitable example.
6. What is Kirchhoff's voltage law? Explain.
7. What is Kirchhoff's current law? Explain.
8. Explain Thevenin's theorem. How is it superior to the KVL or KCL methods for solving variable load problems?
9. How will you convert a star-connected circuit to a delta-connected circuit?
10. What is the efficiency of power transmission at the time of maximum power transfer? Can this efficiency be increased? If yes, how?

Problems

1. In the network shown in Figure P1.8, determine the current I_1 in the 4 Ω resistor.

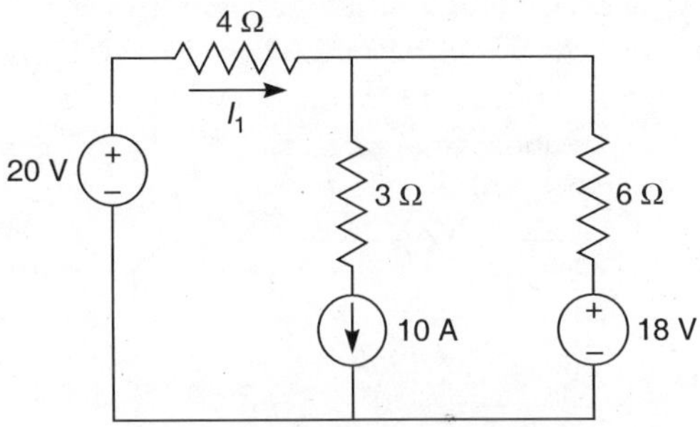

Figure P1.8 Figure for Problem 1.

2. For the network given in Figure P1.9, calculate the value of the current I_1 and I_2. Also, find the value of the unknown resistor R, when the total current drawn in the network is 1.5 A.

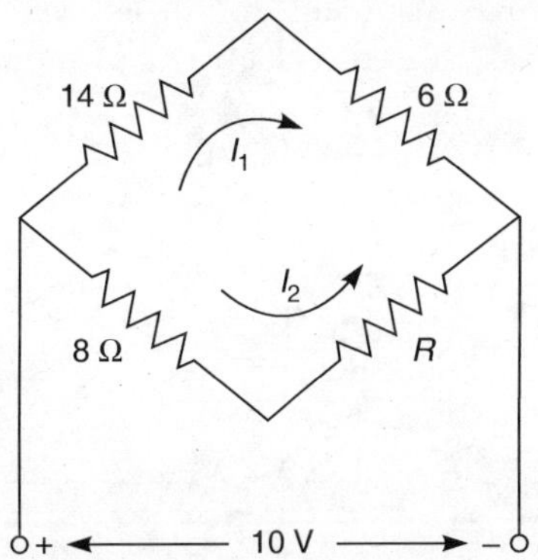

Figure P1.9 Figure for Problem 2.

3. For the network shown in Figure P1.10, find the current I_1 by applying the superposition theorem.

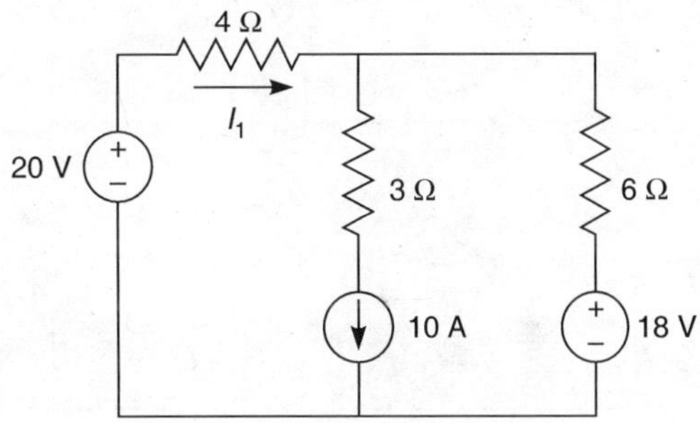

Figure P1.10 Figure for Problem 3.

4. In Figure P1.11, find: (i) the equivalent resistance across a–b, (ii) the reading of the ammeter A and (iii) the power dissipated in each resistor.

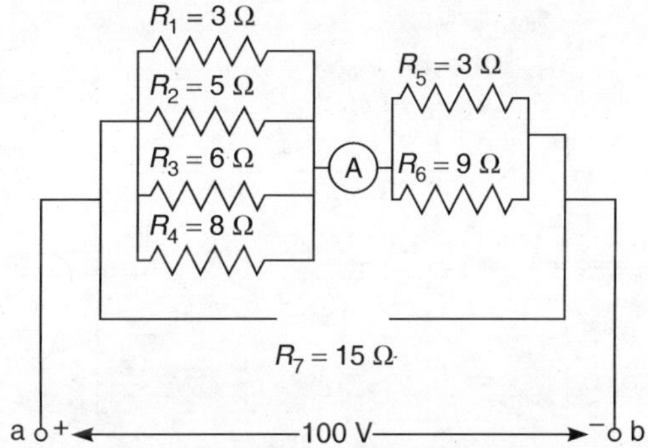

Figure P1.11 Figure for Problem 4.

5. In the network given in Figure P1.12, find the voltage V_{XY} across the terminals X and Y.

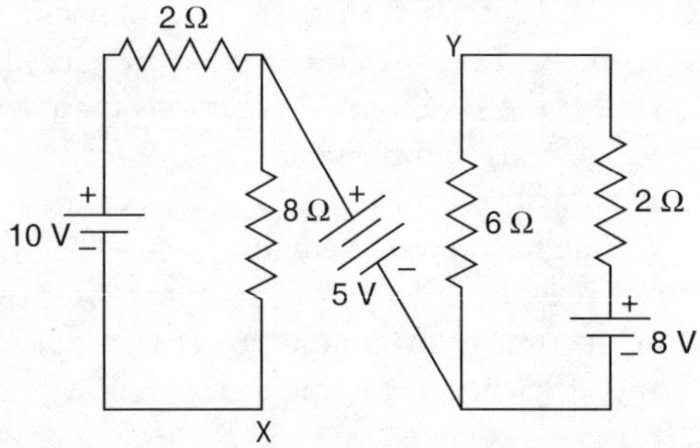

Figure P1.12 Figure for Problem 5.

6. In the network shown in Figure P1.13, find the equivalent resistance between terminals a and b, by using star-delta transformation.

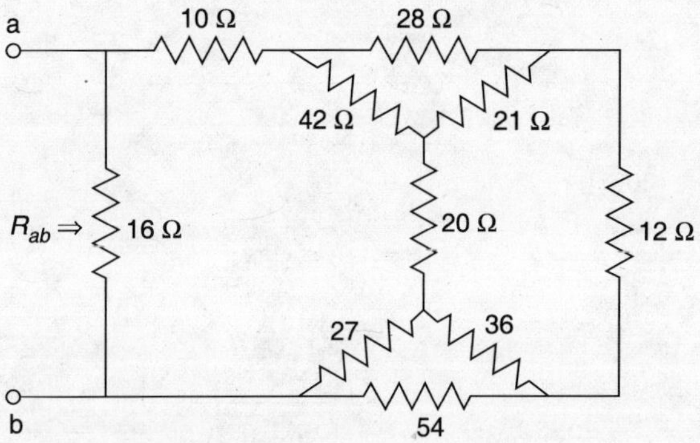

Figure P1.13 Figure for Problem 6.

7. Apply source transformation to calculate the current through the 10 Ω resistor in the circuit shown in Figure P1.14.

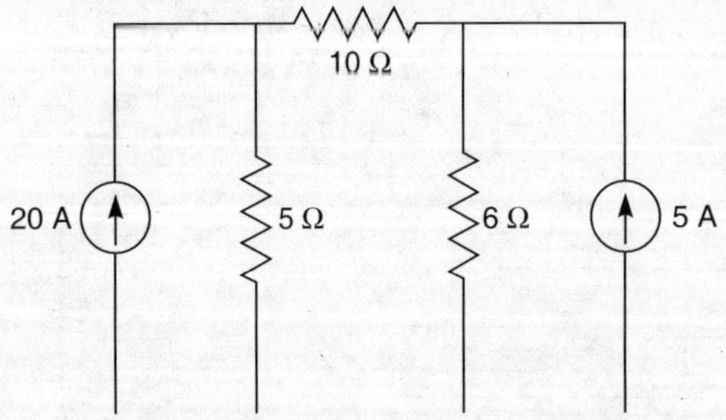

Figure P1.14 Figure for Problem 7.

8. Find the current along with its direction through the ideal galvanometer G in Figure P1.15.

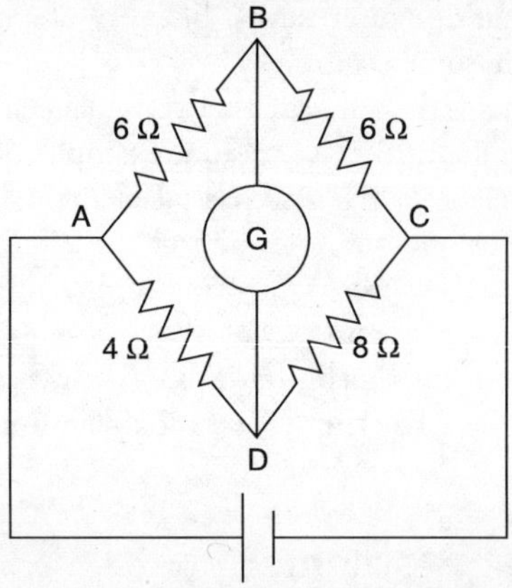

Figure P1.15 Figure for Problem 8.

9. Find the currents in all the branches (i.e. passive elements and sources) of the network shown in Figure P1.16. Use the mesh analysis and confirm the result by solving the same using nodal analysis.

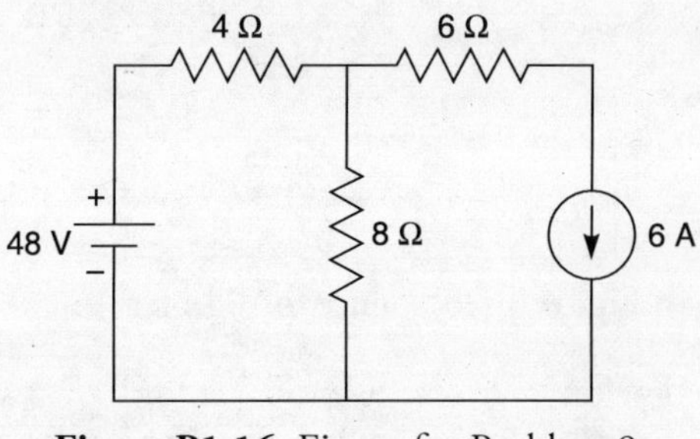

Figure P1.16 Figure for Problem 9.

10. Find the value of R_L required to have the maximum power transferred to it (Figure P1.17). Also, calculate the value of that maximum power.

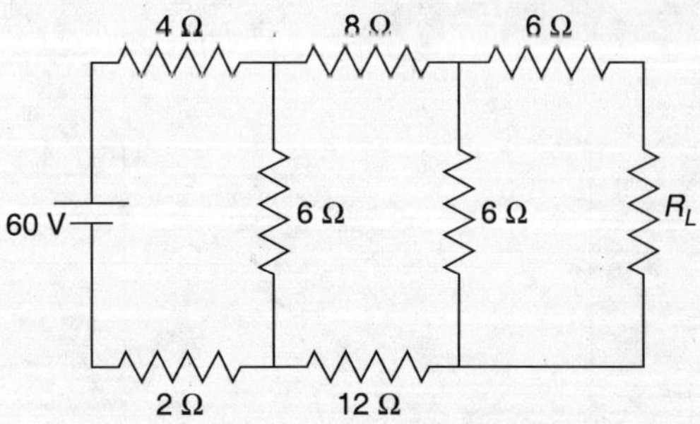

Figure P1.17 Figure for Problem 10.

11. Find the equivalent resistance, as shown in the circuit of Figure P1.18 between the points: (i) A and B, (ii) A and N.
[Hint for R_{AN}: Convert a star made up of 8 Ω, 6 Ω and 12 Ω resistors, having neutral at the point C, to delta.]

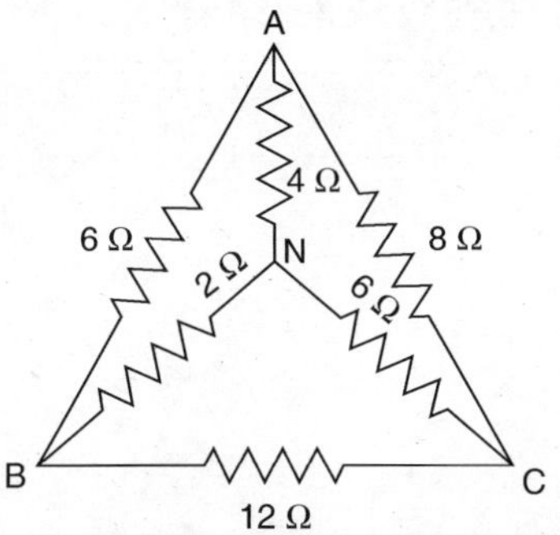

Figure P1.18 Figure for Problem 11.

12. Find the current through the 10 Ω resistor by applying superposition theorem (see Figure P1.19).

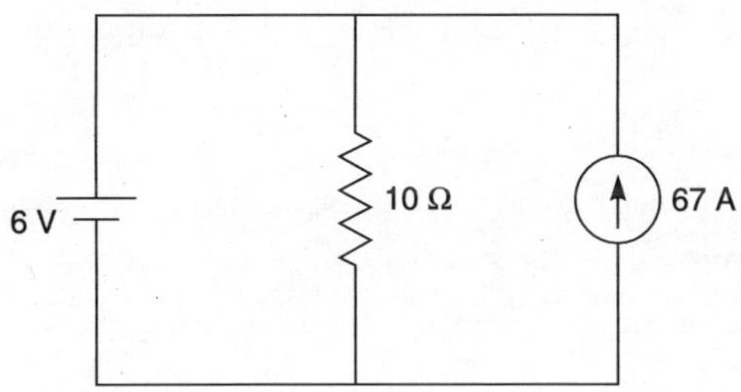

Figure P1.19 Figure for Problem 12.

13. Find the current through the 15 Ω resistance by using Thevenin's theorem (see Figure P1.20).

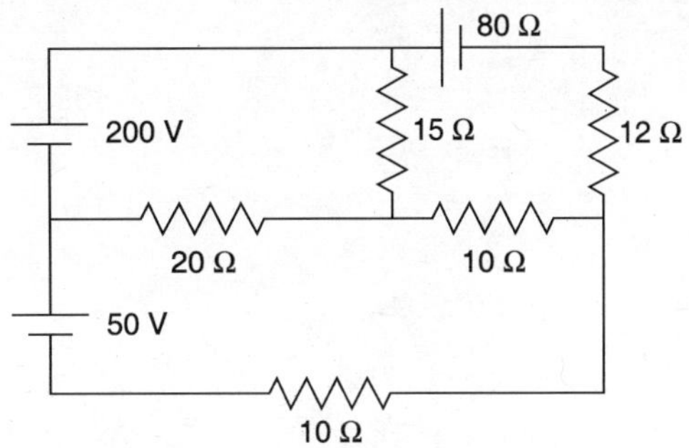

Figure P1.20 Figure for Problem 13.

14. Find the value of R in the circuit shown in Figure P1.21. **(UPTU 2008–09)**
[Hint: Consider the drop across the 14 Ω resistance equal to $4R$.]

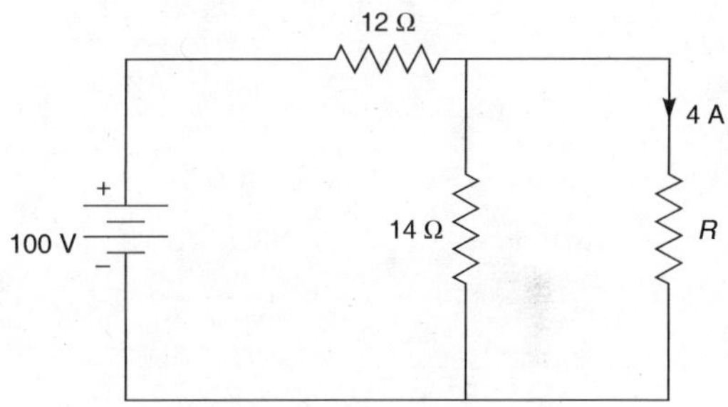

Figure P1.21 Figure for Problem 14.

Answers

Short Answer Questions

1. Charge
2. (i) 2 A, (ii) 5 A
3. 2 A
4. Yes. It is linear because the rate of change of current is directly proportional to the voltage across it. It is bilateral because the magnitude and direction of current do not change on connecting it in the reverse manner. It is passive because it cannot provide energy to drive a circuit for an infinite period of time.
5. 20 A
6. 0 A
7. 4 W
8. 30 Ω
9. 1 A
10. Three star connected resistors of 1 Ω each

Problems

1. 6.2 A
2. I_1 0.5 A, $I_2 = 1$ A, $R = 2\ \Omega$
3. $I_1 = 6.2$ A
4. $R_{ab} = 2.81\ \Omega$, $i = 28.88$ A, $P_{R_1} = 408.8$ W, $P_{R_2} = 245.28$ W, $P_{R_3} = 204.4$ W, $P_{R_4} = 153.3$ W, $P_{R_5} = 1407.47$ W, $P_{R_6} = 469.16$ W, $P_{R_7} = 666.67$ W
5. $V_{XY} = -9$ V
6. $R_{ab} = 12.33\ \Omega$
7. 3.33 A
8. 2.54 A from D to B
9. $I_{(48V)} = 8A\ (\uparrow)$, $I_{(4\Omega)} = 8A\ (\rightarrow)$, $I_{(8\Omega)} = 2A\ (\downarrow)$, $I_{(6\Omega)} = 6A\ (\rightarrow)$, $I_{(6A)} = 6A\ (\downarrow)$
10. $R_L = 10.75\ \Omega$, $P = 0.89$ W
11. $R_{AB} = 2.57\ \Omega$, $R_{AN} = 2.23\ \Omega$
12. 0.6 A
13. 7.78 A
14. 7 Ω

2 Steady-State Analysis of Single-Phase AC Circuits

LEARNING OBJECTIVES

After a careful study of the chapter, you would be able to understand:

- The various types of sinusoidal waveform values (like peak, average, RMS and instantaneous values) and their inter-conversions.
- The concepts of reactance and impedance.
- Types of reactances and impedances.
- Concepts and use of phasors.
- Concepts of lead–lag of currents and voltages in various complex impedance circuits.
- Solution to series, parallel and combined impedances.
- Concepts of resonance (both series resonance and parallel resonance).
- Importance of the terms resonant frequency, quality factor and bandwidth under resonance.

2.1 AC Fundamentals

2.1.1 Sinusoidal Waveform

We know that a sinusoidal waveform is represented by

$$v = V_m \sin \omega t = V_m \sin (2\pi f t)$$

where v is the sinusoidal voltage, V_m is its peak value (also called the *amplitude*), f is the frequency in hertz (see Figure 1).

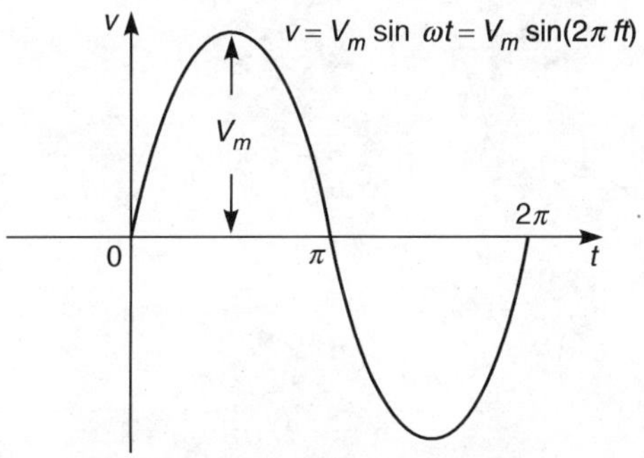

Figure 1 A sinusoidal waveform.

2.1.2 Average and Effective Values

The following are the four values of a sinusoidal waveform. Similar values exist for all other types of (periodic) waveforms:

1. **Average value:** This value denotes the average of the positive and negative areas covered by the sinusoidal waveform. As we see, for a pure sine wave, since the positive and negative areas are exactly equal, the average value is zero. The formula for finding the average value is

$$V_{av} = \frac{1}{2\pi}\int_0^{2\pi} V_m \sin\theta \, d\theta \qquad [\text{where } \omega t = \theta]$$

2. **Root Mean Square value (RMS) or Effective value:** The RMS value is the true representative of a sinusoidal quantity. Its definition goes like "Take an AC circuit. Apply a voltage for a definite amount of time and measure the amount of energy dissipated across a particular resistance of the circuit. Now, replace the AC supply by a DC supply. Now the amount of DC current required to produce the same amount of energy dissipation, when applied for the same time duration, is termed as the RMS value of the AC current." The formula for the RMS value is

$$V_{RMS} = \sqrt{\frac{1}{2\pi}\int_0^{2\pi} V_m^2 \sin^2\theta \, d\theta} \qquad [\text{where } \omega t = \theta]$$

where for a pure sine wave

$$V_{RMS} = \frac{V_m}{\sqrt{2}}$$

3. **Peak value or Amplitude:** In the waveform represented in Figure 1, V_m is the amplitude which occurs twice in every cycle (once each in the positive and negative half).
4. **Instantaneous value:** At any given time instant, the value of the sine wave is its instantaneous value. For example, at $t = 20$ ms, the instantaneous value is

$$V_{inst} = V_m \sin(2\pi f \times 20 \times 10^{-3})$$

For $V_m = 100$ V and $f = 50$ Hz (say)

$$V_{inst} = 100 \sin(2 \times 3.14 \times 50 \times 20 \times 10^{-3})$$

We convert the quantity inside the bracket to degrees, that is,

$$2 \times 3.14 \times 50 \times 20 \times 10^{-3} = 6.28 \text{ radian}$$

Now 3.14 radians − 180°

$$1 \text{ radian} = \frac{180°}{3.14}$$

Then

$$6.28 \text{ radians} = \frac{180°}{3.14} \times 6.28 = 360°$$

So,

$$V_{inst} = 100 \sin 360° = 0 \text{ V}$$

The above calculation show that we can calculate the instantaneous value of a sine wave at any particular instant by putting the value of t at which the value has been desired.

2.1.3 Square and Triangular Waveforms

As shown in Figure 2, a square waveform is having a square shape in both positive (1st) and negative (4th) quadrants. Let the magnitude be $\pm V$ and time period be T, as shown in the figure.

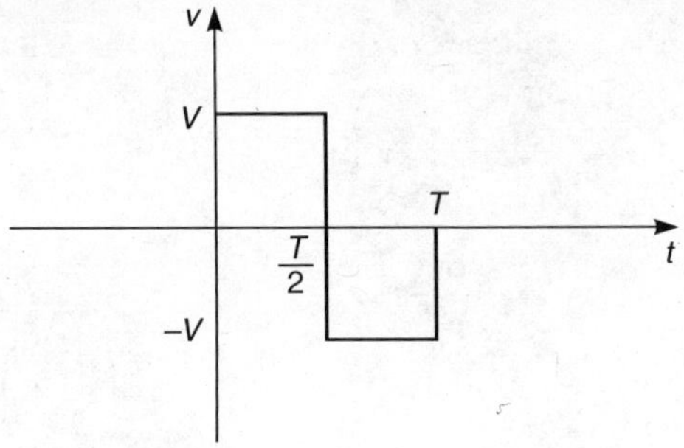

Figure 2 A square waveform.

The effective (or RMS) value can be found as

$$\sqrt{\frac{1}{T}\left[\int_0^{T/2} V^2 dt + \int_{T/2}^{T} (-V)^2 dt\right]} = \sqrt{\frac{1}{T}\int_0^T V^2 dt} = V$$

Since the positive and negative areas are equal, the average value of the square waveform will be zero. For finding the average value of the half-rectified square wave, we have

$$\frac{1}{T}\int_0^{T/2} V\,dt = \frac{V}{T}\frac{T}{2} = \frac{V}{2}$$

Triangular waveform

As shown in Figure 3, a triangular waveform contains two triangles symmetrically located in both positive and negative halves.

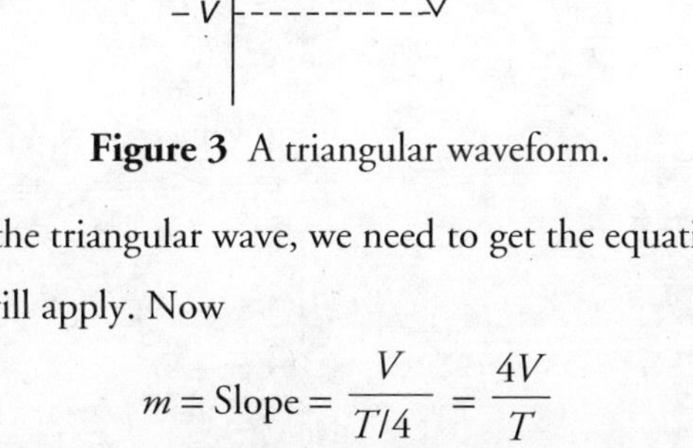

Figure 3 A triangular waveform.

For finding the RMS value of the triangular wave, we need to get the equation of various wave parts:

1. From 0 to $T/4$, $y = mx$ will apply. Now

$$m = \text{Slope} = \frac{V}{T/4} = \frac{4V}{T}$$

So, the equation representing the first part becomes

$$v = \frac{4V}{T}t$$

2. From $T/4$ to $T/2$, we have the second part of the waveform. The equation for this part can be written in the following form:

$$y - y_1 = \frac{y_2 - y_1}{x_2 - x_1}(x - x_1)$$

Taking

$$(x_1, y_1) = \left(\frac{T}{4}, V\right) \quad \text{and} \quad (x_2, y_2) = \left(\frac{T}{2}, 0\right)$$

we get

$$y - V = \frac{0 - V}{(T/2) - (T/4)}\left(x - \frac{T}{4}\right) = \frac{-4V}{T}\left(x - \frac{T}{4}\right)$$

Now with $y = v$ and $x = t$, we have

$$v = V + V - \frac{4V}{T}t = -\frac{4V}{T}t + 2V$$

The effective value will be

$$\sqrt{\frac{1}{T/2}\left[\int_0^{T/4}\left(\frac{4V}{T}\right)^2 t^2 dt + \int_{T/4}^{T/2}\left(-\frac{4V}{T}t + 2V\right)^2 dt\right]} = \frac{V}{\sqrt{3}}$$

2.1.4 Form and Peak (or Crest) Factors

Form factor (FF) is the ratio of RMS value to average value:

$$\text{FF} = \frac{\text{RMS value}}{\text{Average value}}$$

For a sine wave, with RMS value as $V_m/\sqrt{2}$, the average value (taken over half period) is $2V_m/\pi$. So,

$$\text{FF} = \frac{V_m/\sqrt{2}}{2V_m/\pi} = 1.11$$

Peak (or *crest*) *factor* is the ratio of peak value to RMS value. Thus,

$$\text{Peak factor} = \frac{\text{Peak value}}{\text{RMS value}}$$

Hence for a pure sine wave

$$\text{Peak factor} = \frac{V_m}{V_m/\sqrt{2}} = \sqrt{2}$$

Similarly, form factors and peak factors for other waveforms can be calculated.

2.2 Concept of Phasors

Consider Figure 4, where a pure sinusoidal voltage supply is connected across a pure resistance.

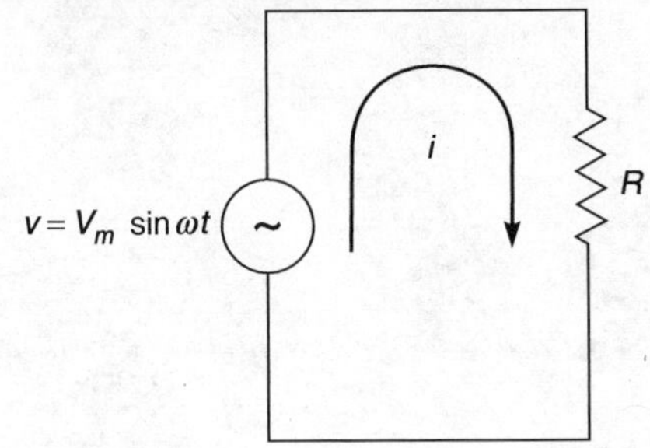

Figure 4 A purely resistive circuit.

The current can be found by using Ohm's law as

$$i = \frac{V_m}{R}\sin\omega t = I_m \sin\omega t \tag{2.1}$$

The current will also be sinusoidal with its maximum value $I_m = V_m/R$ and the frequency ω, same as that of the voltage. If the sine waves of both current and voltage are plotted together, we get Figure 5.

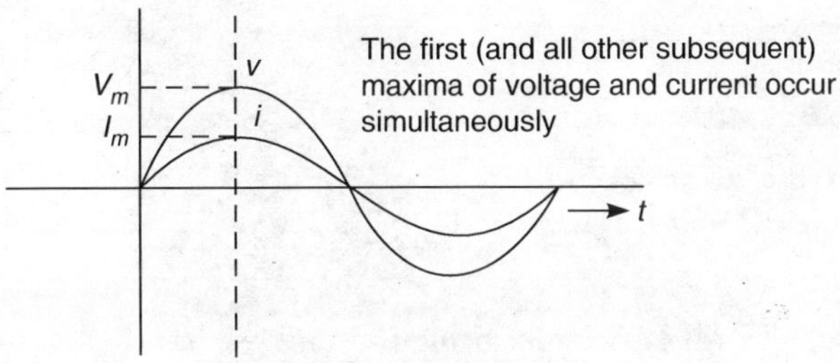

Figure 5 Showing the voltage and current waveforms for a purely resistive circuit.

It can be seen that the current wave attains its positive peak value exactly at the same instant when the voltage wave does, meaning that the first maxima of both the voltage and current waves occur simultaneously. This indicates that there is no phase difference between the two waves. Thus we can say that the phase difference between the two waveforms can be found by observing the difference between the occurrences of their first maxima. The waveform whose first maxima occurs first is said to be leading the second waveform by the difference of their angles. The conclusion in Figure 5 is "The phase difference between the voltage and current waveforms in a purely resistive circuit is zero." Now take a pure inductance and apply the same voltage across it, as shown in Figure 6.

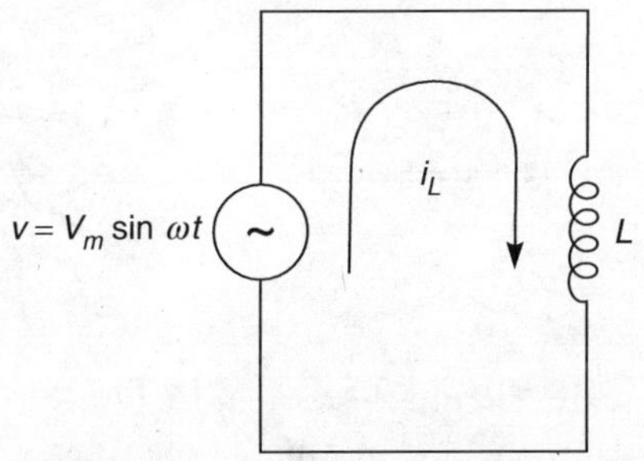

Figure 6 A purely inductive circuit.

Here the relation

$$V_L = L\,\frac{di_L}{dt}$$

holds. So,

$$i_L = \frac{1}{L}\int V_L\,dt$$

But since $V_L = v$ we have

$$i_L = \frac{1}{L}\int V_m \sin\omega t\,dt = \frac{V_m}{\omega L}(-\cos\omega t) = \frac{V_m}{\omega L}\sin\left(\omega t - \frac{\pi}{2}\right) \tag{2.2}$$

Equation (2.2) proves that "In a purely inductive circuit, the current lags behind the voltage by exactly 90°." This also means that the first maxima of voltage occurs exactly 90° before the first maxima of current, as shown in Figure 7. Next, we take a purely capacitive circuit, as shown in Figure 8.

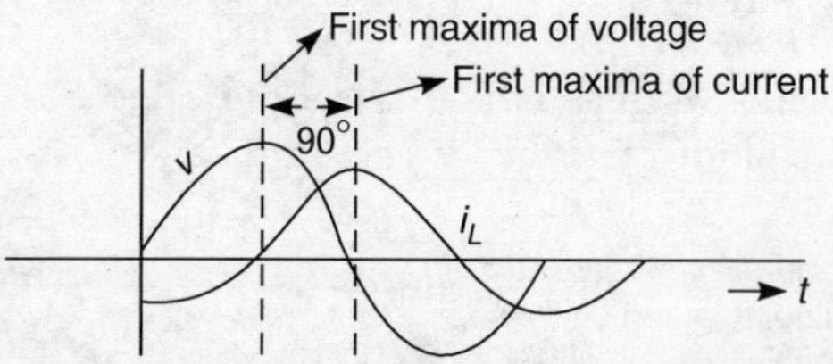

Figure 7 First maxima of the voltage occurs 90° before that of the current in a purely inductive circuit.

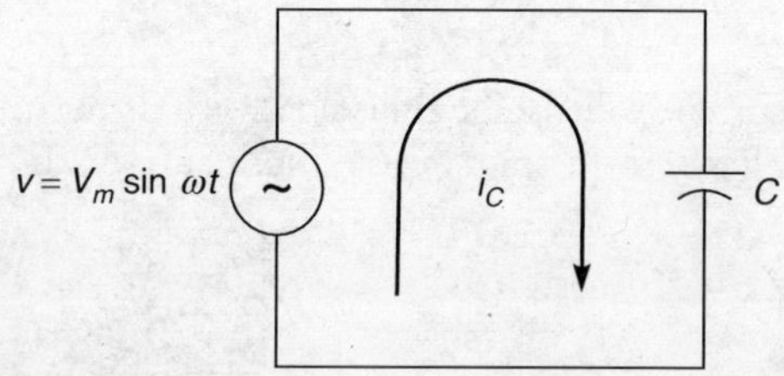

Figure 8 A purely capacitive circuit.

Here we know that $q = cv$, and with capacitance being constant, we have

$$\frac{dq}{dt} = c\,\frac{dv}{dt}$$

Also

$$i = \frac{dq}{dt}$$

That is, current is the rate of flow of charge. So, the expression for capacitive current becomes

$$i_C = C\,\frac{dv}{dt}$$

Substituting for the voltage $v = V_m \sin \omega t$, we get

$$i_C = C\frac{d}{dt}V_m \sin\omega t = V_m \omega C\cos\omega t = \frac{V_m}{1/(\omega C)}\sin\left(\omega t + \frac{\pi}{2}\right) \tag{2.3}$$

This indicates that "In a purely capacitive circuit, the current leads the voltage by exactly 90°." Also, the first maxima of the current occurs 90° before the first maxima of the voltage, as shown in Figure 9.

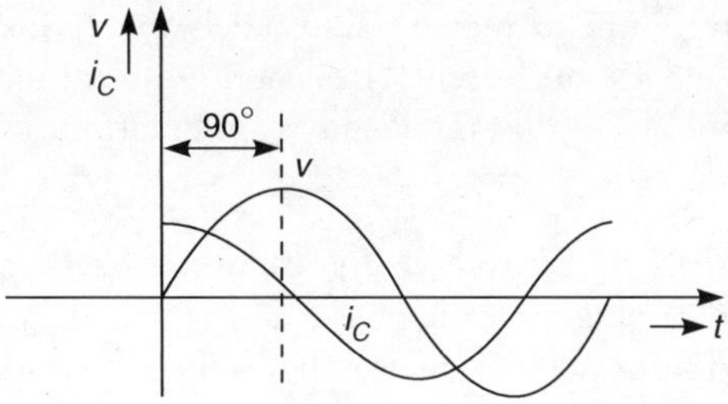

Figure 9 First maxima current occurs 90° before that of the voltage in a purely capacitive circuit.

One more information that we get from Eqs. (2.1)–(2.3) is that according to the Ohm's law, the factor coming in the denominators of the right-hand side of the expression for currents of these equations should be some form of resistance, and its unit should be ohm. In Eq. (2.1), the resistance R has got the unit ohm; in Eq. (2.2), the quantity ωL is called the *inductive reactance* with unit ohm; and in Eq. (2.3), the quantity $1/(\omega C)$ is called the *capacitive reactance* with unit ohm.

Next, we discuss the case when a series combination of resistance and reactance (inductive for example) is supplied by an AC source, as shown in Figure 10.

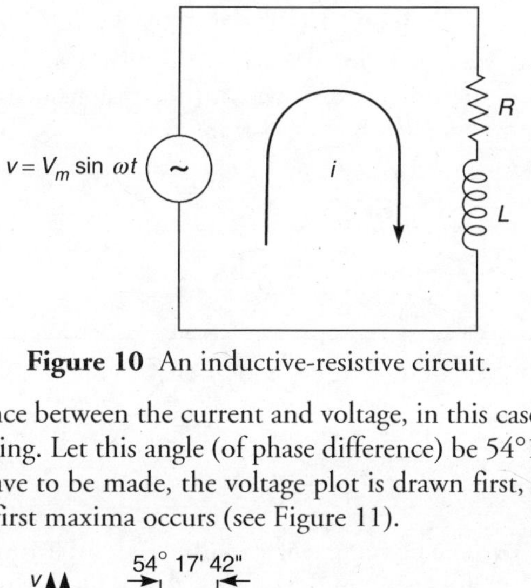

Figure 10 An inductive-resistive circuit.

As expected, the phase difference between the current and voltage, in this case, will be somewhere between zero and 90° with current lagging. Let this angle (of phase difference) be 54°17′42″. Now, if the sinusoidal plots of voltage and current have to be made, the voltage plot is drawn first, and at 54°17′42″, angle from its first maxima, the current's first maxima occurs (see Figure 11).

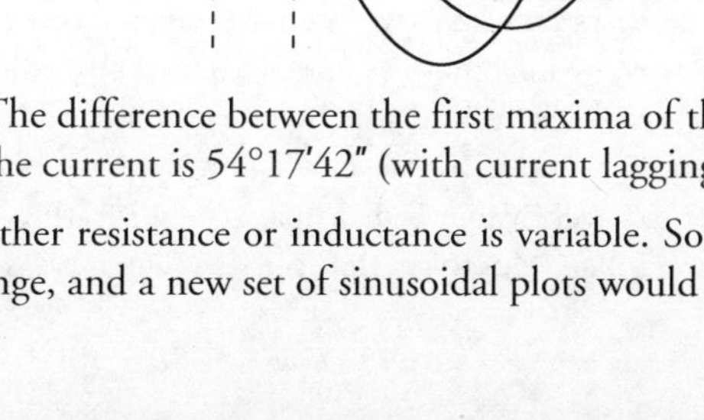

Figure 11 The difference between the first maxima of the voltage and the current is 54°17′42″ (with current lagging).

Next consider the case when either resistance or inductance is variable. So, the moment either of them is varied, the angle of lag will change, and a new set of sinusoidal plots would be required to be made.

Since the above-discussed procedure of making sinusoidal waveforms for showing phase differences between the various quantities (voltages and currents) is a lengthy and cumbersome one, and also calculations for circuit analysis are not possible using instantaneous representation, hence the concept of phasor was introduced.

2.2.1 Phasor Representation of Sinusoidally Varying Voltage and Current

This section explains the method of obtaining the phasor of a sinusoidal quantity. As shown in Figure 12, we take a vector equal to the maximum value (V_m here), that is the amplitude of the sinusoidal quantity, and put it in the negative *x*-axis, pointing towards the origin.

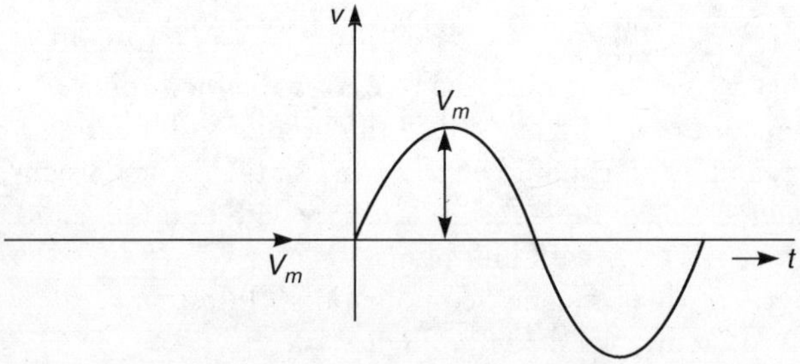

Figure 12 Phasor concept.

Now, as the sinusoidal quantity starts covering its cycle (or oscillating), the arrow tip of the vector is also raised such that the line joining the tip of the arrow with the instantaneous position of the sine wave is parallel to the *x*-axis, as shown in Figure 13 for angles θ and θ'.

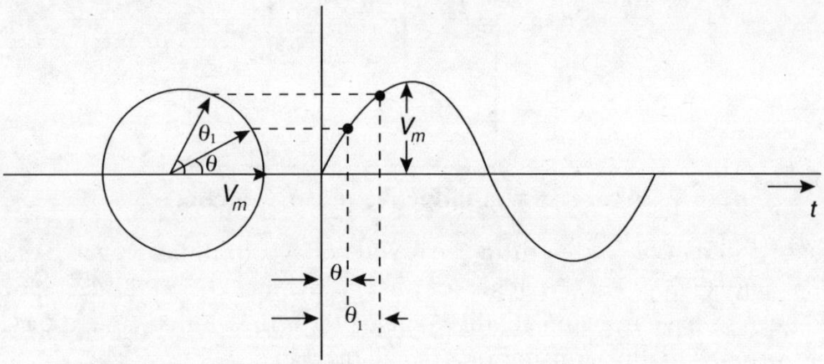

Figure 13 The phasor angular velocity 'ω' is the same as the frequency of the sinusoidal waveform which it represents.

Thus, we can see that when the sinusoidal wave completes one oscillation, the vector V_m completes one rotation in the anti-clockwise direction, and thus the frequency of the sinusoidal wave will be the angular velocity of the rotating vector having radius V_m.

To define a sinusoidal quantity, we require its amplitude and frequency. Since the rotating vector's radius represents the amplitude and the angular velocity represents the frequency, such a rotating vector is successfully able to represent the sinusoidal quantity. Such a rotating vector is called a *phasor*.

The frequency of the current waveform remains same as that of the voltage waveform for various loads (combinations of *R*, *L* and *C*). It is only the phase which varies. Hence for the present case of the *R-L* series load, the current waveform is lagging in phase by 54°17′42″, but its frequency is same as that of the voltage. So the phasor representation of the current waveform will also be a rotating vector (of the radius equal to its amplitude), rotating at the same angular velocity as that of the voltage phasor.

The next step is to superimpose the two phasors. It can be imagined that since the two phasors move at the same angular velocity, any one will appear stationary to the other. If a snapshot (photo) is taken of the two phasors, it will be seen that the voltage phasor will be ahead in the anti-clockwise direction by 54°17′42″ and the current phasor will be behind. If the snapshot is taken when the voltage phasor coincides with the positive *x*-axis, as shown in Figure 14, the current phasor will be seen in the fourth quadrant lagging the voltage phasor by 54°17′42″.

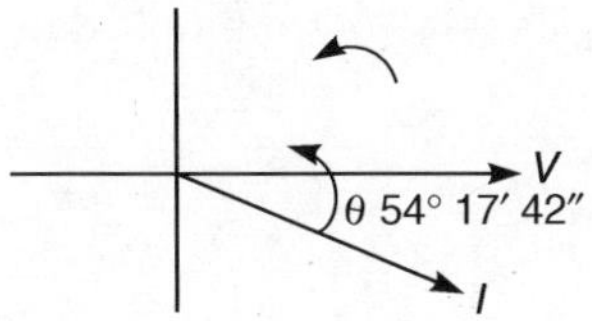

Figure 14 The phasor concept is useful for the comparison of the phases of voltage and current.

Thus, the whole phasor explanation now gets concentrated on drawing two lines: one proportional to the RMS value (the peak value is scaled down to RMS value to represent a phasor according to an internationally accepted convention) of the voltage and the other proportional to the RMS value of the current with the particular angle of lead or lag between them. If the voltage phasor is represented in the positive *x*-axis, the current phasor for a purely resistive load will coincide with the voltage phasor, indicating zero phase difference. For a purely inductive load, the current phasor will be 90° lagging, that is coinciding with the negative *y*-axis. For a purely capacitive load, the current phasor will be 90° leading, that is towards the positive *y*-axis, as shown in Figure 15.

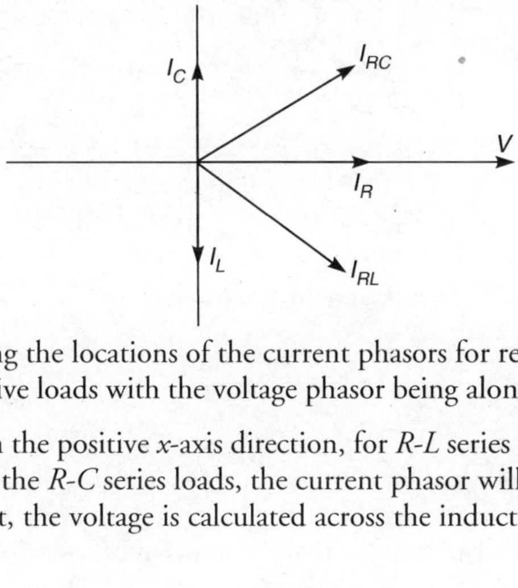

Figure 15 Showing the locations of the current phasors for resistive, inductive and capacitive loads with the voltage phasor being along the positive *x*-axis.

With voltage phasor assumed in the positive *x*-axis direction, for *R-L* series loads, the current phasor will be in the fourth quadrant, and for the *R-C* series loads, the current phasor will be in the first quadrant.

If in a *purely inductive* circuit, the voltage is calculated across the inductance, the expressions for Ohm's law would be

$$v_L = i_L \omega L$$

This equation by no means is able to represent the fact that the inductive voltage v_L leads the current i_L by 90°. Hence to do so, an operator called the "*j* operator" is made use of whose value is $\sqrt{-1}$.

The *j* operator when multiplied to a quantity rotates the quantity by 90° in the anti-clockwise direction. Thus if *j* is multiplied to the RHS of the above equation, we get

$$v_L = j(i_L \omega L) \quad \text{or} \quad v_L = i_L (j\omega L)$$

Thus, now we can say that v_L is leading i_L by 90°. Also, the expression $j\omega L$ is the complete expression for the inductive reactance.

Similarly, for a *purely capacitive* circuit:

$$v_C = i_C \times \frac{1}{\omega C}$$

Here since v_C has to be shown lagging i_C by 90°, $-j$ is multiplied to the RHS. So,

$$v_C = i_C \times \frac{-j}{\omega C}$$

and since $j^2 = -1$, we get

$$v_C = i_C \times \frac{1}{j\omega C}$$

Thus, $1/(j\omega C)$ is the complete expression for the capacitive reactance. Now, we are in a position to define the impedance, that is Z.

2.2.2 Impedance and Impedance Triangle

Impedance is given by

$$\text{Impedance } (Z) = \text{Reactance } (R) \pm j \text{ Reactance } (X)$$

$$Z = R \pm jX$$

$$Z = R \pm jX_L = R + j\omega L \quad \text{(for } R\text{-}L \text{ series)}$$

Also

$$Z = R - jX_C = R - \frac{j}{\omega C} \quad \text{(for } R\text{-}C \text{ series)}$$

Units of impedance, resistance and reactance are ohms. Impedance Z is represented in two forms:

1. Rectangular form (also called Cartesian form).
2. Polar form.

To clarify the difference, let us draw the impedance planes as shown in Figure 16. As can be seen and impedance $Z = (3 + j4)\ \Omega$ is shown as point A in the impedance plane. Similarly, the point B represents and impedance $Z = (3 - j2)\ \Omega$. If we join the impedance point [say A of Figure 16] to the origin as shown in Figure 17 we get the hypotenuse 'r' at an angle θ of the impedance triangle. Here

$$r = \sqrt{3^2 + 4^2} = 5 \quad \text{and} \quad \theta = \tan^{-1}\left(\frac{4}{3}\right)$$

Thus $Z = x + jy$ means

$$\sqrt{x^2 + y^2} \underline{|\tan^{-1}(y/x)}$$

in polar form. Also $x = r\cos\theta$ and $y = r\sin\theta$ where 'r' and θ are as discussed above.

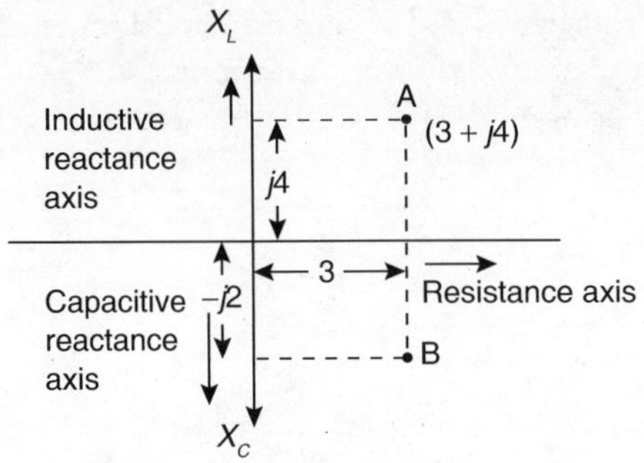

Figure 16 The impedance plane.

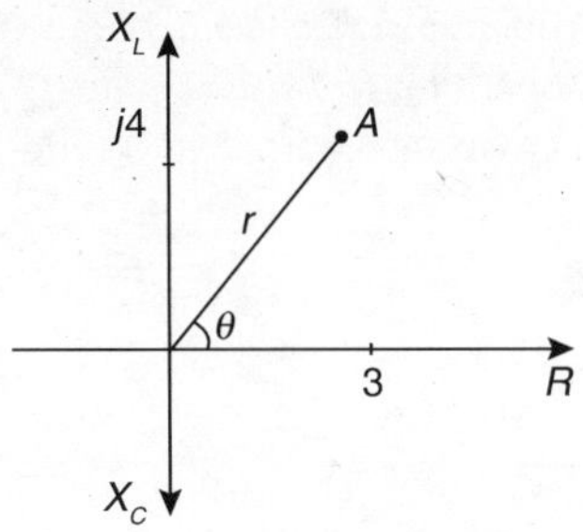

Figure 17 Explanation of polar impedance.

The addition and subtraction of two impedances is always done in rectangular co-ordinates and division and multiplication is usually done in polar co-ordinates.

The impedance triangle is a right-angled triangle whose base is a resistance, altitude is a reactance and hypotenuse is an impedance, as shown in Figure 18. For example, if we take an *R-L* circuit, the impedance will be $Z = R + jX_L$, whose impedance triangle is shown as in Figure 18.

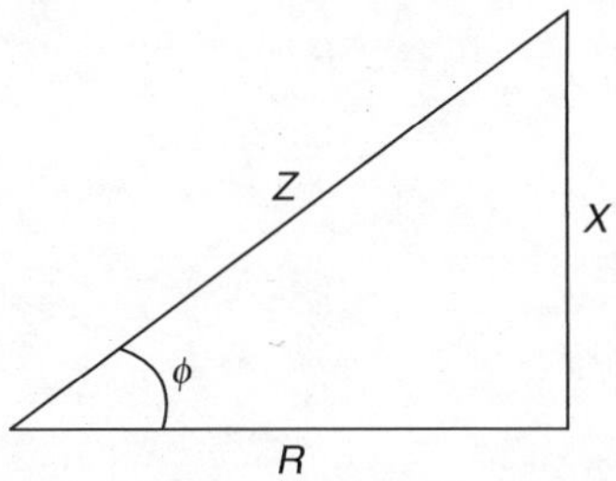

Figure 18 The impedance triangle.

The cosine of the angle between R and Z (i.e. cos ϕ), is called the *power factor*. Thus

$$\text{Power factor} = \frac{R}{\sqrt{R^2 + X^2}}$$

The power factor increases with resistance and becomes unity for $X = 0$ and $Z = R$. Some more discussion on impedance is necessary.

The impedance, reactance and resistance get added when connected in series. A $Z = 3 + j4$ means a resistance of 3 Ω is connected in series with an inductive reactance of 4 Ω. Similarly, $Z = 2 - j5$ means a resistance of 2 Ω is connected in series with a capacitive reactance of 5 Ω. Here $-j5$ means $5/j$. So, comparing this with $1/(j\omega C)$, we get $\omega C = 1/5$, and hence $1/(\omega C) = 5$ Ω.

The inverse of impedance is termed as *admittance* (Y) with unit ℧ (mho), which is the sum of conductance (G) and j times the susceptance (B). Thus $Y = G \pm jB$. Here '+' sign is for capacitive and '−' sign is for inductive susceptance. Thus

$$\text{Inductive susceptance} = \frac{1}{j\omega L} = \frac{-j}{\omega L}$$

and

$$\text{Capacitive susceptance} = \frac{1}{1/(j\omega C)} = j\omega C$$

Admittances connected in parallel get added. Thus, if two admittances, Y_1 and Y_2, are connected in a manner shown in Figure 19, the net admittance is $Y = Y_1 + Y_2$. Similarly, *conductances and susceptances also get added when connected in parallel.* Thus $Y = 3 + j4$ means a conductance of 3 mho (℧) is connected in parallel with a capacitive susceptance of 4 mho (℧).

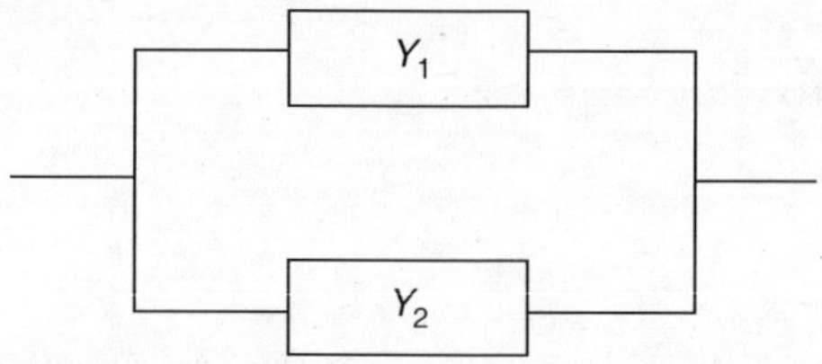

Figure 19 Admittances connected in parallel get added.

The active components in a circuit generate power and the passive elements use that power. For example, in a DC circuit, as shown in Figure 20, the current of 4 A circulates. This results in

1. Power developed by source = VI = 400 W
2. Power dissipated in resistances = $I^2R_1 + I^2R_2 = 320 + 80 = 400$ W

It shows that the power generated is same as the power consumed in the circuit.

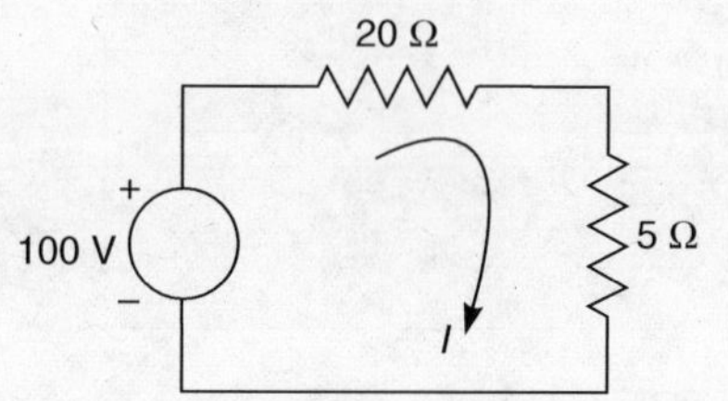

Figure 20 The DC circuit.

The explanation of instantaneous power and its significance are given in the footnote.[1]

---

[1]The instantaneous value of power will simply be the product of v and i, that is

$$vi = V_m \sin \omega t \times I_m \sin(\omega t - \phi)$$
$$= V_m I_m [\sin \omega t \times \sin(\omega t - \phi)]$$
$$= \frac{V_m I_m}{2} [2 \sin \omega t \sin(\omega t \quad \phi)]$$

We know that

$$2 \sin A \sin B = \cos (A - B) - \cos (A + B)$$

So

$$vi = \frac{V_m I_m}{2} [\cos(\omega t - \omega t + \phi) - \cos(\omega t + \omega t - \phi)]$$
$$= \frac{V_m I_m}{2} [\cos \phi - \cos(2\omega t - \phi)]$$

AC power

In AC circuits, the power is represented differently. The total power is represented by "volt-ampere (VA)." Its real and reactive components are represented by "watts" and "volt-ampere-reactive (VAR)." Hence,

$$\text{AC total power, } S = VI^* = P + jQ$$

where

$$P = VI\cos\phi \text{ (watts)}; \quad Q = VI\sin\phi \text{ (VAR)}$$

The power P is called the *real power*, as this component is responsible for doing work. The power Q is called the *reactive* or *imaginary power*, as Q does not contribute any work. The term $\cos\phi$ as already stated is known as the power factor, where the angle ϕ is the angle between the voltage and the current phasors (see Figure 21).

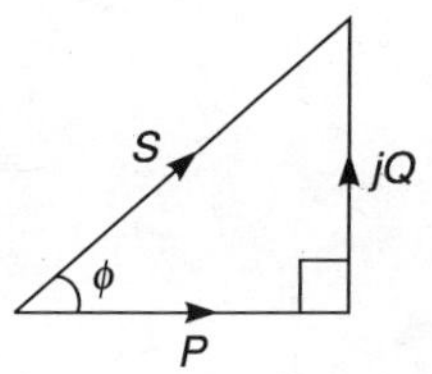

Figure 21 Phasor sum of real and reactive powers is the apparent power.

This can also be written as

$$vi = V_{\text{RMS}}I_{\text{RMS}}\cos\phi - V_{\text{RMS}}I_{\text{RMS}}\cos(2\omega t - \phi) \tag{1}$$

As can be observed from Eq. (1), the instantaneous power comprises two quantities. The first one, $V_{\text{RMS}}I_{\text{RMS}}\cos\phi$, does not depend upon time and is hence constant. The second quantity, $V_{\text{RMS}}I_{\text{RMS}}\cos(2\omega t - \phi)$ varies at twice the frequency of either current or voltage and (as is true for any sinusoidal quantity) its average value over one cycle of voltage is zero.

Hence we can say that the average value of the instantaneous power is given by

$$vi_{(\text{av})} = V_{\text{RMS}}I_{\text{RMS}}\cos\phi$$

where 'ϕ' is the phase difference between voltage and current waveforms. This average value will also represent the ACTUAL amount of (useful or work doing) power which is extracted out of the source and dissipated in the resistive part of the load. We thus also observe that although apparently (or seemingly) vi is the power taken from the source, actually it is only a part of it, i.e. $V_{\text{RMS}}I_{\text{RMS}}\cos\phi$ that is transferred to the load and it is this part which does the useful work.

Some comment on the other (i.e. second) part of the instantaneous power. As can be seen from Eq. (1), this power varies (oscillates) at twice the frequency of voltage or current. It attains negative and positive values and finally has an average value equal to zero.

Thus we can conclude that "If there is phase difference between voltage and current waveforms (due to any reason), the useful power will ALWAYS be less than the product of instantaneous values of voltage and current. The product of the RMS value of component of current **'in'** the direction of voltage with the RMS value of voltage will be the real power (responsible for doing the actual/true or real work) and the product of the RMS value of the **'quadrature'** component of current with the RMS value of voltage will be the respective power (having zero average value).

2.3 Power and Power Triangle

Similar to impedance triangle, the *power triangle* is also a right-angled triangle representing the three types of powers: real power (P), reactive power (Q) and the apparent or total power (S), as shown in Figure 22.

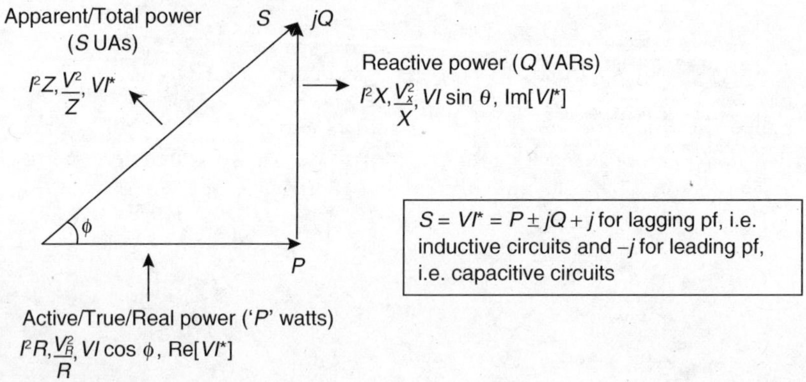

Figure 22 Power triangle. Note: All the values of voltage and current are RMS values.

2.3.1 Real Power

As is clearly evident from Figure 22, the real power or the true power or the active power, is the power dissipated in the pure resistance. Its unit is watts—kilowatts (kW) and megawatts (MW), being the higher units. There are four expressions, as given in Figure 22, of this real power: I^2R, V_R^2/R, $VI\cos\phi$ (with ϕ being the angle between the real power and the apparent power, as shown), real part of V multiplied by the conjugate of I.

2.3.2 Reactive Power

The reactive power has units volt-ampere-reactive (VAR), with higher units being kilovolt-ampere-reactive (kVAR) and megavolt-ampere-reactive (MVAR). The reactive power does not get dissipated or lost forever, because it is a power concerned with the reactance, that is inductance and capacitance. Both these elements are the energy-storing elements, and hence no power is actually consumed by them. So, the power in these two elements is called *reactive power*. The energy associated with this reactive power does not get converted into other usable forms. This is the difference between reactive power and active power. The four formulae for reactive powers are written in Figure 22.

2.3.3 Apparent Power (or Total Power)

The power in impedance is called the *total* or *apparent power*, and its unit is volt-ampere (VA) with higher units being kVA and MVA. The expressions are shown in Figure 22 as $I^2Z, V_Z^2/Z, VI$.

2.4 Power Factor

Power factor is one of the most significant parameters in electrical engineering. Power factor is defined as the cosine of the angle between the voltage and the current. So, if, for example, the voltage phasor is defined as $100\angle{-30}$ and the current phasor as $20\angle 55°$, then the angle between the two phasors is obviously $55° - (-30°) = 85°$, and the cosine of this angle (i.e. cos 85°) is termed as the power factor. Now, if the voltage is $100\angle 30°$ and the current $20\angle{-55°}$, the value of the angle between them will be $-85°$ and the power factor will be cos(−85°). But, as we know $\cos\theta = \cos(-\theta)$, so cos(−85°) = cos 85°. Hence, the complete statement for the power factor is the numerical value of the cosine of the angle between the voltage and current followed by the word "lagging" or "leading" depending upon whether the *current* is lagging or leading the *voltage*. Thus the answer for the first case would be cos 85° leading and for the second case would be cos 85° lagging. The power factor can also be found from the impedance and power triangles as

$$\text{pf} = \cos\phi = \frac{R}{Z} = \frac{\text{Real power}}{\text{Apparent power}}$$

Whenever, in a given circuit, the value of reactance (inductive or capacitive) is more as compared to resistance, the angle between voltage and current increases. This results in lower values of power factor (i.e. cos ϕ), as follows:

1. **Unity power factor:** Refer to Figure 23. Here $V = IR$. Since $\phi = 0$, $\cos\phi = 1$. Thus, there is unity power factor.

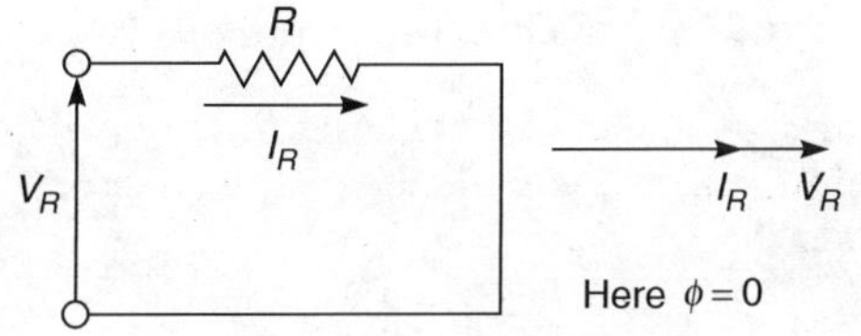

Figure 23 Unit power factor.

2. **Lagging power factor:** Refer to Figure 24. Here $V_L = jI\,X_L$. Since $\phi = -90°$, $\cos\phi = 0$. Thus, there is lagging power factor.

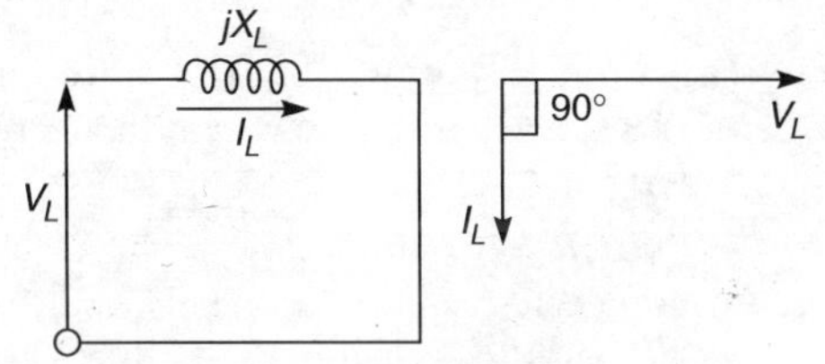

Figure 24 Lagging power factor.

3. **Leading power factor:** Refer to Figure 25. Here $V_C = -jIX_C$. Since $\phi = 90°$, $\cos\phi = 0$. Thus, there is leading power factor.

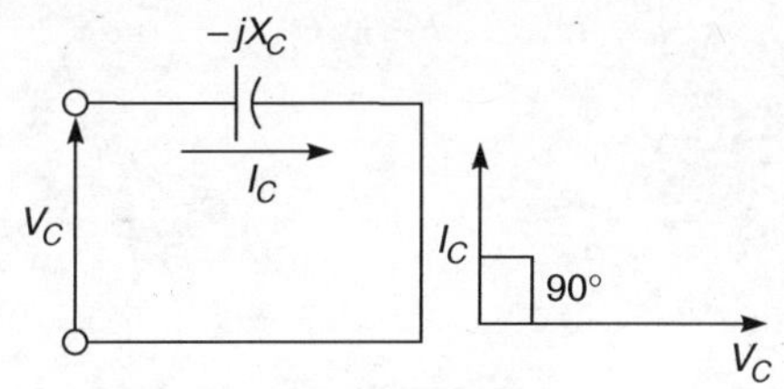

Figure 25 Leading power factor.

Therefore, it is clear from Figures 22–25 that the power factor in an AC circuit can vary from unity (i.e. 1) to zero (from either lagging or leading).

2.4.1 Problems of Low Power Factor

Usually, in a power system all the components (generators, transformers, transmission lines and loads) exhibit inductive property. The resistance of power system components is smaller as compared to inductance. Hence the power factor of a power system becomes very low. The power factor is said to be poor when its value is less.

Suppose a load is supplied by a source of voltage V volts, and it consumes power P watts. If the pf is ϕ_1, it would require S_1 volt-ampere. For the supply voltage V, as shown in Figure 26, the current drawn would be:

$$|I_1| = \frac{S_1}{V}$$

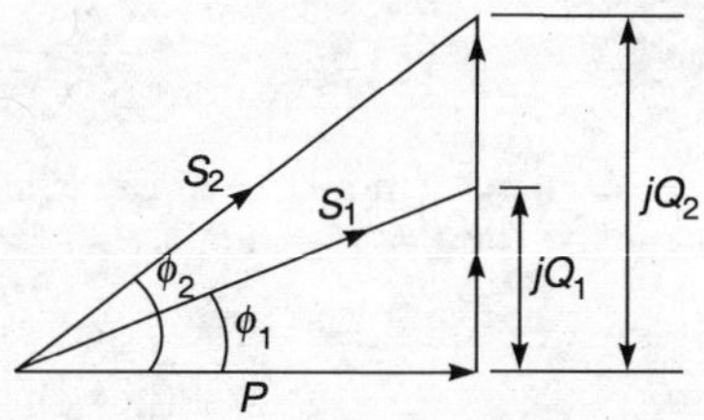

Figure 26 Current drawn in low power factor.

Now if the power factor becomes poor and the angle becomes equal to ϕ_2, then for the same power consumption of P watts, it would require S_2 volt-ampere. So, for the supply voltage V, the current drawn would become

$$|I_2| = \frac{S_2}{V}$$

Now we see that $S_2 > S_1$ in the given example. So it is clear that

$$|I_2| > |I_1|$$

Thus, for the same real power requirement, the circuit would draw more current if the power factor is poor. This will result in more power losses and requirement of more conductor material.

2.4.2 Concept of Power Factor Improvement

Practically, in a power system, circuits are inductive in nature. This results in poor power factor and hence flow of more current through long transmission lines, leading to high I^2R loss in the system. It is, therefore, required to compensate the inductive effect by placing capacitors in the circuits, so that they cancel the inductive effect. Figure 27 shows an R-L circuit for which

$$Z = R + jX_L = \sqrt{R^2 + X_L^2} \;\angle \tan^{-1}\frac{X_L}{R}$$

and

$$I_1 = \frac{V}{\sqrt{R^2 + X_L^2}} \angle \tan^{-1}\left(-\frac{X_L}{R}\right)$$

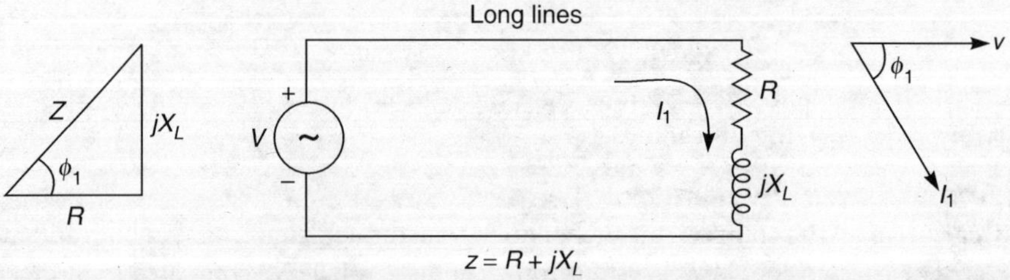

Figure 27 Inductive effects in the circuit.

Let us now place a capacitor for improving the power factor as shown in Figure 28.

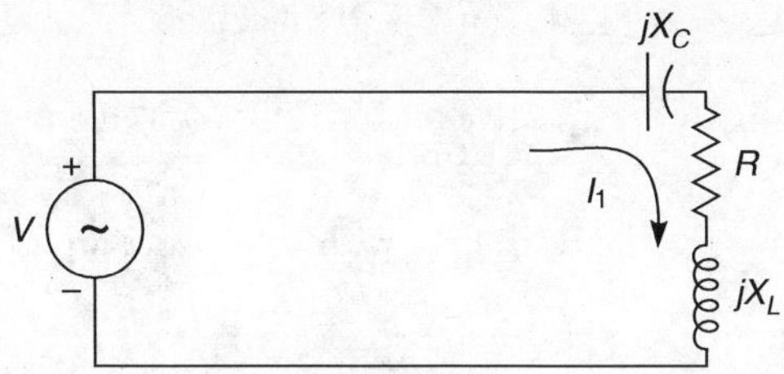

Figure 28 Power factor improvement through capacitor placed in series.

The capacitor X_C is placed in series with an *R-L* circuit, the total impedance would become

$$Z = R + jX_L - jX_C;\ Z = R + j\omega L - \frac{j}{\omega C} \qquad \left[\text{where } X_C = \frac{1}{\omega C}\right]$$

So the angle of the impedance (*Z*) will become

$$\phi_2 = \tan^{-1}\left(\frac{\omega L - \dfrac{1}{\omega C}}{R}\right)$$

As can be seen, $\phi_2 < \phi_1$. So, cos $\phi_2 >$ cos ϕ_1 and hence the power factor gets improved. Thus, the power factor can be improved by using capacitors in the inductive circuits. In power systems, static capacitor banks or synchronous condensers are used for improving the power factor.

The essence of improving the power factor is to nullify/reduce that reactive component of the impedance which is causing a reduction in the power factor. If the presence of excess capacitance is causing a poor power factor, then some inductance will have to be added to nullify or reduce the capacitive effect so that the power factor approaches unity.

2.5 Analysis of Series Circuits

Series *R-L* Circuit

For a series *R-L* circuit, subjected to a sinusoidal voltage $v = V_m \sin \omega t$, the expression for current is given by

$$I = \frac{V\angle 0}{Z\angle \theta} \qquad [\text{where } V = V_m/\sqrt{2} \text{ (the RMS value) and } Z = R + j\omega L]$$

$$= \frac{V}{Z}\angle -\theta$$

$$= \frac{V}{\sqrt{R^2 + (\omega L)^2}}\angle - \tan^{-1}\left(\frac{\omega L}{R}\right)$$

And the instantaneous current expression will be

$$i = \frac{V_m}{\sqrt{R^2 + (\omega L)^2}} \sin\left[\omega t - \tan^{-1}\left(\frac{\omega L}{R}\right)\right]$$

The power factor will be

$$\cos\left(-\tan^{-1}\frac{\omega L}{R}\right) \text{ lagging}$$

The three powers can be found by their respective formulae, as indicated in the example to follow.

Series *R-C* Circuit

For a series *R-C* circuit, the current would be leading with its phasor and its expression is given by

$$I = \frac{V\angle\theta}{Z\angle-\theta} = \frac{V}{2}\angle\theta$$

$$I = \frac{V}{\sqrt{R^2 + \left(\frac{1}{\omega C}\right)^2}}\angle+\tan^{-1}\left(\frac{1}{\omega RC}\right)$$

Thus the instantaneous current expression is

$$i = \frac{V_m}{\sqrt{R^2 + \left(\frac{1}{\omega C}\right)^2}}\sin\left[\omega t + \tan^{-1}\left(\frac{1}{\omega RC}\right)\right]$$

and the power factor will be

$$\cos\left[\tan^{-1}\left(\frac{1}{\omega RC}\right)\right] \text{ leading}$$

Here also the three powers (i.e. the real, reactive and the total powers) can be found as explained in the example below.

Series *R-L-C* Circuit

For series *R-L-C* circuits, the comparison of X_L and X_C is done and the current leads or lags the voltage, depending upon whether $X_C > X_L$ or $X_L > X_C$, respectively (see Figure 29).

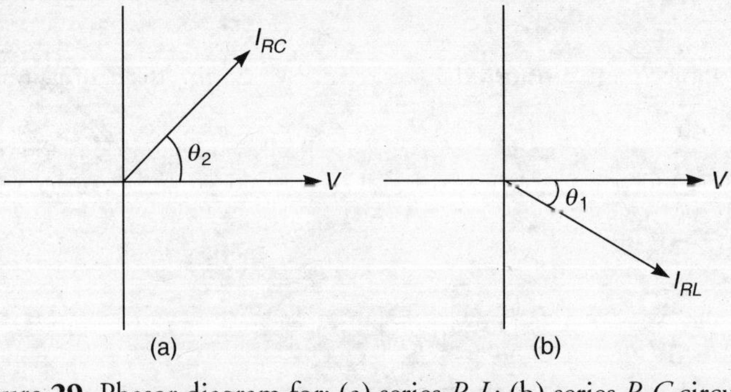

Figure 29 Phasor diagram for: (a) series *R-L*; (b) series *R-C* circuits.

As shown in Figure 30, a resistance in series with an inductive reactance is fed from a 200 V (RMS value), 50 Hz supply. Simple Ohm's law can solve this circuit for current, power factor and various values of powers. We have

$$Z = R + j\omega L = 3 + j4 = 5\angle 53.13°$$

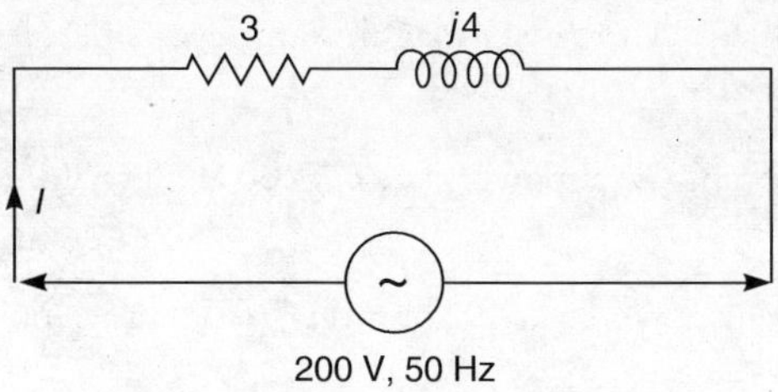

Figure 30 A series *R-L* circuit.

and current

$$I = \frac{200\angle 0}{5\angle 53.13} = 40\angle -53.13$$

Thus, the current is lagging behind the voltage by 53.13° and the pf will thus be cos 53.13° = 0.6 lagging. Hence,

$$\text{Active power} = I^2R = 40^2 \times 3 = 4800 \text{ W}$$

This is the power consumed in the circuit. Also,

$$VI \cos \phi = 200 \times 40 \times 0.6 = 4800 \text{ W}$$

This is the active power developed by the source. Also,

$$\begin{aligned}\text{Active power} &= \text{Re}(VI^*) = \text{Re}(200\angle 0 \times 40\angle -53.13) \\ &= \text{Re}\,[200 \times (40 \cos 53.13 + j\,40 \sin 53.13)] \\ &= 200 \times 40 \cos 53.13 \\ &= 4800 \text{ W}\end{aligned}$$

Similarly, the reactive power can be found using three ways.

$$\text{Reactive power} = I^2X = 40^2 \times 4 = 6400 \text{ VAR}$$

or

$$\begin{aligned}\text{Reactive power} &= VI \sin \phi \text{ (where } \phi = \text{Angle of voltage} - \text{Angle of current)} \\ &= 200 \times 40 \times \sin 53.13 \\ &= 8000 \times 0.8 \\ &= 6400 \text{ VAR}\end{aligned}$$

Also

$$\begin{aligned}\text{Reactive power} &= \text{Im}(VI^*) \\ &= \text{Im}[(200) \times (40 \cos 53.13 + j40 \sin 53.13)] \text{[+sign is due the conjugate of current]} \\ &= j \times 200 \times 40 \sin 53.13 \\ &= j200 \times 40 \times 0.8 \\ &= j6400 \text{ VAR}\end{aligned}$$

The presence of $+j$ in front of the reactive power magnitude suggests that it is the inductive reactive power. For $-j$, it would have been capacitive reactive power.

$$\text{Apparent (total power)} = VI = 200 \times 40 = 8000 \text{ VA}$$

Also

$$\text{Apparent power} = \sqrt{(\text{Real power})^2 + (\text{Reactive power})^2}$$

$$= \sqrt{(4800)^2 + (6400)^2} = 8000 \text{ VA}$$

2.6 Analysis of Parallel AC Circuits

As shown in Figure 31, the voltage across R and $j\omega L$ is same, that is V (RMS $= V_m/\sqrt{2}$). Hence current through R is

$$I_R = \frac{V\angle 0^\circ}{R} = \frac{V}{R}\angle 0^\circ$$

and current through $j\omega L$ is

$$I_L = \frac{V\angle 0}{\omega L\angle 90^\circ} = \frac{V}{\omega L}\angle -90^\circ$$

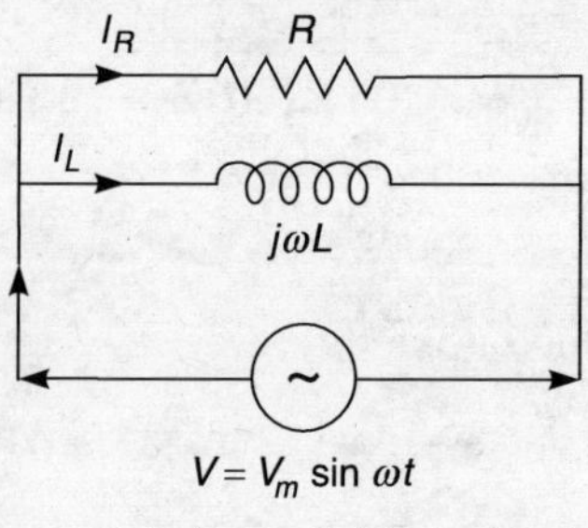

Figure 31 A parallel R-L circuit.

Total supply current is

$$I = I_R + I_L = \frac{V}{R}\angle 0^\circ + \frac{V}{\omega L}\angle -90^\circ \qquad (\text{as } j\omega L = \omega L\angle 90^\circ)$$

$$= V\left(\frac{1}{R} - \frac{j}{\omega L}\right) = V\sqrt{\frac{1}{R^2} + \left(\frac{1}{\omega L}\right)^2}\angle -\tan^{-1}\left[\frac{1/(\omega L)}{1/R}\right]$$

$$= V \times \sqrt{\frac{1}{R^2} + \left(\frac{1}{\omega L}\right)^2}\angle -\tan^{-1}\left(\frac{R}{\omega L}\right)$$

Figure 32 shows a phasor diagram for the parallel R-L circuit.

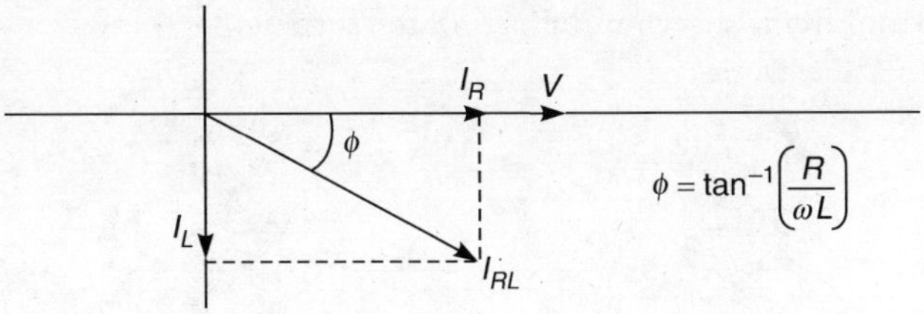

Figure 32 Phasor diagram for the parallel *R-L* circuit.

In the parallel *R-C* circuit shown in Figure 33, we have

$$I_C = \frac{V\angle 0^\circ}{\frac{1}{\omega C}\angle -90^\circ} = V\omega C\angle 90^\circ \quad \left[V = \frac{V_m}{\sqrt{2}};\ \frac{1}{j\omega C} = \frac{-j}{\omega C} = \frac{1}{\omega C}\angle -90^\circ\right]$$

and

$$I_R = \frac{V\angle 0^\circ}{R} = \frac{V}{R}\angle 0^\circ$$

Adding them, we get

$$I = I_R + I_C = \frac{V}{R}\angle 0^\circ + V\omega C\angle 90^\circ = V\left(\frac{1}{R} + \omega C\angle 90^\circ\right)$$

$$= V\left(\frac{1}{R} + j\omega C\right) = V\sqrt{\left(\frac{1}{R}\right)^2 + (\omega C)^2}\angle \tan^{-1}(\omega CR)$$

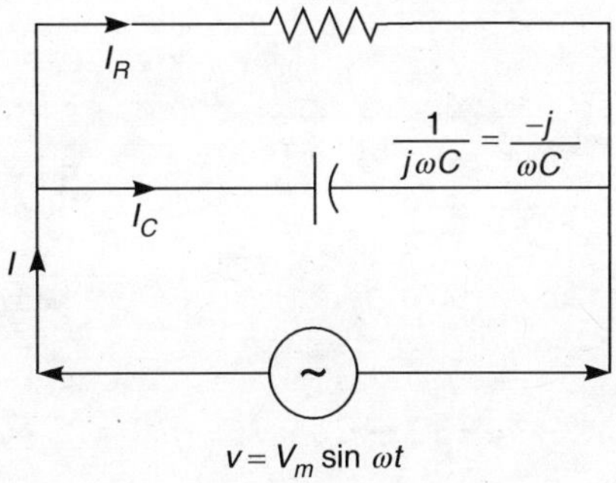

Figure 33 Parallel *R-C* circuit.

Phasor diagram for *R-C* circuit is shown in Figure 34.

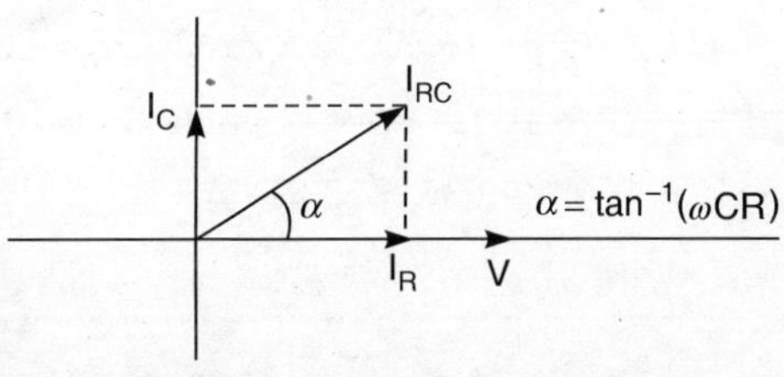

Figure 34 Phasor diagram for parallel *R-C* circuit.

For example, let us solve the circuit shown in Figure 35 for currents, pf, powers and phasor diagram. Here we have (say)

$$Z_1 = 8 - j6 = 10\angle{-36.87^\circ} \quad \text{and} \quad Z_2 = 3 + j4 = 5\angle 53.13^\circ$$

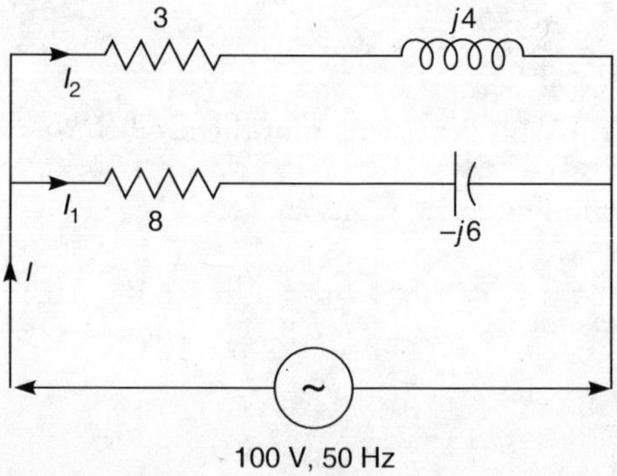

Figure 35 An example of a parallel circuit.

Since the same voltage of 100 V RMS value is applied across both Z_1 and Z_2,

$$I_1 = \frac{V\angle 0}{Z_1} = \frac{100\angle 0^\circ}{10\angle{-36.87^\circ}} = 10\angle 36.87$$

So,

$$I_1 = 10\angle 36.87 = 10(\cos 36.87 + j\sin 36.87) = (8 + j6)\text{ A}$$

Also

$$I_2 = \frac{V\angle 0}{Z_2} = \frac{100\angle 0}{5\angle 53.13} = 20\angle{-53.13}$$

or

$$I_2 = 20\angle{-53.13} = 20(\cos 53.13 - j\sin 53.13) = (12 - j16)\text{ A}$$

Now

$$I = I_1 + I_2 = 8 + j6 + 12 - j16 = 20 - j10 = 22.36\angle{-26.56}$$

Thus, we find that with voltage phasor at zero angle (i.e. along the positive x-axis), the supply current phasor I is having an angle of -26.56°. Hence the overall power factor is $\cos(-26.56) = 0.89$ lagging. The three powers can now be found as

$$P_{\text{active}} = VI\cos\phi = 100 \times 22.36 \times \cos 26.56 = 2000\text{ W}$$

$$P_{\text{reactive}} = VI\sin\phi = 100 \times 22.36 \times \sin 26.56 = 1000\text{ VARs (inductive)}$$

$$P_{\text{apparent}} = VI = 100 \times 22.36 = 2236\text{ VAs}$$

P_{active} can be calculated as

$$I_1^2 R_1 + I_2^2 R_2 = 10^2 \times 8 + 20^2 \times 3 = 2000\text{ W}$$

P_{reactive} also equals to

$$I_1^2 jX_1 - I_2^2 jX_2 = 10^2 \times j6 - 20^2 \times j4 = -j1000 \text{ VARs } = 1000 \text{ VARs (inductive)}$$

As already stated, the inductive reactive power is taken positive and the capacitive reactive power as negative. Phasor diagram for the circuit of Figure 35 is shown in Figure 36.

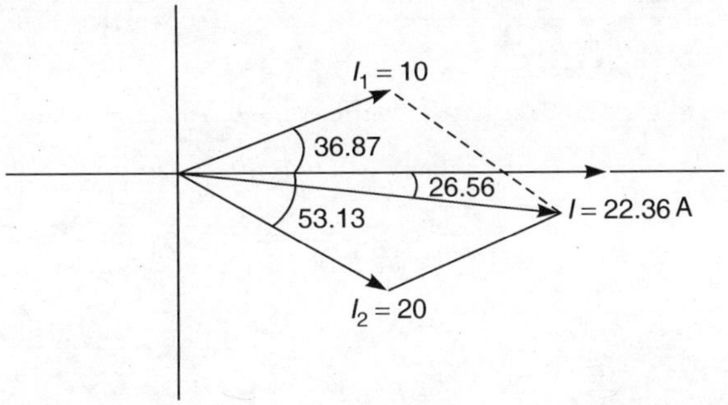

Figure 36 Phasor diagram for the circuit of Figure 35.

2.7 Analysis of Series–Parallel Circuits

Lets us illustrate the analysis of a series–parallel circuit with the help of an example shown in Figure 37. We are required to find I_1, I_2, I, pf, three powers and plot the phasor diagram.

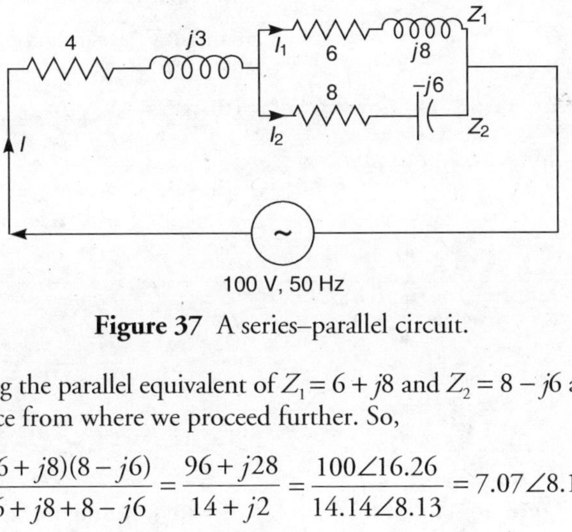

Figure 37 A series–parallel circuit.

Here we can start by finding the parallel equivalent of $Z_1 = 6 + j8$ and $Z_2 = 8 - j6$ and adding it with $(4 + j3)$ to get the overall impedance from where we proceed further. So,

$$Z_1 \parallel Z_2 = \frac{(6+j8)(8-j6)}{6+j8+8-j6} = \frac{96+j28}{14+j2} = \frac{100\angle 16.26}{14.14\angle 8.13} = 7.07\angle 8.13 = 7 + j10$$

Now, $4 + j3$ is in series with $7 + j10$. Adding them, we get

$$Z = 4 + j3 + 7 + j10 = 11 + j13 = 17.03\angle 49.76$$

Current is given by

$$I = \frac{100\angle 0}{17.03\angle 49.76} = 5.87\angle -49.76$$

This current will flow through $(4 + j3)\Omega$ impedance and then divide in Z_1 and Z_2. Thus applying the current divide theorem, we get

$$I_1 = \frac{5.87\angle -49.76 \times (8 - j6)}{6 + j8 + 8 - j6} = 4.15\angle -94.76 = -0.34 - j4.14$$

$$I_2 = \frac{5.87\angle -49.76 \times (6 + j8)}{6 + j8 + 8 - j6} = 4.15\angle -4.76 = 4.13 - j0.344$$

The power factor is the cosine angle of the current I (as the voltage is assumed to be having a zero angle). Hence

$$\text{Power factor} = \cos(-49.76) = 0.64 \text{ lagging}$$

Then

$$\text{Real power} = P_{\text{active}} = VI \cos \phi = 100 \times 5.87 \times 0.64 = 379.2 \text{ W}$$

As before, real power is also equal to

$$I^2 \times 4 + I_1^2 \times 6 + I_2^2 \times 8$$

that is

$$5.87^2 \times 4 + 4.15^2 \times 6 + 4.15^2 \times 8 = 379 \text{ W}$$

Hence

$$\begin{aligned} \text{Reactive power} &= VI \sin \phi \\ &= 100 \times 5.87 \times \sin(-49.76) \\ &= -448.1 \text{ VAR} \\ &= 448.1 \text{ VAR (inductive)} \end{aligned}$$

and

$$\text{Apparent power} = VI = 100 \times 5.87 \text{ VAR}$$

Phasor diagram for the circuit is shown in Figure 38.

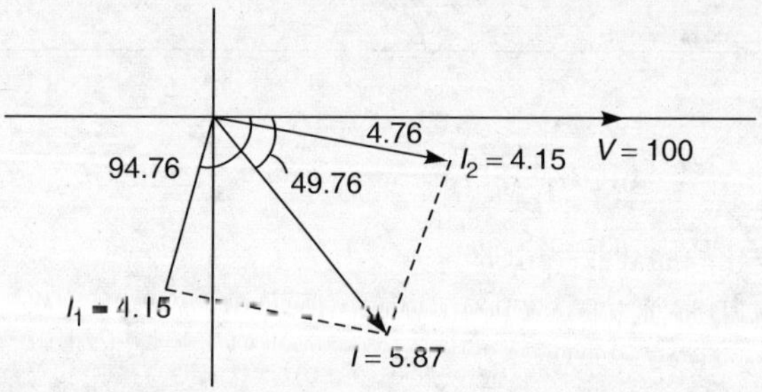

Figure 38 Phasor diagram for the circuit of Figure 37.

2.8 Resonance in Series *R-L-C* Circuit

Resonance, in general, is the behavior of a system under a condition where the frequency of the external system matches with the natural frequency of the system. In electrical circuits, this phenomenon is observed when the frequency of the source matches with the natural frequency of the circuit determined by the values of L and C contained in it. Consider a series R-L-C circuit fed by a variable frequency sinusoidal supply, as shown in Figure 39.

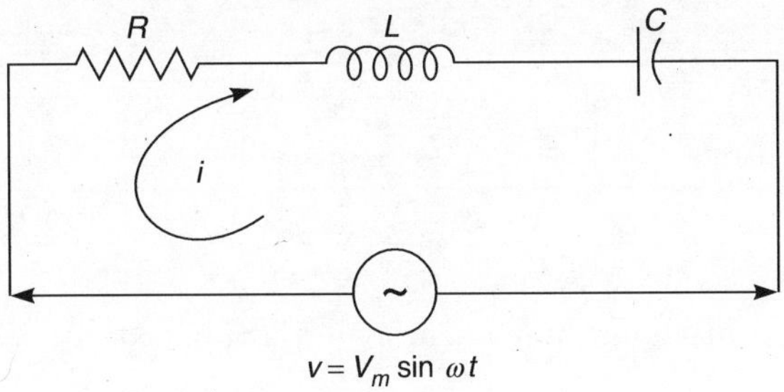

Figure 39 A series R-L-C circuit.

Here

$$Z = R + j\left(\omega L - \frac{1}{\omega C}\right)$$

We get

$$i = \frac{v}{Z} = \frac{V_m \sin(\omega t - \phi)}{\sqrt{R^2 + \left(\omega L - \frac{1}{\omega C}\right)^2}}$$

where

$$\phi = \tan^{-1}\left(\frac{\omega L - \frac{1}{\omega C}}{R}\right) \quad \left[\text{assuming } \omega L > \frac{1}{\omega C}\right]$$

It may so happen that when we vary the frequency of the source, the values of inductive and capacitive reactances may become equal and hence may cancel each other, rendering the impedance as purely resistive. This condition is known as a *series resonance.* Thus, at series resonance

$$\omega L = \frac{1}{\omega C}$$

where

$$\omega = \frac{1}{\sqrt{LC}}, \qquad f = \frac{1}{2\pi\sqrt{LC}}$$

This frequency is called *resonant frequency*, f_r (in hertz) or ω_r (in radians/second). We can say that at the time of series resonance, the imaginary part of the impedance is equated to zero. Circuit is capacitive before resonance because $1/(\omega C) > \omega L$ for $\omega < \omega_r$. The circuit is inductive after resonance because $\omega L > 1/(\omega C)$ for $\omega > \omega_r$.

Now, if the frequency of the input voltage is varied and the RMS values of currents are found at every frequency, a plot between the current and frequency (known as frequency response for the R-L-C series circuit) would look like Figure 40.

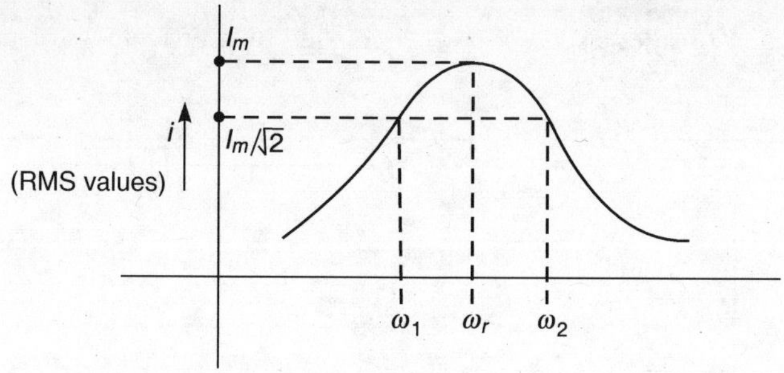

Figure 40 Current is maximum at resonance due to the impedance becoming minimum.

All the current values in Figure 40 are RMS values at various frequencies. I_m is the highest RMS value, which occurs at ω_r. Since there is no reactance present at the time of series resonance, the circuit impedance is purely resistive and also minimum (because the inductive and capacitive reactances completely nullify each other). The power factor at the resonance is thus unity. This causes the current to have its maximum value at the resonant frequency, as shown in Figure 40. Hence,

$$\text{Power dissipated at resonance} = I_m^2 R$$

where I_m is the largest RMS value of the current occurring at resonance. Half of this power will be

$$\frac{I_m^2 R}{2} = \left(\frac{I_m}{\sqrt{2}}\right)^2 R$$

This current value of $I_m/\sqrt{2}$ will occur once before (at ω_1) and once after (at ω_2) the resonance, as shown in Figure 40. The difference of $(\omega_2 - \omega_1)$ is called the *bandwidth* of the *R-L-C* series circuit. This condition (of the current magnitude being $I_m/\sqrt{2}$) will occur when $R = X$, that is the net reactance (in ohms) is equal to the resistance (in ohms). Thus when $R = X$,

$$I_{\text{RMS}} = \frac{V_m/\sqrt{2}}{\sqrt{R^2 + R^2}} = \frac{V_m/\sqrt{2}}{\sqrt{2}R} = \frac{I_m}{\sqrt{2}}$$

It is again reminded that I_m is the RMS value of the current at the time of resonance and not its peak value. Now $R = X$ means

$$R = \frac{1}{\omega_1 C} - \omega_1 L \quad \text{(before resonance)}$$

$$R = \omega_2 L - \frac{1}{\omega_2 C} \quad \text{(after resonance)}$$

The first equation leads to the quadratic equation

$$\omega_1^2 LC + \omega_1 RC - 1 = 0 \tag{2.4}$$

and the second equation leads to another quadratic equation

$$\omega_2^2 LC - \omega_2 RC - 1 = 0 \tag{2.5}$$

Solving for the roots (ω_1 and ω_2) by *Sridharacharya method* and finding the difference gives us

$$\omega_2 - \omega_1 = \frac{R}{L} \tag{2.6}$$

or

$$f_2 - f_1 = \frac{R}{2\pi L} \tag{2.7}$$

Equations (2.6) and (2.7) are the expressions of the bandwidth in rad/second and hertz, respectively.

Quality factor

It is a ratio of two energies and is denoted by the letter Q:

$$Q = 2\pi \frac{\text{Energy stored (by the inductor or the capacitor)}}{\text{Energy dissipated per cycle}}$$

Now, the energy stored by the inductor is

$$\frac{1}{2} L I_{\max}^2$$

and the energy dissipated in the resistance per cycle will be

$$\frac{I_{\text{RMS}}^2 R}{f}$$

Using the relation $I_{\max} = \sqrt{2} I_{\text{RMS}}$, we get

$$Q = 2\pi \frac{(1/2) L I_{\max}^2}{I_{\text{RMS}}^2 R/f} = 2\pi f \frac{L}{R} = \frac{\omega L}{R}$$

If the capacitive energy expression is used, we have

$$Q = 2\pi \frac{(1/2) C V_{c(\max)}^2}{I_{\text{RMS}}^2 R/f}$$

where $V_{c\,(\max)}$ is the maximum voltage across the capacitor. Thus

$$I_{\max} = \frac{V_{c(\max)}}{1/(\omega C)} = V_{c(\max)}\,\omega C \quad \text{or} \quad V_{c(\max)} = \frac{I_{\max}}{\omega C}$$

Putting the above value in the expression for Q, we get

$$Q = 2\pi \frac{\frac{1}{2} C \frac{I_{\max}^2}{\omega^2 C^2}}{I_{\text{RMS}}^2 R/f} = \frac{1}{\omega C R}$$

Now, if we put the expression for the resonant frequency, that is $\omega_r = 1/\sqrt{LC}$ in either of the two expressions obtained for Q we get

$$Q = \frac{\omega L}{R} = \frac{L/\sqrt{LC}}{R} = \frac{1}{R}\sqrt{\frac{L}{C}}$$

which is also an expression for the quality factor. Now, we have seen that the bandwidth is

$$\omega_2 - \omega_1 = \Delta\omega = \frac{R}{L}$$

Multiplying by the resonant frequency in numerator and the denominator, we get

$$\Delta\omega = \frac{\omega_r R}{\omega_r L} = \frac{\omega_r}{\omega_r L/R} = \frac{\omega_r}{Q}$$

which gives

$$Q = \frac{\omega_r}{\Delta\omega} = \frac{\text{Resonant frequency}}{\text{Bandwidth}}$$

Thus, we have four various expressions for the quality factor as follows:

$$Q = \frac{\omega L}{R}, \quad Q = \frac{1}{\omega CR}, \quad Q = \frac{1}{R}\sqrt{\frac{L}{C}}, \quad Q = \frac{\text{Resonant frequency}}{\text{Bandwidth}}$$

Resonance in parallel *R-L-C* circuit

If the inductance, capacitance and the resistance are connected in parallel, the effect of resonance changes. As shown in Figure 41, a resistance, inductance and capacitance in parallel form a simple parallel *R-L-C* circuit.

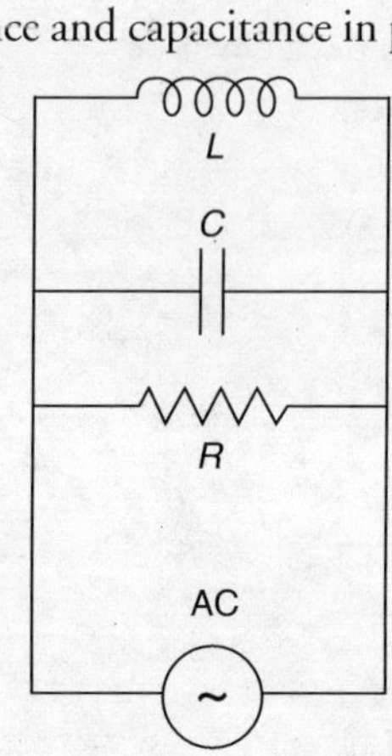

Figure 41 A parallel resonance *R-L-C* circuit.

The net admittance of the circuit is given by

$$Y = \frac{1}{R} + j\omega C + \frac{1}{j\omega L} = \frac{1}{R} + j\omega C - \frac{j}{\omega L} = \frac{1}{R} + j\left(\omega C - \frac{1}{\omega L}\right)$$

In parallel resonance, *the imaginary part of the admittance is equated to zero*. Thus, at resonance

$$\omega C = \frac{1}{\omega L}$$

which gives $\omega = \omega_{rp} = 1/\sqrt{LC}$ (where ω_{rp} is the parallel resonant frequency, which is same as the series resonant frequency). The *energy expression* for the quality factor of a parallel *R-L-C* circuit is same. That is,

$$Q = 2\pi \frac{\text{Energy stored (by the inductor or the capacitor)}}{\text{Energy dissipated per cycle}} = 2\pi \frac{(1/2)CV_{\max}^2}{V_{\text{RMS}}^2/(Rf)} = \omega_{rp} RC$$

In terms of inductive energy stored, the expression for Q looks like

$$Q = 2\pi \frac{(1/2)LI_{\max}^2}{V_{\text{RMS}}^2/(Rf)}$$

where $I_{\max}$ is the maximum current through the inductor. Now

$$I_{\max} = \frac{V_{\max}}{\omega_{rp} L}$$

Putting this value, we get

$$Q = \frac{R}{\omega_{rp} L}$$

As shown in Figure 42, a practical parallel resonance circuit consists of an *R-L* coil in parallel with an *R-C* series branch. The expression for the resonant frequency for this circuit needs to be found out.

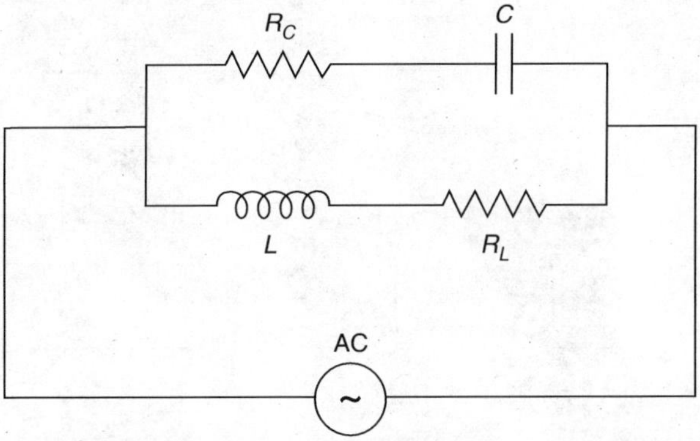

Figure 42 A practical parallel resonance circuit.

From Figure 41, we have

$$\text{Admittance of the } R\text{-}L \text{ branch, } Y_1 = \frac{1}{R_L + j\omega L}$$

$$\text{Admittance of the } R\text{-}C \text{ branch, } Y_2 = \frac{1}{R_C - \dfrac{j}{\omega C}}$$

Hence, admittance of the complete circuit is

$$Y = Y_1 + Y_2 = \frac{1}{R_L + j\omega L} + \frac{1}{R_C - \dfrac{j}{\omega C}}$$

Thus

$$Y = \frac{(R_C - j/\omega C) + (R_L + j\omega L)}{(R_C - j/\omega C)(R_L + j\omega L)} = \frac{R_C + R_L + j\left(\omega L - \dfrac{1}{\omega C}\right)}{\left(R_C R_L + \dfrac{L}{C}\right) + j\left(\omega L R_C - \dfrac{R_L}{\omega C}\right)}$$

Rationalizing the denominator, we get

$$Y=\frac{\left[R_C+R_L+j\left(\omega L-\frac{1}{\omega C}\right)\right]\left[\left(R_CR_L+\frac{L}{C}\right)-j\left(\omega LR_C-\frac{R_L}{\omega C}\right)\right]}{\left(R_CR_L+\frac{L}{C}\right)^2+\left(\omega LR_C-\frac{R_L}{\omega C}\right)^2}$$

Separating the real and imaginary parts of Y, we get

$$Y=\frac{(R_C+R_L)\left(R_CR_L+\frac{L}{C}\right)+\left(\omega L-\frac{1}{\omega C}\right)\left(\omega LR_C-\frac{R_L}{\omega C}\right)}{\left(R_CR_L+\frac{L}{C}\right)^2+\left(\omega LR_C-\frac{R_L}{\omega C}\right)^2}$$

$$+j\frac{\left(R_CR_L+\frac{L}{C}\right)\left(\omega L-\frac{1}{\omega C}\right)-(R_C+R_L)\left(\omega LR_C-\frac{R_L}{\omega C}\right)}{\left(R_CR_L+\frac{L}{C}\right)^2+\left(\omega LR_C-\frac{R_L}{\omega C}\right)^2}$$

As stated before, equating the imaginary (reactive) part of the admittance Y to zero, we get

$$\frac{\left(R_CR_L+\frac{L}{C}\right)\left(\omega L-\frac{1}{\omega C}\right)-(R_C+R_L)\left(\omega LR_C-\frac{R_L}{\omega C}\right)}{\left(R_CR_L+\frac{L}{C}\right)^2+\left(\omega LR_C-\frac{R_L}{\omega C}\right)^2}=0$$

or $$\left(R_CR_L+\frac{L}{C}\right)\left(\omega L-\frac{1}{\omega C}\right)=(R_C+R_L)\left(\omega LR_C-\frac{R_L}{\omega C}\right)$$

or $$R_CR_L\omega L-R_CR_L\frac{1}{\omega C}+\frac{L}{C}\omega L-\frac{L}{C}\frac{1}{\omega C}=(R_C+R_L)(\omega LR_C)-(R_C+R_L)\frac{R_L}{\omega C}$$

or $$\frac{L}{C}\omega L-\frac{L}{C}\frac{1}{\omega C}=R_C^2\omega L-\frac{R_L^2}{\omega C}$$

or $$\frac{L}{C}\omega L-R_C^2\omega L=\frac{L}{C}\frac{1}{\omega C}-\frac{R_L^2}{\omega C}$$

or $$\omega L\left(\frac{L}{C}-R_C^2\right)=\frac{1}{\omega C}\left(\frac{L}{C}-R_L^2\right)$$

or $$\omega^2LC=\frac{(L/C)-R_L^2}{(L/C)-R_C^2}$$

or $$\omega^2LC=\frac{L-CR_L^2}{L-CR_C^2}$$

or
$$\omega^2 = \frac{1}{LC}\frac{L - CR_L^2}{L - CR_C^2}$$

or
$$\omega_{rp} = \sqrt{\frac{1}{LC}\frac{(L - CR_L^2)}{(L - CR_C^2)}}$$

Thus this is the expression for the resonant frequency ω_{rp}. We see that if $R_L = R_C$, then

$$\omega_{rp} = \sqrt{\frac{1}{LC}}$$

which is same as the ideal parallel or series *R-L-C* circuit.

Solved Examples

Example 1

For two phasors, $\bar{A} = a_1 + jb_1$ and $\bar{B} = a_2 - jb_2$, obtain the multiplication and division resultants by using polar form of representation. **(UPTU December 2009)**

Solution

Given

$$\bar{A} = a_1 + jb_1 = \sqrt{a_1^2 + b_1^2}\angle\tan^{-1}\left(\frac{b_1}{a_1}\right) = |A|\angle\theta_1$$

$$\bar{B} = a_2 - jb_2 = \sqrt{a_2^2 + b_2^2}\angle-\tan^{-1}\left(\frac{b_2}{a_2}\right) = |B|\angle-\theta_2$$

Hence

$$\text{Product: } \bar{A}\cdot\bar{B} = \sqrt{a_1^2 + b_1^2}\times\sqrt{a_2^2 + b_2^2}\angle\left(\tan^{-1}\frac{b_1}{a_1} - \tan^{-1}\frac{b_2}{a_2}\right)$$

$$= |A||B|\angle(\theta_1 - \theta_2)$$

and

$$\text{Division: } \frac{\bar{A}}{\bar{B}} = \frac{|A|\angle\theta_1}{|B|\angle-\theta_2}$$

$$= \frac{\sqrt{a_1^2 + b_1^2}\angle\tan^{-1}(b_1/a_1)}{\sqrt{a_2^2 + b_2^2}\angle-\tan^{-1}(b_2/a_2)}$$

$$= \frac{\sqrt{a_1^2 + b_1^2}}{\sqrt{a_2^2 + b_2^2}}\angle\left(\tan^{-1}\frac{b_1}{a_1} + \tan^{-1}\frac{b_2}{a_2}\right)$$

$$= \frac{|A|}{|B|}\angle(\theta_1 + \theta_2)$$

Example 2

Given $v = 200 \sin 377t$ V, $i = 8 \sin (377t - 30°)$A for an AC circuit. Determine: (a) the power factor, (b) true power, (c) apparent power and (d) reactive power. **(UPTU December 2008)**

Solution

Given $v = 200 \sin 377t$, its phasor value is

$$V = \frac{200}{\sqrt{2}}\angle 0°$$

and given $i = 8\sin(377t - 30°)$, its phasor value is

$$I = \frac{8}{\sqrt{2}}\angle -30°$$

(a) The power factor is the cosine of the angle between the current and the voltage. The angle of voltage is 0° and that of current is –30°. So, the mutual angle (between them) is –30°, whose cosine is cos(–30°) = 0.866. Since the current is lagging behind the voltage, the power factor will be 0.866 lagging.

(b) The true power is given by

$$P = V_{\text{RMS}} I_{\text{RMS}} \cos\phi = \frac{200}{\sqrt{2}} \times \frac{8}{\sqrt{2}} \times 0.866 = 692.82 \text{ W}$$

(c) Reactive power is given by

$$Q = V_{\text{RMS}} I_{\text{RMS}} \sin\phi = \frac{200}{\sqrt{2}} \times \frac{8}{\sqrt{2}} \times \sin 30 = 400 \text{ VAR (inductive)}$$

By looking at the numerical value of the reactive power one is not able to tell whether it is inductive or capacitive. So, depending upon whether the current is lagging or leading, we assign inductive and capacitive VAR, respectively.

(d) Apparent power is given by

$$S = V_{\text{RMS}} I_{\text{RMS}} = \frac{200}{\sqrt{2}} \times \frac{8}{\sqrt{2}} = 800 \text{ VA}$$

Alternately

$$S = \sqrt{P^2 + Q^2} = 800 \text{ VA}$$

Example 3

Three voltages are represented by the following equations:

$$e_1 = 15\sin\omega t, \quad e_2 = 5\sin\left(\omega t + \frac{\pi}{6}\right), \quad e_3 = 10\cos\omega t$$

Represent these voltages by phasors, and calculate an expression for the resultant voltage. Check the result obtained graphically. **(UPTU May 2010)**

Solution

For the given voltages, the respective phasors will be

$$E_1 = \frac{15}{\sqrt{2}}\angle 0^\circ, \quad E_2 = \frac{5}{\sqrt{2}}\angle 30^\circ, \quad E_3 = \frac{10}{\sqrt{2}}\angle 90^\circ \qquad \left[\begin{array}{l}\text{The voltage is } e_3 = 10\cos\omega t \\ = 10\sin\left(\omega t + \frac{\pi}{2}\right)\end{array}\right]$$

Then

$$\begin{aligned}\text{Resultant voltage} &= E_1 + E_2 + E_3 \\ &= \frac{15}{\sqrt{2}}\angle 0^\circ + \frac{5}{\sqrt{2}}\angle 30^\circ + \frac{10}{\sqrt{2}}\angle 90^\circ \\ &= \frac{15}{\sqrt{2}} + \frac{5}{\sqrt{2}}(\cos 30^\circ + j\sin 30^\circ) + \frac{10}{\sqrt{2}}(\cos 90^\circ + j\sin 90^\circ) \\ &= \frac{1}{\sqrt{2}}\left(15 + \frac{5\sqrt{3}}{2} + j\frac{5}{2} + j10\right) \\ &= 13.67 + j8.84 \\ &= 16.277\angle 32.89^\circ\end{aligned}$$

whose sinusoidal representation would be

$$16.277\sqrt{2}\sin(\omega t + 32.89^\circ) = 23.02\sin(\omega t + 32.89^\circ)$$

Graphical verification of the above resultant is shown in Figure 43.

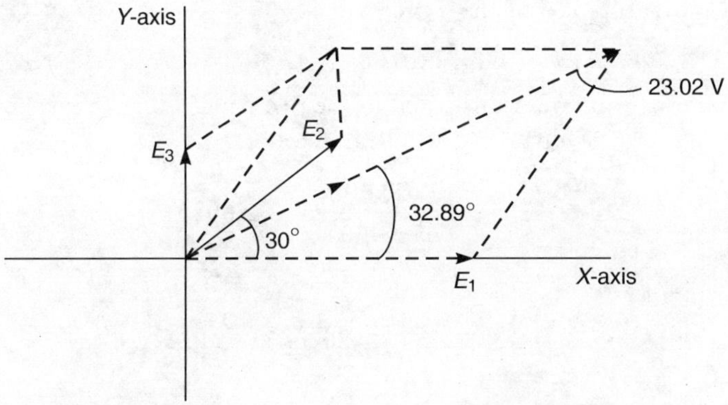

Figure 43 Showing the resultant voltage graphically.

Example 4

A voltage source $e(t) = 141 \sin 377t$ is applied to two parallel branches. The time expression for the current in the first branch is

$$i_1(t) = 7.07\sin\left(\omega t - \frac{\pi}{3}\right)$$

and in the second branch is

$$i_2(t) = 10\sin\left(\omega t - \frac{\pi}{6}\right)$$

Compute the total power supplied by the source. **(UPTU June 2009)**

Solution

Given $e(t) = 141 \sin 377t$, its phasor will be

$$E = \frac{141}{\sqrt{2}}\angle 0° = 100\angle 0°$$

Similarly,

$$i_1(t) = 7.07 \sin\left(\omega t - \frac{\pi}{3}\right) \Rightarrow I_1 = \frac{7.07}{\sqrt{2}}\angle{-60°} = 5\angle{-60°}$$

and

$$i_2(t) = 10\sin\left(\omega t - \frac{\pi}{6}\right) \Rightarrow I_2 = \frac{10}{\sqrt{2}}\angle{-30°} = 7.07\angle{-30°}$$

Now since the two branches in which I_1 and I_2 are flowing are connected in parallel, the total supply current I can be found by summing I_1 and I_2. That is,

$$I = I_1 + I_2 = 5\angle{-60°} + 7.07\ \angle{-30°} = 2.5 - j4.33 + 6.12 - j3.535$$
$$= 8.62 - j7.865 = 11.67\angle{-42.38°}$$

Now the total active power supplied by the source will be

$$VI \cos \phi = 100 \times 11.67 \cos(-42.38°) = 859.5 \text{ W}$$

Second method: The total real power dissipated in the two parallel branches is obtained by adding the individual I^2R losses. Resistance of branch 1 will be the real part of

$$Z_1 = \frac{E}{I_1} = \frac{100\angle 0°}{5\angle{-60°}} = 20\angle 60°$$

or

$$R_1 = \text{Re}(20\angle 60°) = 20\cos 60° = 10\ \Omega$$

Hence

$$I_1^2 \times 10 = 5^2 \times 10 = 250 \text{ W}$$

Similarly, the resistance of branch 2 will be the real part of

$$Z_2 = \frac{E}{I_2} = \frac{100\angle 0°}{7.07\angle{-30°}} = 14.14\angle 30°\ \Omega$$

or

$$R_2 = \text{Re}(14.14\angle{-30°}) = 14.14\cos(-30) = 12.25\ \Omega$$

Hence

$$I_2^2 \times 12.25 = 7.07^2 \times 12.25 = 612.28 \text{ W}$$

Sum of the two powers (250 + 612.28 = 862.28 W) is same as the value of power found before with minor calculation differences.

Example 5

Show that average power demand in a purely inductive ac circuit is zero (see Figure 44).

(UPTU 2002–03)

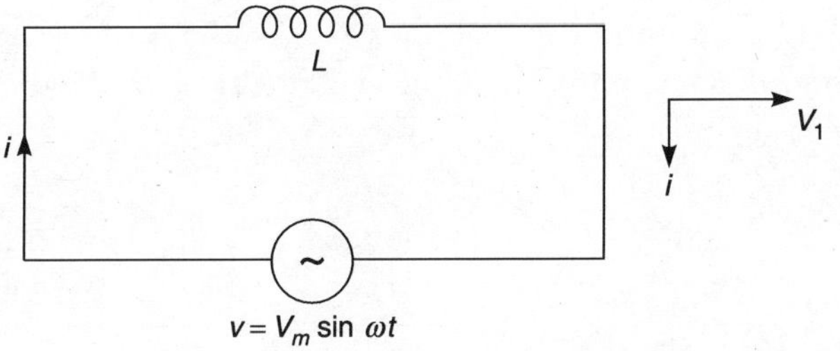

Figure 44 Figure for Example 5.

Solution

Let an alternating voltage v is applied to a purely inductive coil. Because of self-inductance of the coil, a back emf is produced. The back emf at every step opposes the rise or fall of current through the coil. Since there is no ohmic voltage drop, the applied voltage will compensate only with self-induced emf. So, at every step

$$V = L\frac{di}{dt}$$

Now

$$v = V_m \sin \omega t \;\Rightarrow\; V_m \sin \omega t = L\frac{di}{dt} \;\Rightarrow\; di = \frac{V_m}{L}\sin \omega t\, dt$$

On integrating, we get

$$i = \frac{V_m}{L}\int \sin \omega t\; dt = -\frac{V_m}{\omega t}\cos \omega t \quad \text{or} \quad i = I_m \sin\left(\omega t - \frac{\pi}{2}\right) \quad \left[\text{Let } I_m = \frac{V_m}{\omega t}\right]$$

Hence the phase difference between the two is $\pi/2$ with voltage leading instantaneous power given by

$$p = V_m I_m \sin \omega t \sin\left(\omega t - \frac{\pi}{2}\right) = V_m I_m \sin \omega t \cos \omega t = \frac{V_m I_m}{2}\sin 2\omega t$$

Then the power flow for whole cycle is

$$p = \frac{V_m I_m}{2}\int_0^{2\pi} \sin 2\omega t \; dt = 0$$

Hence for the whole cycle, the average power will be zero.

Example 6

An iron choke coil takes 4 A when connected to a 20 V DC supply and takes 5 A from 65 V, 50 Hz supply. Determine:

(a) The resistance and inductance of the coil.
(b) The power drawn by coil.
(c) The power factor. **(UPTU 2000–01)**

Solution

(a) The iron choke coil contains resistance and inductance. When the coil is connected by a DC source, then the inductance acts as a short circuit and only resistance will present in the circuit. Then,

$$R = \frac{20}{4} = 5\ \Omega$$

When the coil is connected by an AC source, then both resistance and inductance take part in circuit. Then the total impedance is

$$Z = \frac{65}{5} = 13\ \Omega$$

Now

$$Z = \sqrt{R^2 + X_L^2} \Rightarrow X_L = \sqrt{Z^2 - R^2} \Rightarrow X_L = \sqrt{13^2 - 5^2} = 12\ \Omega$$

$$\Rightarrow \omega L = 12 \Rightarrow L = \frac{12}{2\pi 50} = 0.038\,\text{H}$$

(b) The power drawn by coil is

$$P = I^2R = 5^2 \times 5 = 125\ \text{W}$$

(c) Power factor is given as

$$\text{pf} = \cos\phi = \frac{R}{Z} = \frac{5}{13} = 0.384$$

Example 7

A coil having a resistance of 6 Ω and an inductance of 0.0255 H is connected across a 230 V, 50 Hz AC supply. Calculate:

(a) Current
(b) Power factor
(c) Active power
(d) Reactive power
(e) Apparent power

It is desired to improve power factor to 0.8. What is the value of capacitance to be connected in series R and what is reduction in reactive power? **(UPTU December 2009)**

Solution

Given the resistance of the coil = 6 Ω, inductance = 0.0255 H. So,

$$jX_L = j2\pi fL = j2 \times 3.14 \times 50 \times 0.0255 = j8\ \Omega$$

Thus the coil impedance, $Z = 6 + j8$ (see Figure 45).

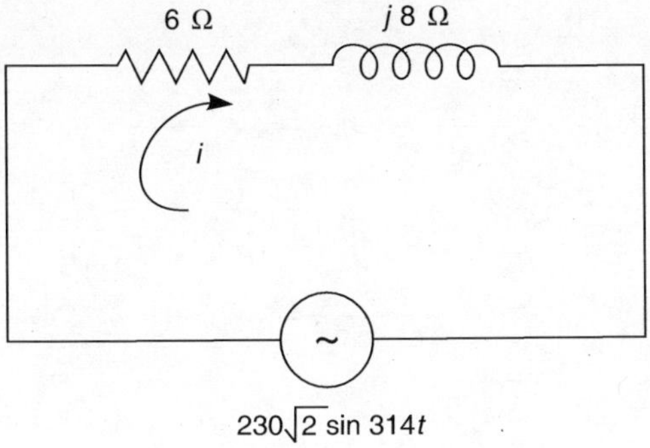

Figure 45 Figure for Example 7.

(a) The RMS value of current is

$$\frac{V_{\text{RMS}}}{Z} = \frac{230\angle 0}{6 + j8} = \frac{230\angle 0}{10\angle 53.13} = 23\angle -53.13 \text{ A}$$

Thus the current is $i = 23\sqrt{2}\sin(314t - 53.13)$.

(b) Power factor = cos(53.13) = 0.6 lagging; lagging because the circuit is inductive.

(c) Active power = $VI\cos\phi$; ϕ is the angle between voltage and current. Here $V = 230\angle 0°$, $I = 23\angle -53.13$, $\phi = 53.13$. So,

$$P = 230 \times 23 \times \cos 53.13 = 3174 \text{ W}$$

Also, $I^2R = P$. So, $23^2 \times 6 = 3174$ W.

(d) Reactive power = $VI\sin\phi = 230 \times 23 \times \sin 53.13 = 4232$ VAR. Also, $I^2X = Q$. So, $23^2 \times 8 = 4232$ VAR.

(e) Apparent power = $VI = 230 \times 23 = 5290$ VA

Refer to Figure 46.

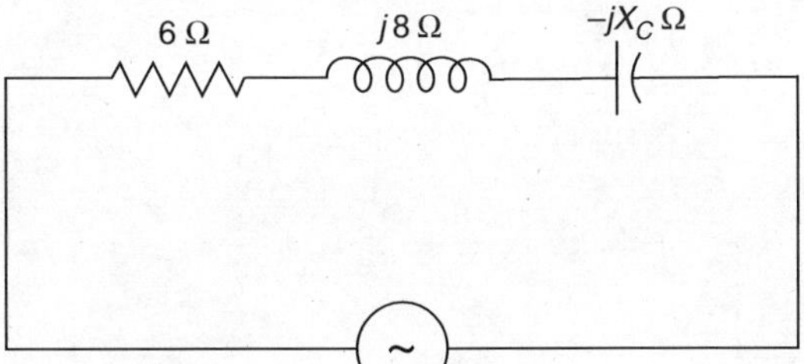

Figure 46 Solution figure for Example 7.

On connecting a capacitor in series as shown the total reactance becomes

$$j(X_L - X_C) = j\left(8 - \frac{1}{\omega C}\right)$$

Now the resistance being constant at 6 Ω, the angle of the impedance is

$$\tan^{-1}\left(\frac{8 - \dfrac{1}{\omega C}}{6}\right)$$

Now the new power factor (desired value) in 0.8, so the new impedance angle will be $\cos^{-1} 0.8 = 36.87$. Thus

$$\tan^{-1}\left(\frac{8-\dfrac{1}{\omega C}}{6}\right) = 36.87 \Rightarrow \frac{8-\dfrac{1}{\omega C}}{6} = 0.75 \Rightarrow 8-\frac{1}{\omega C} = 4.5 \Rightarrow \frac{1}{\omega C} = 8-4.5 = 3.5$$

Thus $X_C = -j3.5\ \Omega$ and total $X = +j8 - j3.5 = +j4.5\ \Omega$. Hence

$$C = \frac{1}{314 \times 3.5} = 909.9\ \mu\text{F}$$

Therefore, new reactive power = $VI \sin \phi$. This gives $V = 230$ V. $\sin \phi = 0.6$, as $\cos = \phi\ 0.8$. Net reactance value (in ohms) = $j4.5\ \Omega$. So,

$$\text{Current} = \frac{230}{\sqrt{6^2 + (4.5)^2}} = 30.67\ \text{A}$$

or

$$Q = 230 \times 30.67 \times 0.6 = 4232\ \text{VAR}$$

Example 8

A 46 mH inductive coil has a resistance of 10 Ω. Determine

(a) The amount of current drawn by the inductive coil when connected across a 100 V, 60 Hz source.

(b) The value of the capacitance that must be connected across the coil to make the power factor of the overall circuit unity.(UPTU December 2008)

Solution

Given a 46 mH inductance in series with a 10 Ω resistance. Then

$$X_L = 2 \times \pi \times f \times L = 2 \times 3.14 \times 60 \times 46 \times 10^{-3} = 17.34\ \Omega$$

This coil is fed from a 100 V, 60 Hz supply (see Figure 47).

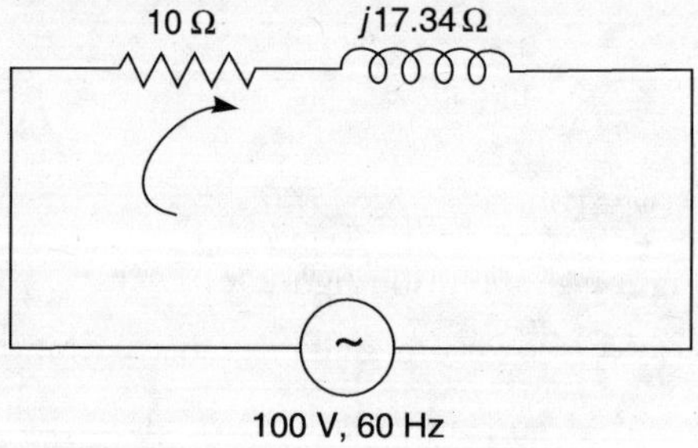

Figure 47 Figure for Example 8.

(a) The current drawn by the coil will be

$$I = \frac{V}{Z} = \frac{100}{\sqrt{10^2 + (17.34)^2}} = 5\angle{-60^\circ}$$

(b) A capacitance has to be connected across, that is parallel to the coil, to make the overall power factor unity. The resultant circuit looks as shown in Figure 48.

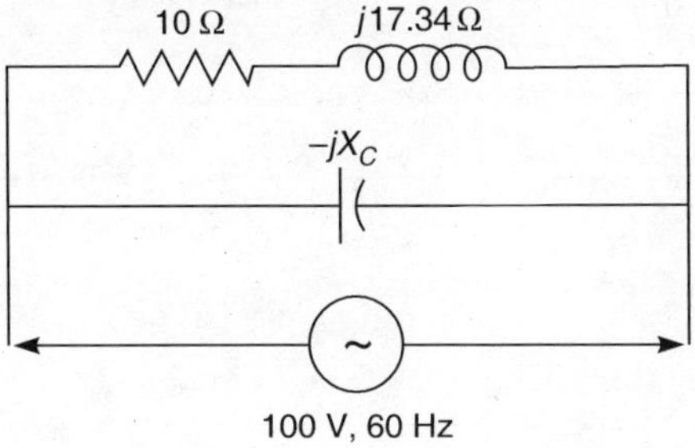

Figure 48 A capacitor connected parallel to a coil.

There are two methods of finding the value of the capacitance.

First method: We calculate the overall impedance of Figure 48 and equate its imaginary (or reactive) part to zero to get the unity power factor, so that the impedance must be purely resistive. Thus,

$$Z = \frac{(10 + j17.34) \times (-jX_c)}{10 + j17.34 + (-jX_c)} = \frac{17.34X_c - j10X_c}{10 + j(17.34 - X_c)} \quad [\text{as } j^2 = 1]$$

Rationalizing the denominator, we multiply by its conjugate [i.e. $10 - j(17.34 - X_c)$] both the numerator and the denominator. We get

$$Z = \frac{(17.34X_c - j10X_c)[10 - j(17.34 - X_c)]}{10^2 + (17.34 - X_c)^2}$$

Now since we are interested in equating the imaginary part of Z to zero, we pick only that part. That is,

$$\frac{-j17.34X_c(17.34 - X_c) - j100X_c}{10^2 + (17.34 - X_c)^2} = 0$$

or $$17.34X_c(17.34 - X_c) = -100X_c$$

or $$17.34 - X_c = \frac{-100}{17.34} = -5.76$$

or $$X_c = 17.34 + 5.76 = 23.11\ \Omega$$

or $$\frac{1}{2\pi fC} = 23.11$$

or $$\frac{1}{2 \times 3.14 \times 60 \times C} = 23.11$$

which gives $C = 114.7\ \mu$F.

Second method: We can also use the phasor diagram method to find the value of C. As shown in the phasor diagram in Figure 49, the coil (or R-L) current phasor, $I_{R\text{-}L}$, lags behind the voltage phasor V (which is along the positive x-axis) by 60°. Now, if the overall power factor has to be made unity, the supply current should make zero degree with V. This will happen only if I_C (the capacitive current phasor) is equal and opposite to the negative y-axis component of $I_{R\text{-}L}$, that is $I_{R\text{-}L}$ sin 60°.

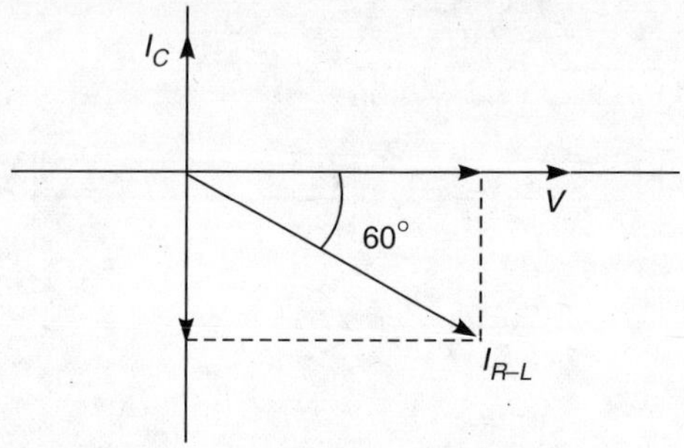

Figure 49 Phasor diagram of the circuit of Figure 47.

Thus $I_{R\text{-}L}$ sin 60° = I_C. Now, we know $I_{R\text{-}L}$ = 5A. So,

$$5 \sin 60^\circ = I_C \quad \text{or} \quad I_C = 4.33 \text{ A}$$

Also,

$$I_C = \frac{V}{1/(\omega C)} = V\omega C$$

So,

$$C = \frac{I_C}{V\omega} = \frac{4.33}{100 \times 2 \times \pi \times 60} = 114.7\ \mu\text{F}$$

Example 9

Refer to the circuit shown in Figure 50. Find: (a) RMS line current I, (b) power dissipated and (c) power factor.

(UPTU June 2009)

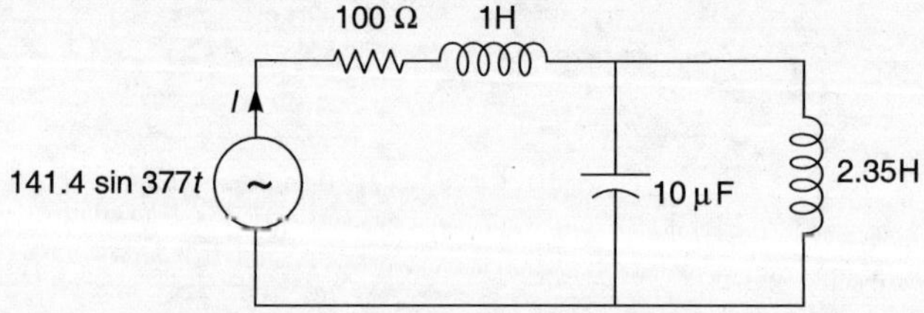

Figure 50 Figure for Example 9.

Solution

Calculating the reactance value (in ohms) with the frequency being 377 rad/sec (given), we have

$$X_{L1} = j\omega L_1 = j \times 377 \times 2.35 = j885.95\ \Omega$$

$$X_{L2} = j\omega L_2 = j \times 377 \times 1 = j377\ \Omega$$

$$X_C = \frac{1}{j\omega C} = \frac{1}{j \times 377 \times 10 \times 10^{-6}} = -j265.25\Omega$$

The circuit becomes as shown in Figure 51.

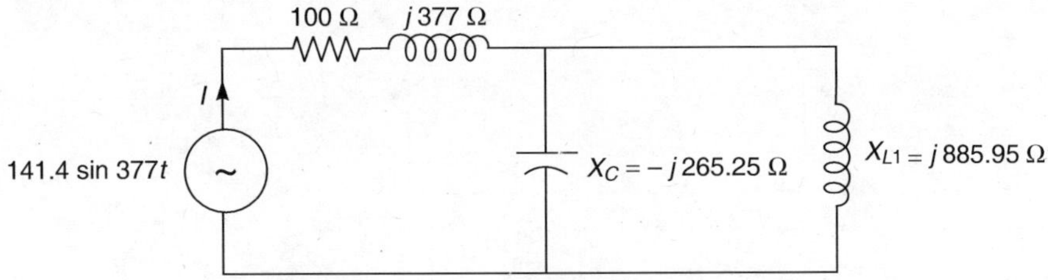

Figure 51 Resultant circuit for Example 9.

Taking the parallel combination of X_c and X_{L1}, we get

$$\frac{X_C \times X_{L1}}{X_C + X_{L1}} = \frac{-j265.25 \times j885.95}{-j265.25 + j885.95} = -j378.61\Omega$$

The circuit becomes as shown in Figure 52.

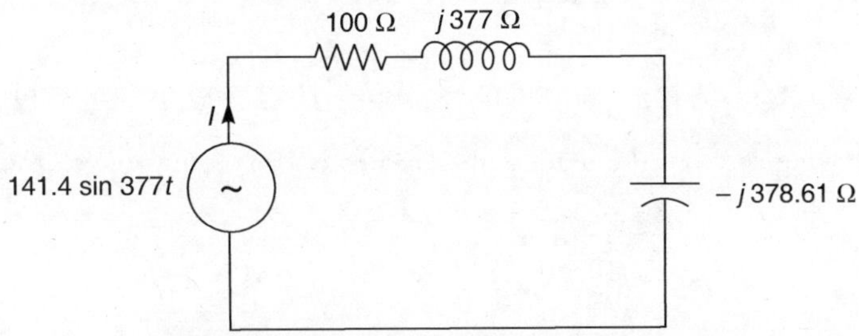

Figure 52 Final circuit for Example 9.

To calculate the RMS line current, $Z = 100 + j377 - j378.61 = 100 - j1.61$

$$I = \frac{\left(141.4/\sqrt{2}\right)\angle 0^\circ}{100 - j1.6} = 0.99\angle 0.916^\circ$$

Hence, the RMS line current will be 0.99 A. So, the power dissipated is

$$I_{\text{RMS}}^2 R = 0.99^2 \times 100 = 99.97 \text{ W}$$

and $$\text{Power factor, } \frac{R}{Z} = \frac{100}{\sqrt{100^2 + (1.61)^2}} = 0.99 \text{ lagging}$$

Example 10

In the circuit shown in Figure 53, determine the voltage at 50 Hz to be applied across AB in order that a current of 10 A flows in the capacitor. **(UPTU 2000–01)**

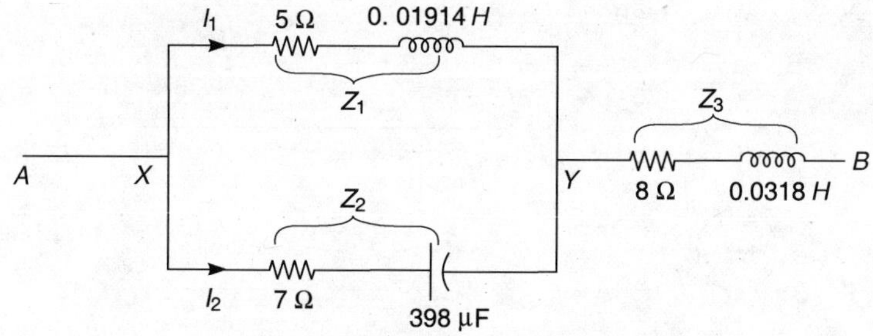

Figure 53 Figure for Example 10.

Solution

The circuit diagram of the problem is shown in Figure 53. Now,

$$Z_1 = R_1 + jX_{L_1} = 5 + j(2\pi \times 50 \times 0.0191) = 5 + j6\ \Omega = 7.8\angle 50.2°\ \Omega$$

Similarly,

$$Z_2 = 7 - j\left(\frac{1}{2\times\pi\times 50\times 398\times 10^{-6}}\right) = 7 - j8 = 10.6\angle -48.8°\ \Omega$$

and

$$Z_3 = 8 + j(2\pi \times 50 \times 0.0318) = 8 + j10\ \Omega = 12.8\angle 51.3°\ \Omega$$

Since the current flowing through the branch of capacitor is 10 A, the voltage drop across the capacitor branch is

$$V_{XY} = I_2 Z_2 = (10\angle 0°)(10.6\angle -48.8°) = 106\angle -48.8°\ \text{V}$$

Now

$$I_1 = \frac{V_{XY}}{Z_1} = \frac{106\angle -48.8°}{7.8\angle 50.2°} = 13.59\angle -99°\ \text{A}$$

The total current flowing in the circuit is

$$I = I_1 + I_2 = 13.6\angle -99° + 10\angle 0° = -2.13 - j13.4 + 10 = 7.87 - j13.4 = 15.54\angle -59.57°$$

Now voltage drop across YB is

$$V_{YB} = IZ_3 = (15.54\angle -59.57°)(12.8\angle 51.3°) = 198.9\angle -8.27°$$

Then voltage across AB is

$$\begin{aligned}\overline{V}_{AB} &= \overline{V}_{XY} + \overline{V}_{YB} = 106\angle -48.8° + 198.9\angle -8.27° \\ &= 69.8 - j79.76 + 196.9 - j28.4 \\ &= 266.7 - j108.16 = 287.8\angle -22.1°\end{aligned}$$

Example 11

Derive the response of *R-L-C* series circuit to sinusoidal input. Also, derive the condition of resonance.

(UPTU June 2011)

Solution

As shown in Figure 54, a sinusoidal input voltage $v = V_m \sin \omega t$ is applied across a series *R-L-C* circuit. The current i can be found by a simple application of Ohm's law. We have the impedance:

$$z = R + j\omega L - \frac{j}{\omega C} = R + j\left(\omega L - \frac{1}{\omega C}\right) \text{ohms}$$

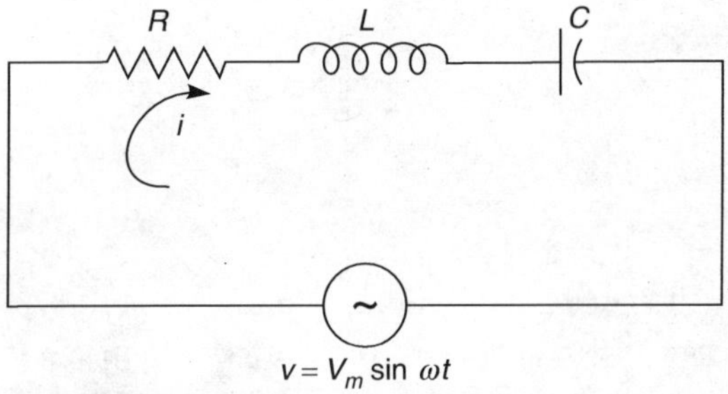

Figure 54 Figure for Example 11.

If the current phasor, that is the RMS value of the current i, is I and if the voltage phasor is assumed to be the reference, we have

$$V = \frac{V_m}{\sqrt{2}} \angle 0° \quad \text{and} \quad I = \frac{I_m}{\sqrt{2}} \angle \pm\theta$$

$+\theta$ will be there if $\omega L < 1/(\omega C)$ (current leading) and $-\theta$ will be there if $\omega L > 1/(\omega C)$ (current lagging). Thus,

$$V = IZ$$

or

$$I = \frac{V}{Z} = \frac{V_m \angle 0°/\sqrt{2}}{\sqrt{R^2 + \left(\omega L - \dfrac{1}{\omega C}\right)^2} \angle \tan^{-1}\left(\dfrac{\omega L - \dfrac{1}{\omega C}}{R}\right)}$$

or

$$I = \frac{V_m \angle -\tan^{-1}\left(\dfrac{\omega L - \dfrac{1}{\omega C}}{R}\right)}{\sqrt{2}\sqrt{R^2 + \left(\omega L - \dfrac{1}{\omega C}\right)^2}}$$

If $\omega L < 1/(\omega C)$, the angle of current would be positive and if $\omega L > 1/(\omega C)$, the angle of current would be negative. Also, maximum (or peak) value of the current is

$$I_m = \sqrt{2}\,I = \frac{V_m}{\sqrt{R^2 + \left(\omega L - \dfrac{1}{\omega C}\right)^2}}$$

The expression for the instantaneous value of current will be

$$i = I_m \sin(\omega t \pm \theta)$$

or

$$i = \frac{V_m}{\sqrt{R^2 + \left(\omega L - \dfrac{1}{\omega C}\right)^2}} \sin\left[\omega t \pm \tan^{-1}\left(\frac{\omega L - \dfrac{1}{\omega C}}{R}\right)\right]$$

As stated before ± sign will be there depending upon the nature of the circuit (capacitive or inductive respectively). When the frequency of the source voltage is varied, the values of inductive and capacitive reactances [i.e., $X_L = \omega L$ and $X_C = 1/(\omega C)$] change. As the frequency is increased, the capacitive reactance reduces and the inductive reactance increases. At a particular value of the frequency, $X_L = X_C$. That is,

$$\omega L = \frac{1}{\omega C} \quad \text{or} \quad \omega^2 = \frac{1}{LC} \quad \text{or} \quad \omega = \frac{1}{\sqrt{LC}}$$

So

$$f = \frac{1}{2\pi\sqrt{LC}}$$

This is called the *series resonant frequency*.

Example 12

A 10 mH coil is connected in series with a loss-free capacitor to a variable frequency source of 20 V. The current in the circuit has the maximum value of 2 A at a frequency of 10 kHz. Calculate: (a) resistance of the coil, (b) capacitance, (c) half-power frequencies and (d) Q factor of the coil.

Solution

Given the coil, that is the *R-L* circuit connected in series with a capacitor C makes the case of a series *R-L-C* circuit. Under the resonant condition, the circuit impedance becomes purely resistive. Hence the value of the resistance can be obtained by

$$R = 20/2 = 10\ \Omega$$

The inductive reactance at resonance:

$$\omega L = j2\times\pi\times f\times L = j2\times 3.14\times 10\times 10^3\times 10\times 10^{-3} = j628\ \Omega$$

For nullifying this reactance, an equal (and opposite) capacitive reactance is required. Thus,

$$X_C = -j628\ \Omega$$

or
$$\frac{1}{\omega C} = 628$$

or
$$C = \frac{1}{628\times 2\times 3.14\times 10\times 10^3} = 25\text{ nF}$$

The bandwidth is given by *R*/*L*, which will here be equal to

$$\frac{10}{10^{-2}} = 1000\text{ rad/second} = \frac{1000}{2\times 3.14} = 159.14\text{ Hz}$$

The two half-power frequencies will be

$$f_1 = \text{Resonant frequency} - \frac{\text{Bandwidth}}{2} = 10^4 - \frac{159.14}{2} = 9920.43\text{ Hz}$$

$$f_2 = \text{Resonant frequency} + \frac{\text{Bandwidth}}{2} = 10^4 + \frac{159.14}{2} = 10079.57\text{ Hz}$$

Hence

$$\text{Quality factor, } Q = \frac{\omega L}{R} = \frac{2\times 3.14\times 10^4\times 10^{-2}}{10} = 62.8$$

Example 13

Derive the quality factor *Q* of the series *R-L-C* circuit at resonance. Discuss the expression of *Q* in terms of energy. **(UPTU June 2009)**

Solution

The quality factor for a series *R-L-C* circuit is defined as

$$Q = 2\pi\frac{\text{Energy stored}}{\text{Energy dissipated/cycle}}$$

Thus, the quality factor is a measure of the ratio of the energy stored by the inductor, or the capacitor, to the energy dissipated in the resistance in a series *R-L-C* circuit per cycle. In an input cycle, the energy is stored and exchanged between the inductor and the capacitor, and some part of it is dissipated (lost forever) in a resistor. Taking the energy stored in the inductor, we get

$$Q = 2\pi \frac{(1/2)LI_m^2}{I^2_{\text{RMS}} R/f}$$

where I_m = peak value and I_{RMS} = RMS value of the current. Now

$$Q = \frac{\omega_r L}{R}$$

with ω_r the resonant frequency. Similarly, if the capacitive energy is considered, then

$$Q = 2\pi \frac{(1/2)CV_{\text{cm}}^2}{I_{\text{RMS}}^2 R/f}$$

where V_{cm} = peak value of the voltage across the capacitor. Now,

$$V_{\text{cm}} = I_m \times X_c = I_m \times \frac{1}{\omega C}$$

Hence,

$$Q = 2\pi \frac{\frac{1}{2} C \times I_m^2 \times \frac{1}{\omega^2 C^2}}{I_{\text{RMS}}^2 R/f} = \frac{2\pi f \times I_{\text{RMS}}^2 \times 1}{I_{\text{RMS}}^2 R \omega^2 C}$$

Here $\omega = \omega_r$ at resonance. Therefore,

$$Q = \frac{1}{\omega_r RC}$$

Example 14

Two impedances, $z_1 = (10 + j15)\ \Omega$ and $z_2 = (6 - j8)\ \Omega$, are connected in parallel. The total current supplied is 15 A. What is the power taken by the impedances? **(UPTU May 2010)**

Solution

Given impedances can be represented graphically as shown in Figure 55.

Applying the current divide theorem to find I_1 and I_2, we get

$$I_1 = \frac{15 \times (6 - j8)}{10 + j15 + 6 - j8} = \frac{90 - j120}{16 + j7} = 8.59\angle{-76.76^\circ}$$

and

$$I_2 = \frac{15(10 + j15)}{10 + j15 + 6 - j8} = \frac{150 + j225}{16 + j7} = 15.48\angle 32.68$$

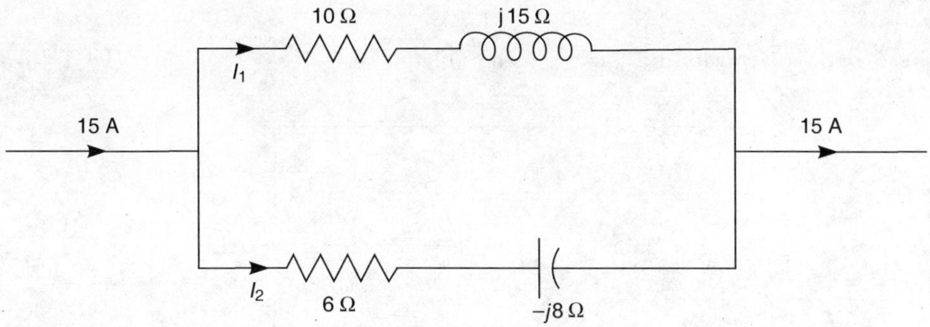

Figure 55 Circuit for Example 14.

Now, power taken by z_1 and z_2 can be found, respectively, as

$$I_1^2 R_1 = 8.59^2 \times 10 = 737.88 \text{ W}$$

$$I_2^2 R_2 = 15.48^2 \times 6 = 1437.78 \text{ W}$$

The sum of the two powers in Z_1 and Z_2 is 737.88 + 1437.78 = 2176 W

Second method: We can also find the power across any one of the branches by using the relation $P = VI \cos\phi$, where V is the RMS value of the voltage found, I is the RMS value of the total current (which is given as 15 A) and ϕ is the angle between them. Now, we have

Voltage across the $10 + j\,15\ \Omega$. impedance $= I_1 \times (10 + j15) = 8.59\angle{-76.76} \times 18.027\angle 56.31°$

Hence

$$V = 154.85\angle{-20.45}$$

We could have also found the voltage across the parallel branches as

$$I_2 Z_2 = 15.48\angle 32.68 \times (6 - j8) = 154.8\angle{-20.45°}$$

and

$$P_1 = VI_1 \cos\phi_1 = 154.85 \times 8.59 \times \cos(-56.31) = 737.8 \text{ W}$$

where (−56.31°) is the angle between the voltage V and I_1. We also have

$$P_2 = VI_2 \cos\phi_2 = 154.85 \times 15.48 \times \cos(53.013) = 1438 \text{ W}$$

where (53. 13°) is the angle between the voltage V and I_2. So

$$VI \cos\phi = 154.85 \times 15 \times \cos(-20.45) = 2176.36 \text{ W}$$

This is same as the sum of the two powers in Z_1 and Z_2 obtained by the first method.

Example 15

Determine the power factor for the circuit shown in Figure 56. **(UPTU June 2011)**

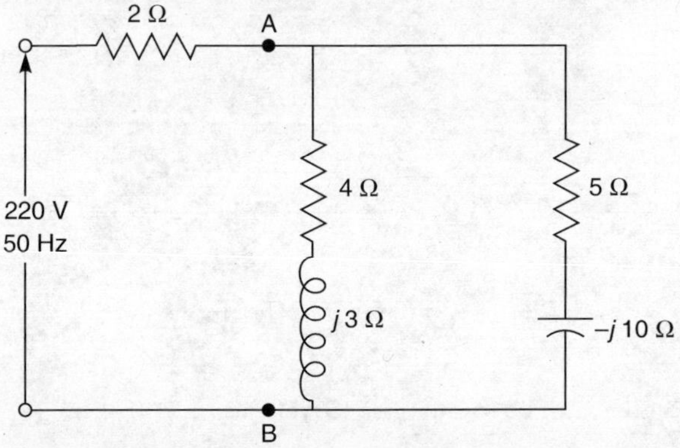

Figure 56 Figure for Example 15.

Solution

For finding the power factor, the equivalent impedance across the voltage source is calculated. Then the cosine of the angle of the impedance, that is

$$\tan^{-1}\left(\frac{\text{Reactance}}{\text{Resistance}}\right)$$

gives the power factor. The $(5 - j\,10)\ \Omega$ branch is in parallel with the $(4 + j\,3)\ \Omega$ branch. Their equivalence will be

$$Z_{\text{AB}} = \frac{(5-j10)\times(4+j3)}{(5-j10)+(4+j3)} = \frac{11.18\angle{-63.43}\ \times 5\angle 36.87}{9-j7} = \frac{55.9\angle{-26.56}}{11.4\angle{-37.95}} = 4.9\angle 11.39 = 4.8 + j\,0.97$$

Now,

$$\begin{aligned}\text{Equivalent } Z = Z_{AB} + 2\ \Omega &= 4.8 + j\,0.97 + 2\ \Omega \\ &= 6.8 + j\,0.97 = 6.87\ \angle 8.11\end{aligned}$$

Thus

$$\text{Power factor} = \cos(8.11°) = 0.99 \text{ lagging}$$

Example 16

Two impedances, $Z_1 = 7 + j5\ \Omega$ and $Z_2 = 3 - j10\ \Omega$, are connected in parallel. The total current supplied is 20 A. What is the power taken by each impedance? Also find the reactive powers. **(UPTU 2001–02)**

Solution

As the diagram of circuit is shown in Figure 57, by the current divider rule we have

$$I_1 = I\frac{Z_2}{Z_1+Z_2} = 20\times\frac{3-j10}{7+j5+3-j10} = \frac{20\times 10.44\ \angle{-73.28}}{11.18\ \angle{-26.56}} = 18.67\ \angle{-46.72}$$

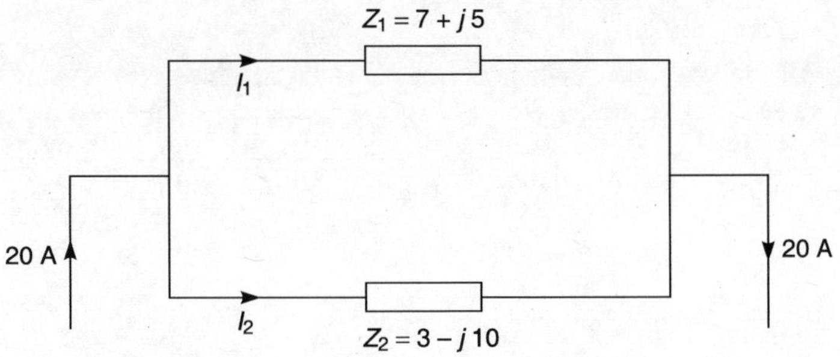

Figure 57 Figure for Example 16.

and

$$I_2 = I\frac{Z_1}{Z_1 + Z_2} = 20 \times \frac{7 + j5}{11.18\angle{-26.56}} = \frac{20 \times 8.6\angle 35.53}{11.18\angle -26.56} = 15.38 \angle 62.1$$

Now $R_1 = 7\ \Omega$ and $R_2 = 3\ \Omega$. Hence

$$\text{Power absorbed by } Z_1 = I_1^2 R_1 = 2440 \text{ W}$$

$$\text{Power absorbed by } Z_2 = I_2^2 R_2 = 709.64 \text{ W}$$

Example 17

Differentiate among real, reactive and apparent powers. Obtain the power factor of a two-branch parallel circuit, where the first branch has $Z_1 = 2 + j4$ and the second branch $Z_2 = 6 + j0\ \Omega$. To what value must $6\ \Omega$ resistor be changed to result in overall pf 0.90 lagging? **(UPTU 2000–01)**

Solution

Since the two branches are in parallel, the equivalent resistance is

$$Z = \frac{Z_1 Z_2}{Z_1 + Z_2} = \frac{(2 + j4)(6 + j0)}{2 + j4 + 6 + j0} = \frac{12 + j24}{8 + j4} = 2.4 + j1.8 = 3\angle 36.86^\circ$$

Hence pf will be

$$\text{pf} = \cos\phi = \cos 36.86 = 0.8$$

Now, let the new value of resistance for pf equal to 0.9 be R. Then the equivalent impedance is

$$Z = \frac{(2 + j4)R}{R + 2 + j4} = \frac{2R + j4R}{(R + 2) + j4} \times \frac{(R + 2) - j4}{(R + 2) - j4}$$

$$= \frac{2R^2 + 20R + j4R^2}{(R + 2)^2 + 16} = \frac{2R^2 + 20R}{(R + 2)^2 + 16} + j\frac{4R^2}{(R + 2)^2 + 16}$$

Now,

$$|Z| = \frac{1}{(R + 2)^2 + 16}\sqrt{(2R^2 + 20R)^2 + (4R^2)^2}$$

Then

$$\text{pf} = \cos\phi = \frac{\text{Real part}}{\text{Total impedance}} = \frac{\dfrac{2R^2 + 20R}{(R+2)^2 + 16}}{\dfrac{\sqrt{20R^4 + 400R^2 + 80R^3}}{(R+2)^2 + 16}} = 0.9 \text{ (given)}$$

Hence

$$4R^4 + 400R^2 + 80R^3 = 16.2R^4 + 324R^2 + 64.8R^3 \Rightarrow 12.2R^4 - 76R^2 - 15.2R^3 = 0$$
$$\Rightarrow 12.2R^2 - 15.2R - 76 = 0$$

On solving, the value of $R = 3.19\,\Omega$.

Example 18

Derive the expression for Q-factor in the R-L-C parallel circuit. **(UPTU June 2011)**

Solution

As shown in Figure 58, the parallel R-L-C circuit is given an AC voltage input V. The definition of quality factor or Q-factor is

$$Q = 2\pi \frac{\text{Energy stored}}{\text{Energy dissipated/cycle}}$$

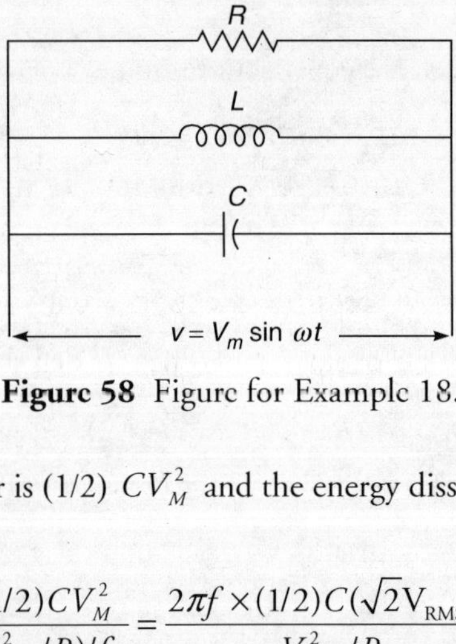

Figure 58 Figure for Example 18.

The energy stored in the capacitor is (1/2) CV_M^2 and the energy dissipated in the resistor is V_{RMS}^2/R. So, putting the values, we get

$$Q = 2\pi \frac{(1/2)CV_M^2}{(V_{\text{RMS}}^2/R)/f} = \frac{2\pi f \times (1/2)C(\sqrt{2}\text{V}_{\text{RMS}})^2}{V_{\text{RMS}}^2/R} = \omega RC$$

This expression is inverse of the expression for quality factor of a series R-L-C circuit.

Example 19

In the circuit shown in Figure 59, find the value of R such that the impedance of whole circuit should be independent of the frequency of supply. If voltage = 200 V, $L = 0.16$ H and $C = 100$ μF, calculate power loss in the circuit. **(UPTU 2001–02)**

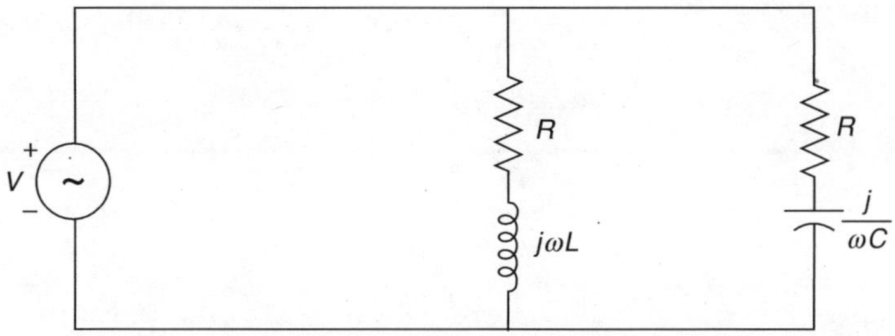

Figure 59 Figure for Example 19.

Solution

We have to find the value of R such that the impedance of whole circuit should be independent of the frequency of supply, then it will be the condition of resonance only. Hence, the impedance of the circuit is

$$Z = \frac{(R + j\omega L)(R - j/\omega C)}{2R + j(\omega L - 1/\omega C)} = \frac{R^2 + (L/C) + jR(\omega L - 1/\omega C)}{2R + j(\omega L - 1/\omega C)}$$

Now for Z to be equal to R under resonant condition (i.e. when $\omega L = 1/\omega C$)

$$Z = \frac{R^2 + L/C}{2R}$$

For Z to be equal to R, R^2 has to be equal to L/C. Thus

$$Z = R = \sqrt{\frac{L}{C}}$$

becomes purely resistive. Now

$$R^2 = \frac{L}{C} \Rightarrow R = \sqrt{\frac{L}{C}}$$

Then

$$Z = \frac{R^2 + R^2 + jR\left(\omega L - \frac{1}{\omega C}\right)}{2R + j\left(\omega L - \frac{1}{\omega C}\right)} = R\left[\frac{2R + j\left(\omega L - \frac{1}{\omega C}\right)}{2R + j\left(\omega L - \frac{1}{\omega C}\right)}\right] = R$$

Hence for the impedance of the whole circuit, to be independent of supply, the value of R should be $\sqrt{L/C}$. This implies

$$R = \frac{\sqrt{0.16}}{\sqrt{100 \times 10^{-6}}} = 40\ \Omega$$

and the power loss in the circuit is

$$P = \frac{V^2}{R} = \frac{200 \times 200}{40} = 1000\ \text{W}$$

Example 20

Show that the condition for resonance in parallel R-L-C circuit is same as that in a series R-L-C circuit. State the applications of series and parallel resonance. **(UPTU May 2010)**

Solution

In a series resonance, the *imaginary part of impedance* of the circuit becomes zero (see Figure 60).

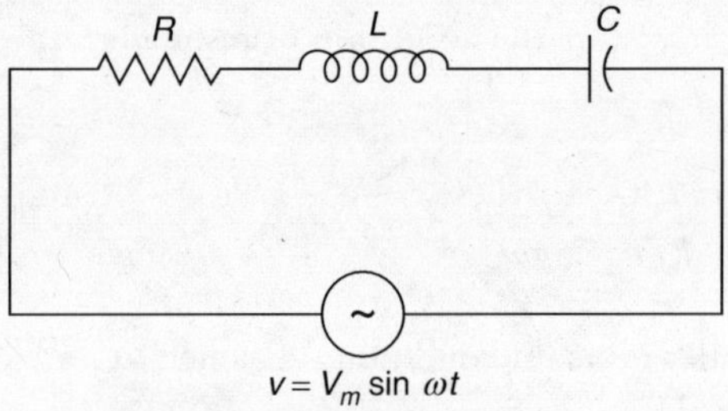

Figure 60 Series resonance.

As shown in the series R-L-C circuit in Figure 60, the impedance is

$$Z = R + j\left(\omega L - \frac{1}{\omega C}\right)$$

Now at the resonance, ohmic values of the inductive and capacitive reactances become equal. So, the net reactance becomes zero. The resonant frequency is derived on the basis of this information. That is,

$$\omega_r C - \frac{1}{\omega_r L} = 0$$

$$\omega_r L = \frac{1}{\omega_r C}$$

gives

$$\omega_r = \frac{1}{\sqrt{LC}} \quad \text{or} \quad f_r = \frac{1}{2\pi\sqrt{LC}}$$

For parallel resonance, as shown in Figure 61, the *imaginary part of the admittance* becomes zero.

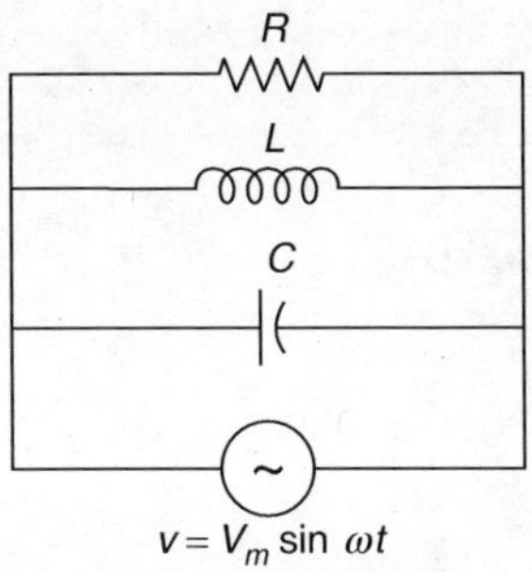

Figure 61 Parallel resonance.

Thus,

$$Y = G + jB = \frac{1}{R} + j\left(\omega C - \frac{1}{\omega L}\right)$$

where G is the conductance (reciprocal of the resistance) and B is the susceptance (reciprocal of the reactance). Now at parallel resonance, $B = 0$. That is,

$$\omega_r C - \frac{1}{\omega_r L} = 0$$

$$\omega_r L = \frac{1}{\omega_r C}$$

Hence

$$\omega_r = \frac{1}{\sqrt{LC}} \quad \text{or} \quad f_r = \frac{1}{2\pi\sqrt{LC}}$$

which is same as that of a series resonance.

Applications:

1. *Series resonance*: Here at the resonance frequency, the impedance is minimum, so the current is maximum. Hence it can be said that the strongest signal is transmitted at a resonant frequency. Hence this circuit is also used as an *acceptor circuit*. Also, the voltage across each of the inductance and capacitance becomes magnified by Q times, where Q is the quality factor of the series R-L-C circuit.
2. *Parallel resonance*: Here, the value of admittance is minimum at the resonant frequency, and hence (for a given voltage magnitude) the current is minimum (as $I = VY$). Hence this circuit is also used as a *rejecter circuit*.

Example 21

A coil has impedance of 30 Ω with the voltage–current phasor angle through it being 60°. This coil is connected in series with a resistance so that the current through it is 4 A when a 200 V (RMS), 50 HZ voltage is applied across the coil–resistance combination. Find the value of the series resistance and the circuit phase angle.

Solution

See Figure 62.

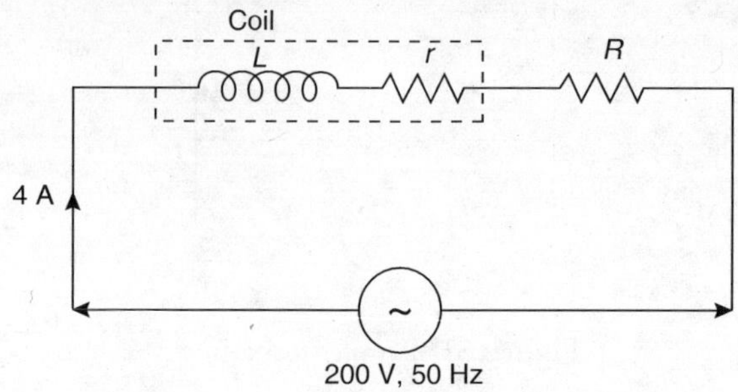

Figure 62

The overall impedance of the circuit will be

$$Z = \frac{200}{4} = 50\ \Omega$$

The resistance and reactance of the coil can be found because the ohmic value of the coil impedance and also the angle of impedance are given as 30 Ω and 60°, respectively. Thus

$$R_{(\text{coil})} = 30 \cos 60° = 15\ \Omega$$

and

$$X_{(\text{coil})} = 30 \sin 60° = 26\ \Omega$$

Now

$$\sqrt{(R+15)^2 + 26^2} = 50\ \Omega \Rightarrow (R+15)^2 = 50^2 - 26^2 \Rightarrow R = 27.72\ \Omega$$

Thus complete resistance of the circuit is

$$R_{(\text{coil})} + R = 15 + 27.72 = 42.72\ \Omega$$

and circuit phase angle is

$$\tan^{-1}\frac{X}{R} = \tan^{-1}\frac{26}{42.72} = 31.32°$$

Summary

The chapter explains steady-state analysis of single-phase AC circuits in details. The concepts of phasor is discussed in the chapter, wherein it is observed that a sinusoidal quantity can be represented by its equivalent phasor whose value is the RMS value of the sinusoidal quantity which it is representing. Relationships between the current and voltage phasors have been established for (a) purely resistive, (b) purely inductive, (c) purely capacitive, (d) *R-L* series, (e) *R-C* series and finally (f) *R-L-C* series and parallel circuits. The phase difference between current and voltage for a purely resistive circuit is zero degree. Here, we have $V = IR$. The phase difference between current and voltage for a purely inductive circuit is 90° with the current lagging behind the voltage. The inductive reactance is $X_L = j\omega L = 2\pi f L$ and the voltage is $V = jX_L I = j\omega LI$. Similarly, the phase difference between current and voltage for a purely capacitive circuit is 90° with the current leading the voltage. The capacitive reactance is

$$X_C = \frac{1}{j\omega C} = -j\frac{1}{2\pi fC}$$

and the voltage is

$$V = -jX_c I = -j\frac{1}{\omega C}I$$

Power factor is the cosine of the angle between voltage and current (say, θ). This angle is same as that between R and Z in impedance triangle and also as that between the real power and the apparent power in the power triangle. The apparent (or total power) is given by V_{RMS}, in RMS (the unit is VA, i.e. volt-ampere). The real (or active power) is given by $V_{RMS} \cdot I_{RMS} \cos\theta$ (the unit is watt). The imaginary (or reactive power) is given by $V_{RMS} \cdot I_{RMS} \sin\theta$ (the unit is VAR, i.e. volt-ampere-reactive).

When a variable frequency source supplies a circuit containing a series or parallel combination of R, L and C, the natural frequency of the circuit is given by

$$f_r = \frac{1}{2\pi\sqrt{LC}}$$

Now, let us say that the frequency of the source is varied. The moment it matches the above natural frequency of the circuit, a phenomenon, called resonance, takes place.

The inductive and capacitive current components, in case of series circuit, and the inductive and capacitive voltage components, in case of parallel circuit, nullify each other, making the circuit purely resistive. In series *R-L-C* circuits: quality factor is given by

$$Q = \frac{\omega L}{R} = \frac{1}{\omega RC} = \frac{1}{R}\sqrt{\frac{L}{C}} = \frac{\text{Resonant frequency}}{\text{Bandwidth}}$$

In parallel *R-L-C* circuits:

$$\text{Quality factor, } Q = \frac{R}{\omega L} = \omega RC$$

The resonant frequency and the quality factors are the key parameters of a resonant circuit. A lower bandwidth circuit gives a better quality factor, making the circuit highly selective. High quality factor circuits can generate responses (voltage, in the case of series *R-L-C* circuits and current, in the case of parallel *R-L-C* circuits), which are greater in magnitudes than that produced by their respective sources.

Short Answer Questions

1. What is the phase angle between voltage and current phasors for a purely inductive load?
2. The power factor of a purely capacitive circuit will be
3. If the power factor of a series *R-L-C* circuit is unity, what conclusion can be drawn regarding the circuit?
4. What happens to the bandwidth of a series *R-L-C* circuit if the resistance is increased?
5. What is the explanation of $Y = (2 - j4)$ mho?
6. What are the units of the three types of powers in AC circuits?
7. A circuit has a voltage $v = 200\sin(\omega t - 60°)$ and current $i = 50\cos(\omega t - 60°)$. The real power consumed will be equal to
8. What is the nature of a series *R-L-C* circuit before resonance?
9. What is the power factor of a series *R-L* circuit whose inductive reactance is equal to the resistance?
10. What is the relationship between the quality of a series *R-L-C* circuit and its bandwidth?

Review Questions

1. Distinguish between the instantaneous, RMS, average and peak values of a sinusoidal quantity.
2. What are form factor and peak factor? Derive their values for a sinusoidal wave.
3. What is the difference between a vector and a phasor? What is the significance of phasor in the sinusoidal steady state circuit analysis? Explain.
4. Derive the phasor relationships between voltage and current in a/an
   - **(a)** Purely inductive circuit
   - **(b)** Purely capacitive circuit
   - **(c)** Purely resistive circuit
   - **(d)** *R-C* series circuit
   - **(e)** *R-L* parallel circuit
5. Derive the expression for the half power frequencies in a series *R-L-C* circuit.
6. What is the physical significance of the half-power frequencies?
7. Derive the expression for the quality factor in a
   - **(a)** Series *R-L-C* circuit
   - **(b)** Parallel *R-L-C* circuit
8. How is bandwidth related to the quality factor? Derive the relationship between them.
9. What is power factor? Why is it desired that it remains high?
10. What are the methods to improve the power factor?

Problems

1. As shown in Figure P2.1, a voltage of 200 V, 50 Hz is applied across a series *R-C* circuit having a phase angle of 45°. The real power consumed by the circuit is 250 W at 100 V. Find the values of *R* and *C*.

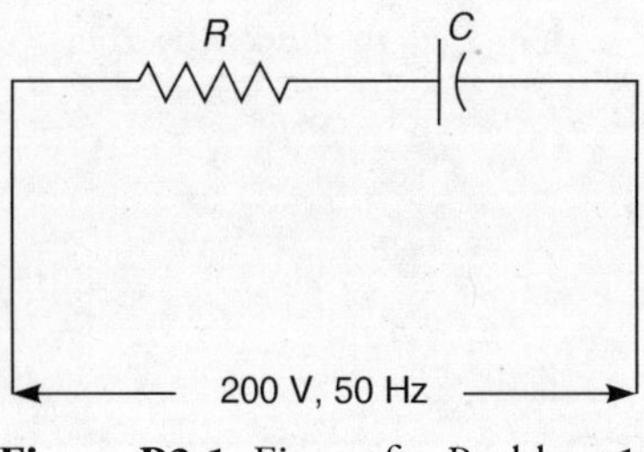

Figure P2.1 Figure for Problem 1.

2. A voltage wave, $e(t) = 141.4 \sin 314t$, produces a current $i(t) = 7.07 \sin(314t + 60°)$ in a particular circuit. Find the element values and the power factor of the circuit.
3. A coil (*R-L* series circuit) has a resistance of 10 Ω and draws a current of 8 A when fed with a 200 V, 50 Hz supply. Find the inductance and power factor of the coil.
4. Two impedances, $Z_1 = 6 + j8$ and $Z_2 = 4 - j10$, are connected in parallel across a 50 Hz voltage supply. The supply current is 20 A.
   - **(a)** Find the real, reactive and apparent powers drawn by each impedance.
   - **(b)** Find the source voltage (complete expression).
   - **(c)** Also draw the complete phasor diagram.
5. If a resistance of 10 Ω, capacitance of 200 μF and inductance of 50 mH are connected in series, find
   - **(a)** Resonance frequency in hertz
   - **(b)** Bandwidth in hertz
   - **(c)** Quality factor
   - **(d)** Voltage across the capacitance at resonance
   - **(e)** Value of current at resonance

   Consider that the above circuit is fed by a 200 V variable frequency AC supply.
6. A series *R-L-C* circuit has $R = 50\ \Omega$, $L = 0.8$ H. The power factor of the circuit at a frequency of 35 Hz is 0.6 lagging. Find the resonant frequency.
   [Hint: Find X_C with the help of the angle given.]
7. Prove that $f_r = \sqrt{f_1 f_2}$, where f_r = resonant frequency, f_1, f_2 are half power point frequencies for indicating the bandwidth.
8. Prove that the parallel resonant frequency for the circuit, as shown in Figure P2.2, is given by

$$f_r = \frac{1}{2\pi}\sqrt{\frac{1}{LC} - \frac{R^2}{L^2}}\ \text{Hz}$$

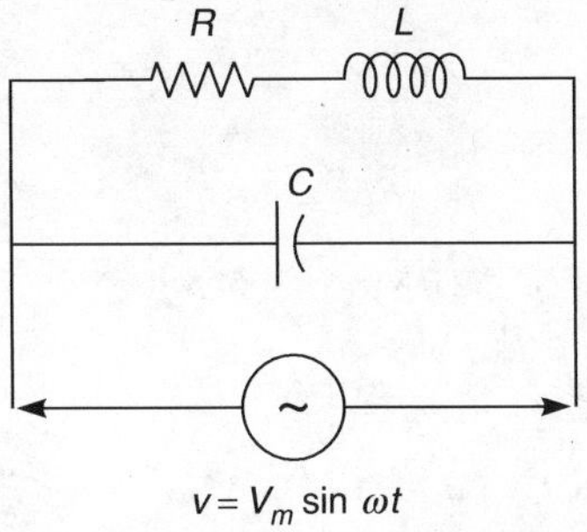

Figure P2.2 Figure for Problem 8.

9. An *R-L-C* series circuit is connected across a variable frequency, 25 V, AC supply with $R = 50\ \Omega$. When the frequency is 400 Hz, the current attains its maximum value at 0.5 A, and the potential difference across the inductor is 150 V. Find the values of L and C.
10. A sinusoidal current having RMS value of $8\angle 0°$ is added to another sinusoidal current of RMS value $6\angle 90°$. What is the RMS value of the resultant current?
11. A coil (*R-L* series) when connected across a 250 V, 50 Hz supply takes a current of 25 A. The phase angle of the coil is 45°. When a capacitance is connected in parallel with the coil (across the same supply), the magnitude of supply current becomes 40 A. Find the supply current and the overall power factor when the same capacitance is connected in series with the coil and the whole circuit is fed by the same supply.

Answers

Short Answer Questions

1. The voltage and current phasors have a phase difference of 90° with current lagging the voltage.
2. In a capacitive circuit, the voltage and current phasors have a phase difference of 90° with current leading the voltage, hence the power factor will be zero leading.
3. The unity power means that the series *RLC* circuit is under resonance where the inductive and capacitive reactances neutralize each other and the circuit becomes purely resistive giving a unity power factor.
4. The bandwidth is given by *R*/*L*. So if the applied frequency and inductance are held constant, the bandwidth increases as the resistance increases.
5. The unit mho indicates that the quantity is an admittance. Since the admittances connected in parallel get added, hence we can (for the present case) say that, a conductance of 2 mho is connected in parallel with an inductive susceptance of 4 mho.
6. Real power has watt, reactive power VAR and apparent power VA.
7. If *x*-axis is taken as reference, the voltage phasor lies at 60° clockwise in the fourth quadrant. Now current $i = 50\cos(\omega t - 60°)$. We know that $\cos\alpha = \sin(90 + \alpha)$. Hence current $i = 50\sin(90° + \omega t - 60°) = 50\sin(\omega t + 30°)$. Hence we see that the current phasor is at 30° from the positive *x*-axis. A close look will reveal that the angle between voltage and current phasors is 90° and thus the real power consumed by such a load will be zero.
8. Since the capacitive reactance is greater than the inductive reactance before resonance, hence the circuit is capacitive before resonance.
9. The phase angle in the case mentioned will be 45° (tan inverse of unity) and hence its cosine (which will be the power factor) will be 0.707 lagging.
10. The quality factor is given by $Q = f_r/\Delta f$, where Δf is the bandwidth of the circuit. Hence it is clear that if the bandwidth is increased, the quality of signals received will be poorer.

Problems

1. $R = 80\ \Omega$, $C = 39.8\ \mu F$
2. **(a)** $R = 10\ \Omega$, $C = 183.8\ \mu F$
   (b) 0.5 leading
3. 72.97 mH, Pf = 0.4
4. **(a)** $P_{z1} = 2676.32$ W, $Q_{z1} = 3568.43$ VARs (inductive), $S_{z1} = 4460.54$ VAs, $P_{z2} = 1538.1$ W, $Q_{z2} = -3845.45$ VARs (capacitive), $S_{z2} = 4141.64$ VAs
   (b) $298.7 \sin(\omega t - 3.75°)$
   (c) Phasor diagram is shown below:

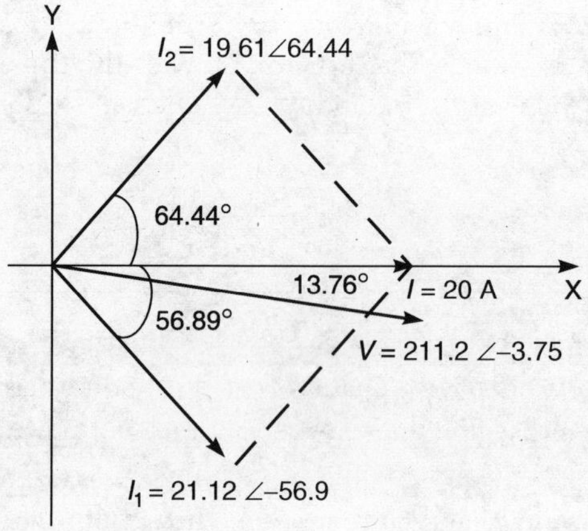

5. **(a)** 50.33 Hz
   (b) 31.83 Hz
   (c) 1.58
   (d) 316.23 (RMS value) volts
   (e) 20 A (RMS)
6. 27.58 Hz
9. $L = 119.36$ mH, $C = 1.33\ \mu F$
10. $8 + j6 = 10\angle 36.87°$
11. 33.48 A and 0.947 lagging

3 Three-Phase System

LEARNING OBJECTIVES

After a careful study of the chapter, you would be able to understand:

- Need, importance, advantages and performance of a three-phase system.
- Meaning of phase sequence in a three-phase system.
- Voltage and current relationships in star and delta connected loads.
- Balanced supply and balanced loads.
- Three-phase power and its measurement.
- Types of measuring instruments along with their construction and working principles.
- Extension of instrument range through shunts and multipliers.

3.1 Three-Phase System: Necessity and Advantages

Three-phase AC systems are used for generation, transmission and distribution of bulk electrical power. Smaller machines, requiring less power, can be operated by single-phase supply. The larger machines perform smoothly and economically by using three-phase supply. Thus, wherever larger power is required, it is found that use of three numbers of single phases is uneconomical and cumbersome as compared to the use of three-phase systems. Besides, the following are specific advantages of three-phase power over the single-phase power:

1. The magnitude of the three-phase power is constant unlike the pulsating single-phase power. This results in quieter operation of various machines, lesser vibrations and thus an overall increased efficiency and lifetime of the machine.
2. The resultant magnetic field produced by a balanced three-phase supply is a steady rotating field which enables a smooth self starting of three-phase induction motors.
3. For same sending and receiving end voltages in a transmission system, the three-phase transmission requires lesser conductor material thus enhancing the efficiency of the power system.

3.2 Meaning of Phase Sequence

A balanced three-phase voltage is shown in Figure 1. V_{RN}, V_{BN} and V_{YN} are (the phasors of) the phase to neutral (the meeting point of the three-star phases) voltages of exactly equal magnitude and each having a progressive phase displacement of 120°, as shown in Figure 1. The corresponding sinusoidal waveforms are shown in Figure 2. It can clearly be seen that the voltage v_{RN} leads the voltage v_{YN} by 120° and v_{YN} leads v_{BN} by 120° in the phase sequence RYB. As phasor value is the RMS value, hence in the sinusoidal waveform, peak (or maximum) value is taken to show the phase differences.

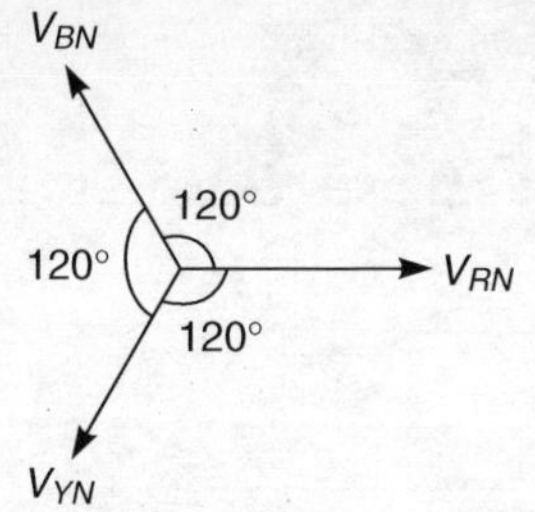

Figure 1 Balanced three-phase supply (phasor diagram).

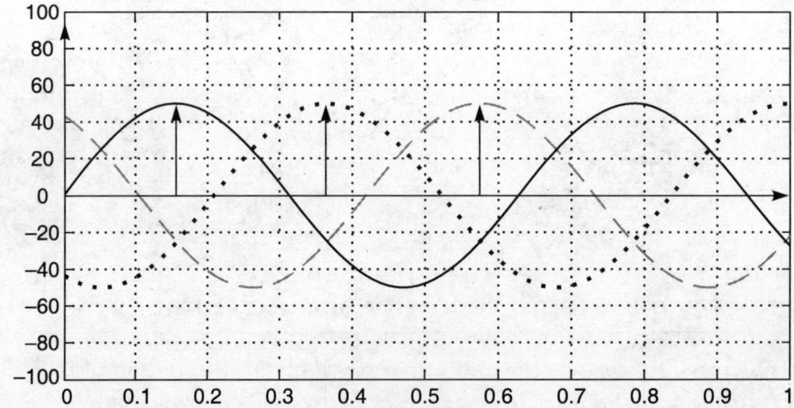

Figure 2 Balanced three-phase supply (sinusoidal diagram).

The instantaneous values of these voltages can be represented by

$$v_{RN} = \sqrt{2}\, V_{RN} \sin \omega t$$
$$v_{YN} = \sqrt{2}\, V_{YN} \sin(\omega t - 120°)$$
$$v_{BN} = \sqrt{2}\, V_{BN} \sin(\omega t - 240°)$$

In a balanced three-phase system, $|V_{RN}| = |V_{YN}| = |V_{BN}|$ and they are displaced by 120° from each other in such a way that if v_{RN} is considered as the reference voltage, then v_{YN} lags v_{RN} by 120°. Further v_{BN} lags behind v_{YN} by 120°. Thus, if a person walks in clockwise direction, he will come across the voltages in sequence of *R*, *Y* and *B*. In other words, the maxima of the voltages appear in order of *R*, *Y*, *B*. This is called the *phase sequence* in three-phase AC system. Phase sequence is usually represented by RYB.

3.3 Star and Delta Connections

A three-phase supply is fed to a three-phase load. Three-phase loads can be connected only in two ways: either in a star (Y) connection or in a delta (Δ) connection. Figures 3(a) and (b) show a balanced star-connected three-phase load and a balanced delta-connected load.

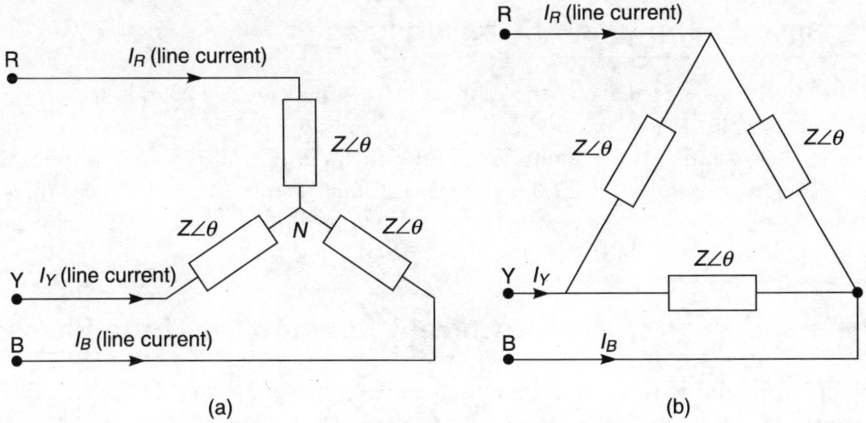

Figure 3 (a) Star connection; (b) delta connection.

They can be obtained as described below. Let there be three independent phases given by a_1a_2, b_1b_2 and c_1c_2, as shown in Figure 4.

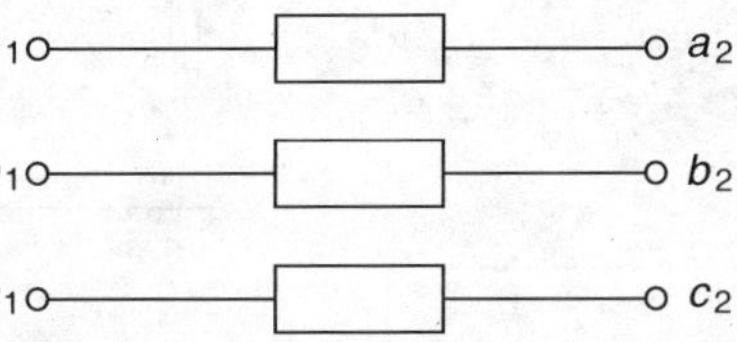

Figure 4 Three independent phases in connected stage.

If these three independent phases are required to form a three-phase system, then they can be combined to form either a star or a delta, as shown in Figures 5(a) and (b).

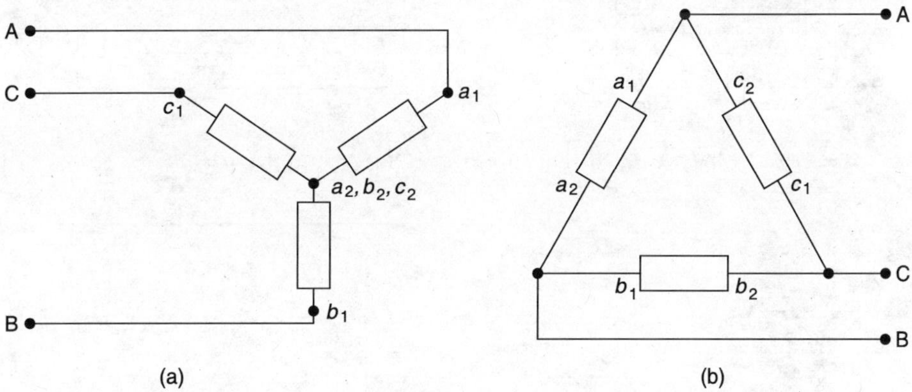

Figure 5 (a) Star (Y) connection; (b) delta (Δ) connection.

As it is clear from Figure 5 that in Y connection, three terminals (one from each phase, i.e. a_1, b_1, c_1) are brought out while the other three ends are shorted together (a_2, b_2, c_2). This common point is termed as *neutral point*. A Δ connection is formed by connecting the start of one phase (e.g. a_1) to the end of any of the two remaining phases (say c_2), as shown in Figure 5(b). In other words, a_1c_2, c_1b_2 and b_1a_2 form the three terminals of Δ connection. The Y and Δ connections are interconvertible, as already described in Chapter 1.

3.4 Balanced Supply and Balanced Load

If the voltage magnitudes of the three phases of a three-phase supply are equal, and the only difference is in their phases (which are 120° as explained before), it is said to be a *balanced three-phase supply*. Similarly, if the three-phase load is such that the magnitudes and the angles of each of the three phases are exactly the same, it is said to be a *balanced three-phase load*. Figures 3(a) and (b) represent the balanced three-phase loads connected in star and delta, respectively. A balanced supply, feeding a balanced load, circulates balanced three-phase currents in the three-phase system.

3.5 Line and Phase Voltage/Current Relations in Three Phases

The voltages and currents in three-phase star and delta connections can be described as shown in Figures 6(a) and (b). As shown in Figure 6(a), I_R, I_Y, I_B are the phase currents (I_{ph}), and I_R, I_Y, I_B are also the line currents (I_L). Similarly, V_{RN}, V_{YN}, V_{BN} are the phase voltages (V_{ph}) and V_{RY}, V_{YB}, V_{BR} are the line voltages (V_L). In delta connection, as shown in Figure 6(b), I_R, I_Y, I_B are the line currents (I_L) and I_{RY}, I_{YB}, I_{BR} are the phase currents (I_{ph}). V_{RY}, V_{YB}, V_{BR} are the phase voltages (V_{ph}) and also the line voltages V_L. In Y connection, it is obvious that the line and phase currents are the same. In Δ connection, it is obvious that the line and phase voltages are the same, that is in star $I_{ph} = I_L$ and in delta $V_{ph} = V_L$. The line and phase currents and voltages in star and delta connections are shown in Figure 7.

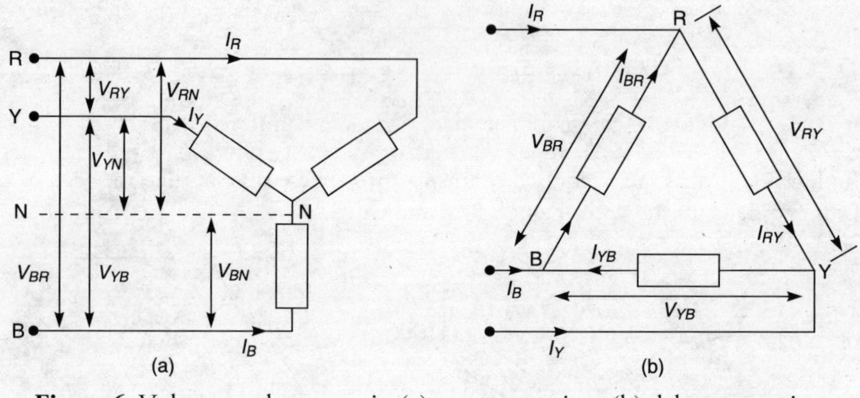

Figure 6 Voltages and currents in (a) star connection; (b) delta connection.

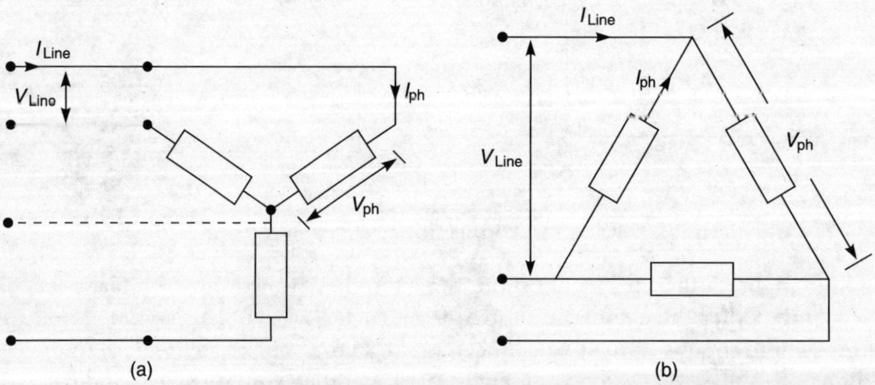

Figure 7 Phase and line voltages and currents in (a) star connection; (b) delta connection.

The relationship between line and phase quantities can be obtained by the simple geometry, as follows. A set of balanced three phases (voltage or current) can be represented by an equilateral triangle ABC, where $AN = BN = CN$ = phase quantity and $AB = BC = CA$ = line quantity (N is the neutral point). Hence,

$$AO = AB \sin 60^\circ = \frac{\sqrt{3}}{2} AB$$

Also,

$$AN = \frac{2}{3} AO \quad \text{or} \quad AN = \frac{2}{3}\frac{\sqrt{3}}{2} AB \quad \text{or} \quad AN = \frac{1}{\sqrt{3}} AB$$

In other words,

$$\text{Phase quantity} = \frac{1}{\sqrt{3}} \text{(Line quantity)}$$

The actual phase voltages in a star-connected load are shown in Figure 8(b), where a phase sequence of RYB has been adopted. The phase voltages are V_{RN}, V_{YN} and V_{BN}, as shown in Figure 8. The line voltages are V_{RY}, V_{YB} and V_{BR}. V_{RY} has been shown as the vector sum of V_{RN} and V_{YN}, whose numerical value comes out to be $V_{RY} = \sqrt{3}\, V_{RN}$ as explained in Figure 8(b).

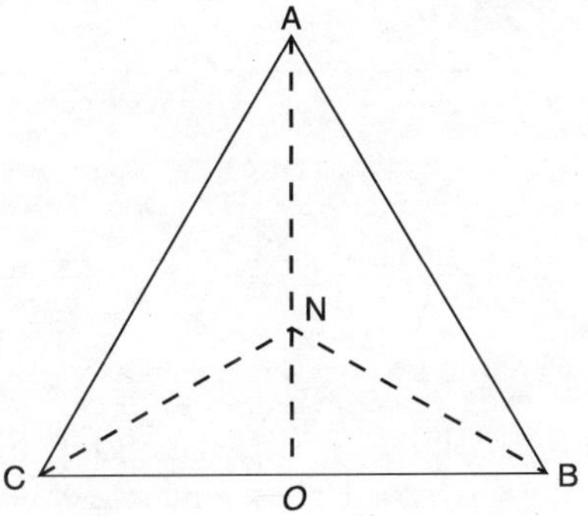

(a)

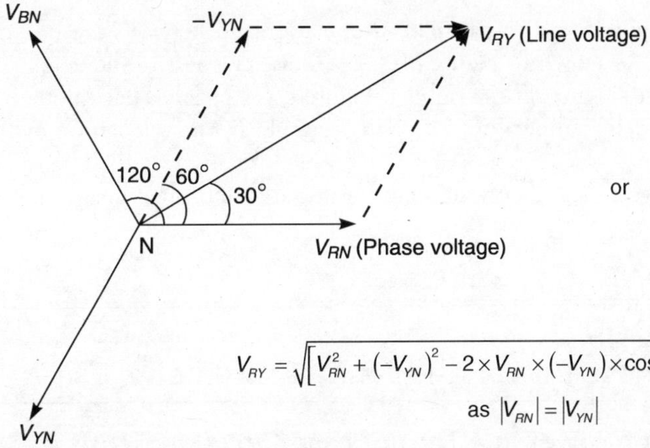

$$\overline{V}_{RY} = \overline{V}_{RN} + \overline{V}_{NY}$$

$$\text{or} \quad \overline{V}_{RY} = \overline{V}_{RN} - \overline{V}_{YN}$$

$$V_{RY} = \sqrt{\left[V_{RN}^2 + (-V_{YN})^2 - 2 \times V_{RN} \times (-V_{YN}) \times \cos 60^\circ\right]} = \sqrt{3} \times V_{RN}$$

$$\text{as } |V_{RN}| = |V_{YN}|$$

(b)

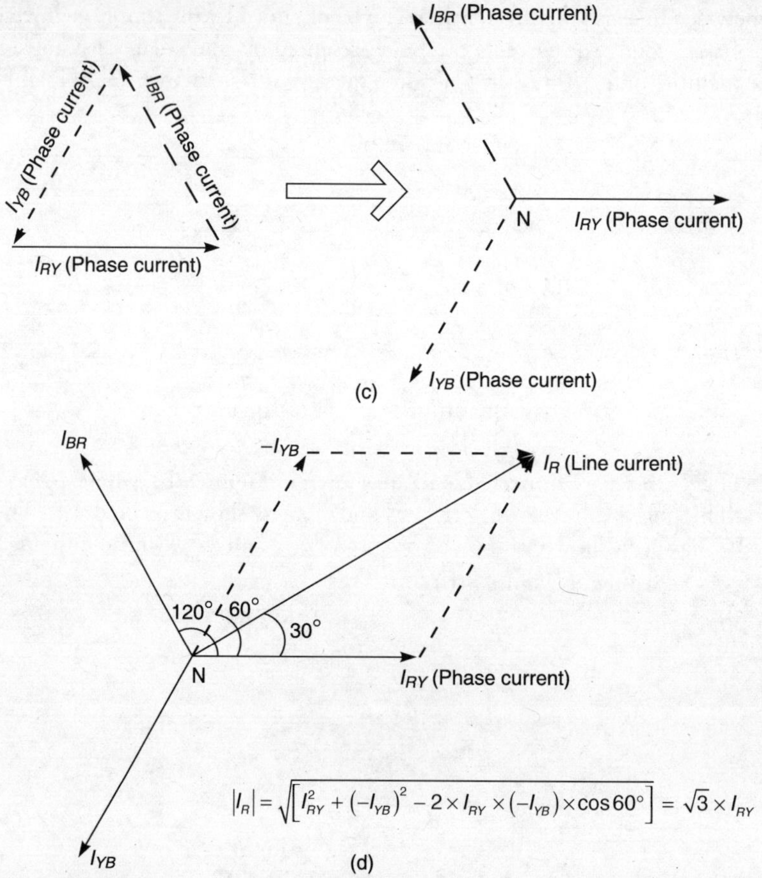

Figure 8 (a) Finding phase and line voltage relationship in a star connection (in a mathematical way); (b) actual phase and line voltages in a star connection; (c) actual phase currents in a delta connection can be represented in a parallel manner; (d) calculating the line currents in a delta connection.

Similar reasoning can be adopted to calculate the relationship between the line and phase currents in a delta-connected circuit. As shown in Figure 8(c), the phase currents in the delta connection can be shown in a parallel manner by the figure at the right in Figure 8(c). This has been done to ease the calculation of the line currents in the delta connection. Next, in Figure 8(d), the calculation of the line currents in delta connection is shown.

Thus, the following voltage and current relationships are established in a balanced three-phase system:

1. In Y connection; $I_{\text{ph}} = I_L$ and $V_{\text{ph}} = V_L/\sqrt{3}$.
2. In Δ connection, $I_{\text{ph}} = I_L/\sqrt{3}$ and $V_{\text{ph}} = V_L$.

3.6 Three-Phase Power and Its Measurement

3.6.1 Expression for Power in a Three-Phase Circuit in Terms of Line Quantities

Let us connect a balanced load of $z\angle\theta$ ohms/phase in star connection, as shown in Figure 9.

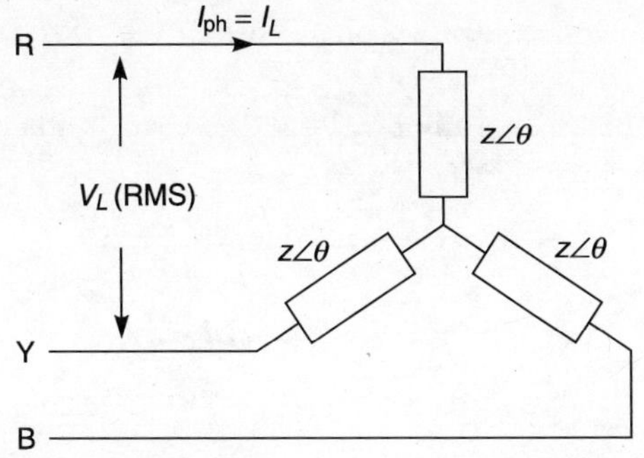

Figure 9 Balanced three-phase star-connected load.

The magnitude of phase current is given by

$$\frac{V_L/\sqrt{3}}{z} = I_{ph} \quad \text{(let)}$$

Now, the expression of power in each phase is given by

$$\text{Apparent power} = \frac{V_L}{\sqrt{3}} \times I_{ph} \text{ (VA)}$$

$$\text{Real power} = \frac{V_L}{\sqrt{3}} \times I_{ph} \cos\theta \text{ (Watts)}$$

$$\text{Reactive power} = \frac{V_L}{\sqrt{3}} \times I_{ph} \sin\theta \text{ (VAR)}$$

Hence

$$\text{Three-phase real power} = 3 \times \frac{V_L}{\sqrt{3}} \times I_{ph} \cos\theta \text{ VA} = \sqrt{3}\, V_L I_{ph} \cos\theta$$

$$= \sqrt{3}\, V_L I_L \cos\theta \text{ (as } I_{ph} = I_L \text{ in star)} \tag{3.1}$$

Similarly, the other two power expressions for three-phase can be calculated.

Next, let us connect the same load in delta, as shown in Figure 10, with the same line-to-line voltage, V_L.

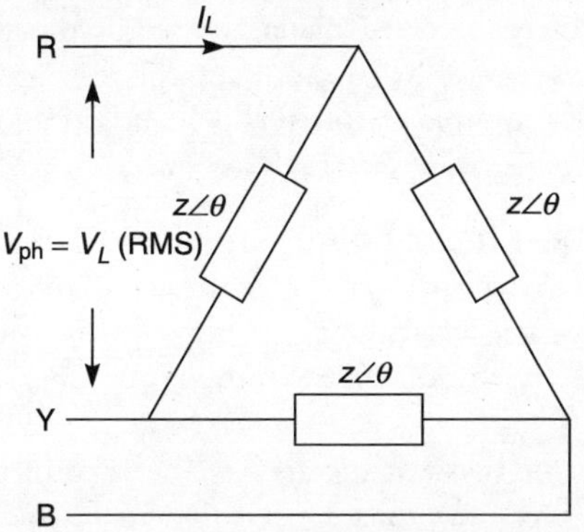

Figure 10 Same load (of Figure 9) connected in three-phase delta.

The per phase voltage is V_L and magnitude of per phase current is $I_L/\sqrt{3}$. So,

$$\text{Per phase real power} = V_L \frac{I_L}{\sqrt{3}} \cos\theta \text{ Watts}$$

$$\text{Reactive power} = V_L \frac{I_L}{\sqrt{3}} \sin\theta \text{ VAR}$$

$$\text{Apparent power} = V_L \frac{I_L}{\sqrt{3}} \text{ VA}$$

Then

$$\text{Three-phase real power} = 3 \times \frac{V_L I_L}{\sqrt{3}} \cos\theta = \sqrt{3}\, V_L I_L \cos\theta$$

which is same as Eq. (3.1).

The other two power expressions will also be similar to those in star. Thus, the conclusion that can be drawn is: The expressions for powers (active, reactive and apparent) in terms of *line quantities* remain *unchanged* on connecting the load either in star or delta.

3.6.2 Measurement of Three-Phase Power

The instrument to measure the real power in an electrical circuit is known as the *wattmeter*. It consists of two coils, namely the *voltage coil* (also called the *potential coil* or the *pressure coil*) carrying large number of thin turns and the *current coil* carrying less number of thick turns. The voltage coil (which represents the voltmeter part of the wattmeter) is connected in parallel and the current coil (which represents the ammeter part of the wattmeter) is connected in series to the load in which the power is to be measured. The reading given by the wattmeter is the product of the RMS values of the voltage (across the voltage coil), RMS value of the current (in the current coil) and the cosine of the angle between them.

There are various methods of measurement of three phase power using the wattmeters:

1. **One-wattmeter method:** In this method, the voltage coil of the wattmeter is connected across the phase (line to neutral for star and across the phase in delta), and the current coil is connected in series with the phase. Thus, at a time, the wattmeter gives power in one phase only. So after the measurement in one phase, the wattmeter is disconnected and then reconnected in each of the two remaining phases, and finally the three readings are added to get the total power. Thus, it consumes a lot of time to measure the three-phase power.
2. **Three-wattmeter method:** Here three wattmeters are required with one connected across each phase. This gives the total three-phase power by summing up the readings of the three wattmeters. Thus, the simultaneous requirement of the three wattmeters is mandatory in this method, which may be quite expensive.

3.6.3 Measurement of Three-Phase Power by Using the Two-Wattmeter Method

This method illustrates that if two wattmeters are connected in an appropriate manner, as shown in Figure 11, the sum of the two wattmeter readings gives the total three-phase real power. A balanced star-connected load (with per-phase impedance of $z\angle\theta$) is considered, as shown in Figure 11.

As can be seen from Figure 11, the two wattmeters have been connected in the two lines (R and B, for example), and their potential (or pressure) coils are terminated in the line, where there is no wattmeter (i.e. line Y in the present case). Now since we know that a wattmeter measures real power, which is given

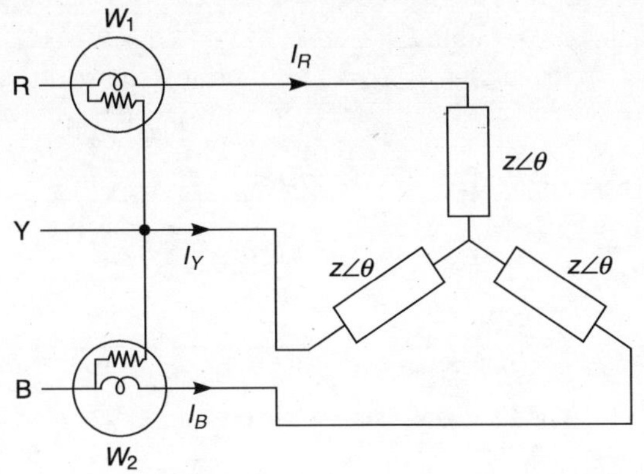

Figure 11 Measurement of three-phase power by two-wattmeter method.

by the product of the voltage across its potential coil, the current through its current coil and the cosine of the angle between them, we have

$$W_1 = V_{RY}\, I_R \cos \angle V_{RY}, I_R \tag{3.2}$$

$$W_2 = V_{BY}\, I_B \cos \angle V_{BY}, I_B \tag{3.3}$$

Figure 12 shows the phasor diagram of the three-phase voltages, two line voltages and the three-phase (or line) currents. The load has been assumed to be *R-L* (resistive-inductive), so that the phase current lags the phase voltage by θ degree, as shown. Also, the sum of the phase voltages V_{RN} and V_{YN} gives the line voltage V_{RY} and the phasor sum of the phase voltages V_{BN} and V_{YN} gives V_{BY}, as shown in Figure 12. It is also clearly visible that the angle between V_{RY} and I_R is $(30° + \theta)$ and the angle between V_{BY} and I_B is $(30° - \theta)$.

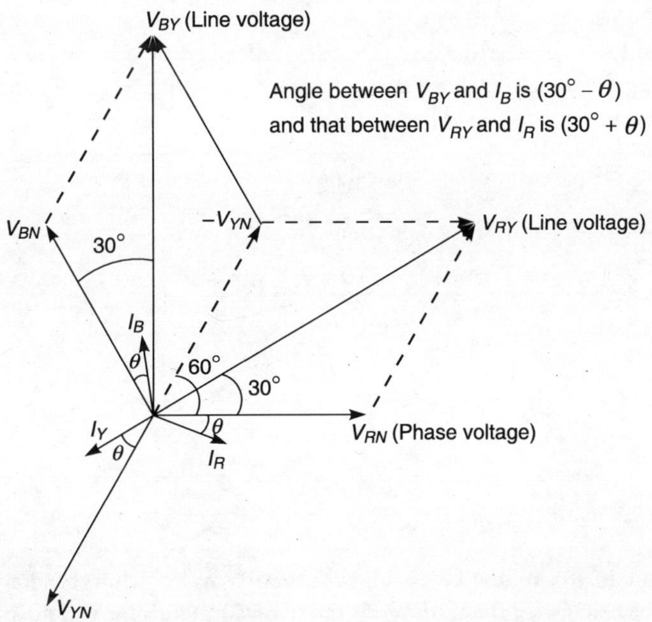

Figure 12 Phasor diagram showing the phase voltages, line voltages and line (or phase) currents.

Also, since the supply is balanced, the line voltages $|V_{RY}| = |V_{BY}| = V_L$ (say), and since the load is also balanced, so the line (or phase currents) $I_R = I_Y = I_B = I_L$ (say). Substituting the above information in Eqs. (3.2) and (3.3), we get

$$W_1 = V_L \, I_L \cos(30° + \theta) \tag{3.4}$$

$$W_2 = V_L \, I_L \cos(30° - \theta) \tag{3.5}$$

Now adding Eqs. (3.4) and (3.5), we get

$$\begin{aligned} W_1 + W_2 &= V_L I_L [\cos(30° + \theta) + \cos(30° - \theta)] \\ &= V_L I_L (\cos 30° \cos\theta - \sin 30° \sin\theta + \cos 30° \cos\theta + \sin 30° \sin\theta) \\ &= V_L I_L (2\cos 30° \cos\theta) \end{aligned}$$

Then

$$W_1 + W_2 = \sqrt{3} V_L I_L \cos\theta \tag{3.6}$$

Now from Eq. (3.1), we know that Eq. (3.6) is indeed the total three-phase real power (in terms of lines quantities V_L and I_L). Thus the sum of the two wattmeter readings gives us the three-phase real power. Further, subtracting Eq. (3.4) from Eq. (3.5), we get

$$W_2 - W_1 = V_L I_L [\cos(30 - \theta) - \cos(30 + \theta)] = V_L I_L (2 \sin 30 \sin\theta)$$

Hence

$$W_2 - W_1 = V_L I_L \sin\theta \tag{3.7}$$

If the right-hand side of Eq. (3.7) is multiplied by $\sqrt{3}$, we will get the three-phase reactive power:

$$\sqrt{3}\,(W_2 - W_1) = \sqrt{3} V_L I_L \sin\theta \tag{3.8}$$

Thus, if the difference of the two wattmeter readings is multiplied by $\sqrt{3}$, we get the three-phase reactive power in VAR. Knowing the three-phase real and reactive powers (in terms of W_1 and W_2), we can calculate the three-phase apparent power (in VA). That is,

$$\text{Apparent power} = \sqrt{(\text{Real power})^2 + (\text{Reactive power})^2}$$

Now dividing Eq. (3.7) by Eq. (3.6), we can get the value of angle θ:

$$\frac{V_L I_L \sin\theta}{\sqrt{3} V_L I_L \cos\theta} = \frac{W_2 - W_1}{W_2 + W_1}$$

or

$$\tan\theta = \frac{W_2 - W_1}{W_2 + W_1}\sqrt{3}$$

or

$$\theta = \tan^{-1}\left(\sqrt{3}\,\frac{W_2 - W_1}{W_2 + W_1}\right)$$

Thus, θ (the angle of the load) can also be calculated in terms of the two wattmeter readings: W_1 and W_2. Now, we will discuss the changes in the readings of the two wattmeters for various values of θ, the angle of the load (see Table 1).

Table 1 Changes in reading of two wattmeters based on load angle θ

θ	*Nature of the load*	*Wattmeter readings*
0 ($pf = 1$)	Purely resistive	$W_1 = W_2$ [from Eqs. (3.4) and (3.5)]
$0 < \theta < 60°$	Complex (i.e. combination of resistance and inductive or capacitive reactance)	$W_1 \neq W_2$ (both positive)
$\theta = 60°$ ($pf = 0.5$)	Complex (i.e. combination of resistance and inductive or capacitive reactance)	$W_1 = 0,\ W_2 =$ positive
$60° < \theta < 90°$	Complex (i.e. combination of resistance and inductive or capacitive reactance)	$W_1 =$ negative, $W_2 =$ positive
$\theta° = 90$ ($pf = 0$)	Purely reactive (inductive or capacitive reactance)	$W_1 = -W_2$, i.e. $W_1 + W_2 = 0$

As can be seen from Table 1, for purely resistive load, the readings W_1 and W_2 are equal. For $\theta = 60°$ (point 3 in the table), one of the wattmeters (here W_1) shows zero reading. This problem arises because the angle between the potential coil voltage (i.e. V_{RY}) and current coil current (i.e. I_R) become 90° when $\theta = 60°$. Thus with cos 90° = 0, the wattmeter shows zero value.

For $60° < \theta < 90°$, by similar reasoning, the angle between V_{RY} and I_R becomes greater than 90° (with cos 91° and so on being negative values). The wattmeter (W_1 here) starts showing negative reading. Finally with $\theta = 90°$, the load is purely reactive, and hence there is no question of any active power getting dissipated in it. So the sum of the two wattmeter readings is $W_1 + W_2 = 0$ (and thus, $W_1 = -W_2$).

Now, since the wattmeter is not a "centre zero" instrument (unlike a galvanometer), it cannot show negative reading. Thus, if the pointer hits the left end of the scale (tries to move anticlockwise), it is indicative of $\theta > 60°$, and the wattmeter is said to be showing negative reading. In this case, either the potential coil or the current coil connection is reversed. This causes the direction of torque in the wattmeter to reverse and hence the wattmeter starts displaying positive reading. But, the value used for numerical calculations (of power) is taken to be minus of that positive reading (i.e. if the reading is 200 W after reversing either potential coil or current coil connection, it is taken as $W = -200$ W).

3.7 Types of Measuring Instruments

There are mainly two types of measuring instruments:

1. **Primary instruments:** They measure the absolute physical quantity directly, for example tangent galvanometer.
2. **Secondary instruments:** They measure the actual quantity in terms of other proportional quantity. Such instruments are calibrated to read the actual quantity, for example voltmeter, energy meter and ECG recorder.

The secondary instruments are more popular, and also their cost is less. They are further classified as:

1. **Indicating type:** These instruments measure the quantity and indicate its value/variation on a graduated scale through a pointer, for example voltmeter, ammeter and wattmeter.
2. **Integrating type:** These instruments integrate the quantity with respect to time and show the integrated numerical value on their panel, for example energy meter and ampere-hour meter.
3. **Recording type:** These instruments produce on record (such as on a paper strip), the variation of one quantity with respect to the other, for example *X-Y* plotter, ECG, EEG and EMG.

3.7.1 Various Torques Acting on Instruments

Before proceeding to learn about various types of instruments individually, it is essential to know about the various torques that work in any measuring instrument. These are: deflecting torque, controlling torque and damping torque.

1. **Deflecting torque (or operating torque):** The magnetic, electrostatic, chemical or thermal effects, produced by the quantity to be measured, develop the deflecting torque (T_D) such that $T_D \propto$ operating quantity.
2. **Controlling torque:** Arrangement is made in the instruments such that a controlling torque acts on its moving part. The controlling torque (i) controls/stops the movement of the pointer beyond the desired reading and (ii) brings the pointer back to its zero position, when the operating quantity is removed. The controlling torque, T_C, is usually obtained either by a spring control or by gravity control mechanism. Hence,

$$T_C \propto \text{deflection } \theta \text{ (in spring control)}$$

and $$T_C \propto wl\sin\theta \text{ (in gravity control)}$$

where w is the weight put at a distance l from spindle. Instruments using spring control have uniformly graduated scale, while the instruments using gravity control have cramped scale as shown in Figures 13(a) and (b), respectively.

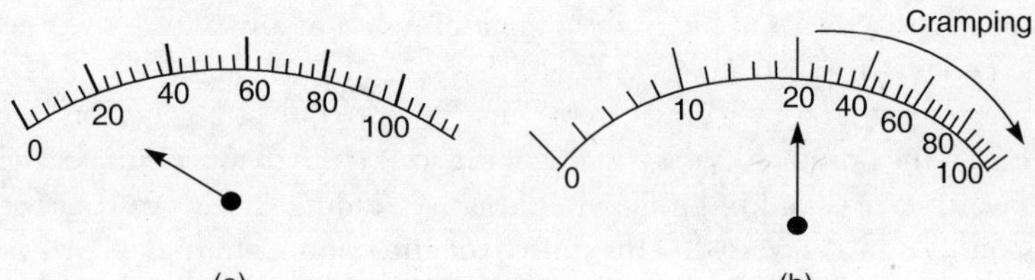

Figure 13 (a) Uniform scale; (b) cramped scale.

3. **Damping torque:** Simultaneous operation of T_D and T_C on movement of the pointer makes it oscillate near its final position to indicate the reading. Arrangement is made in the instruments to prevent such oscillations by using damping mechanism. Damping is done by generating air or fluid friction or by action of eddy currents. The effect of damping on deflection (θ) is shown in Figure 14.

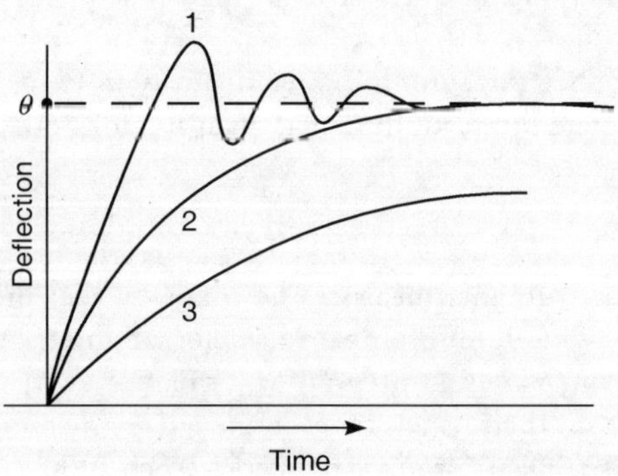

Figure 14 Damping responses in instruments.

In Figure 14, "1" represents under-damped instrument, "2" critically damped instrument (known as dead beat instrument) and "3" over-damped instrument.

3.7.2 Types of Indicating Instruments

Table 2 gives a broad comparison of various types of indicating instruments, which would be studied later at length.

Table 2 Comparison of different indicating instruments

Type	*Deflecting quantity*	*Control*	*Damping*	*General features*
Moving iron (M.I.) type	Voltage/current (RMS value of AC, DC)	Spring/gravity control	Air friction	Cheaper, accurate
Moving coil (M.C.) types				
1. Permanent magnet moving coil (PMMC) type	Voltage/current (DC/Average values)	Spring control	Eddy current	Very accurate, popular
2. Dynamometer type	Voltage/current/power (AC/DC)	Spring control	Air friction	Mostly used for wattmeter
Induction type	Voltage/current/power/ energy (AC only)	Spring control	Eddy current	Costly, not very accurate, mostly used for wattmeter and energy meter

3.8 Permanent Magnet Moving Coil Instruments

Figure 15 shows a permanent magnet moving coil (PMMC) instrument. It has the following components: C, moving coil; P_1, pivot to support spindle; P_2, pointer; S_1, control spring; S_2, spindle to which the coil C and the pointer P are rigidly attached; S_3, graduated scale; N, S are North and South poles of a permanent magnet; and R, cylindrical magnetic core on which the coil is placed.

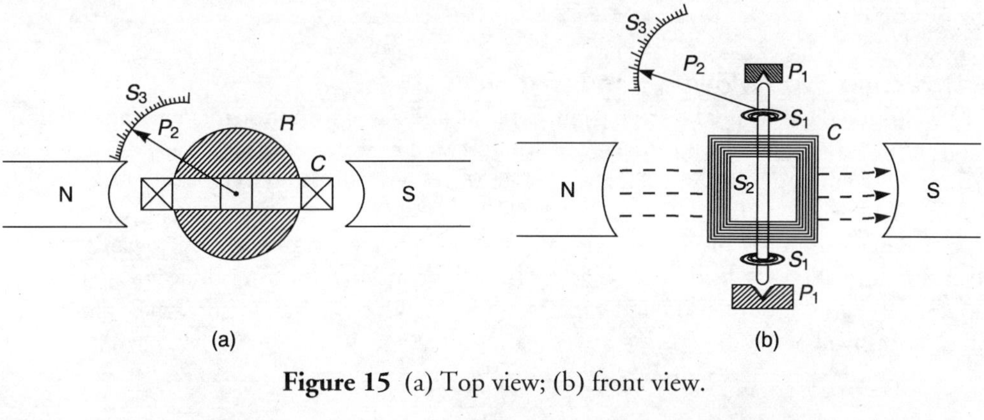

Figure 15 (a) Top view; (b) front view.

A current, proportional to the quantity to be measured, is allowed to flow through the moving coil C. This coil is placed in a constant magnetic field, developed by the North and South poles of a permanent magnet.

The current carrying coil placed in the magnetic field experiences a torque, which is proportional to the current through the coil. The coil, and hence the pointer, gets deflected to read on scale the amount of quantity to be measured. The magnetic core, on which the coil is wound, provides eddy-current damping. The reason for movement can be easily understood by Figure 16.

By Flemings left-hand rule, coil side "ab" (carrying current from "a" to "b") will be pushed backward (into the paper) by the force, while the coil side "cd" (carrying current from "c" to "d") will be pushed forward (out of the paper) by the same amount of force. Thus, looking from top, the coil will appear to rotate in clockwise direction. Since the forces developed are proportional to the current through the coil, the pointer (which is rigidly connected to the spindle carrying the coil) will move proportionately to read the unknown quantity on the calibrated scale.

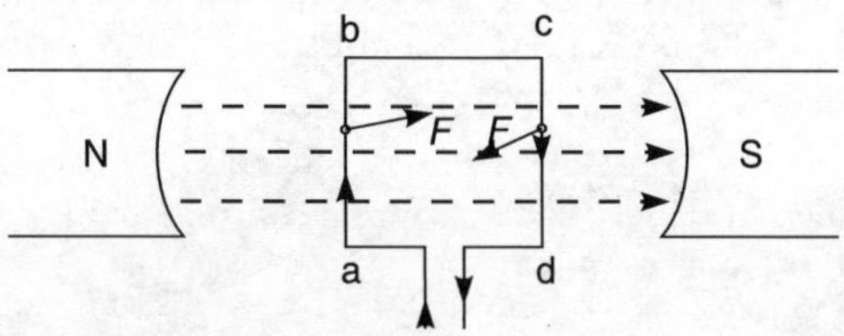

Figure 16 Explaining the torque production in a PMMC.

A PMMC instrument thus measures the average value of the quantity, which is input to its moving coil. Thus, if a pure sine wave (or any alternating quantity) is input to this instrument, it will show zero reading. In other words, a PMMC can be used only for measuring average values of mixed signals and DC signals.

3.9 Moving Iron Instruments

In PMMC instruments, the current to be measured is allowed to flow through the coil, which is placed in the constant magnetic field developed by a permanent magnet. In moving iron instruments, the current to be measured is allowed to flow through a stationary coil, which produces a magnetic field proportional to the quantity to be measured. A piece of iron, rigidly fixed with the moving system (carrying spindle and pointer), gets attracted/repelled proportionately and gives reading on the calibrated scale. There are two types of moving iron instruments based on principle of attraction or repulsion. We will discuss them in the following subsections.

3.9.1 Attraction Type Moving Iron Instruments

The coil "C" wound on a solenoid, carrying current to be measured, is stationary. The moving iron (made of soft iron) is placed near it, which is free to get attracted and move towards the coil. The spindle is rigidly connected to the pointer, to the controlling weight, to the moving iron and to the piston that moves inside the damping chamber (see Figure 17).

Due to the magnetic field inside the solenoid, the moving iron tries to move towards the solenoid. While doing so, it makes a deflection of angle $d\theta$ (say). The mechanical work done = $T_D\, d\theta$, where T_D is the deflecting (operating) torque. The inductance of iron changes by dL and hence, the energy stored by it during this process will be

$$E = \frac{1}{2} dL \times I^2$$

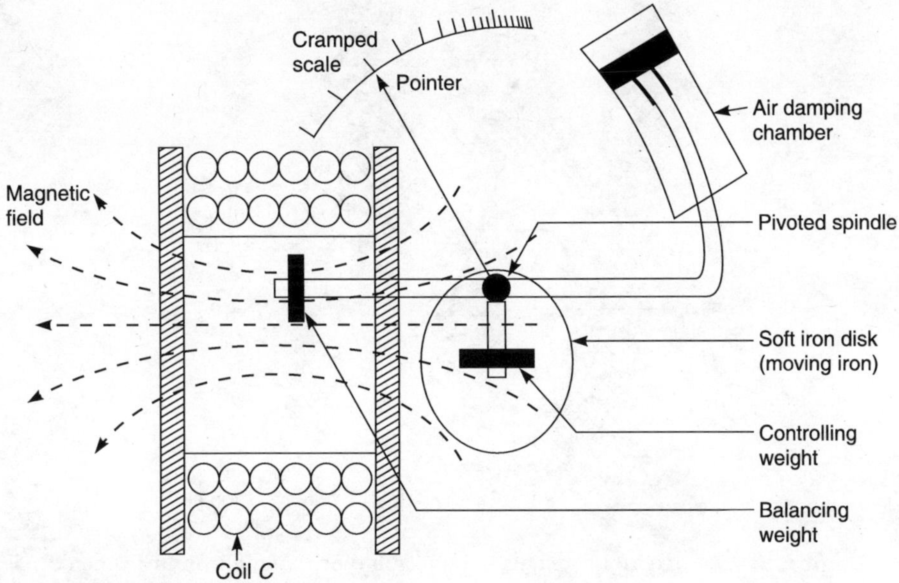

Figure 17 Constructional details of moving-iron attraction-type instrument.

At the verge of operation, the energy stored and the work done are equal. Hence,

$$T_D\, d\theta = \frac{1}{2} dLI^2 \quad \text{or} \quad T_D = \frac{1}{2}\frac{dL}{d\theta} I^2$$

It shows that the deflecting (operating) torque, $T_D \propto I^2$. Also, the controlling torque, $T_C \propto \theta$. On the verge of operation, under steady state,

$$T_C = T_D \quad \text{or} \quad \theta \propto I^2 \;\; \text{[assuming that } dL/d\theta \text{ is constant]}$$

Since the deflection is proportional to the square of the current, the moving iron instruments read the RMS value of the signal input to them. Thus they read both the AC as well as DC currents/voltages. (It needs to be noted that there is no difference between the RMS value and the average value of a pure DC quantity.) Since this instrument uses gravity control for the controlling torque, its scale is cramped.

3.9.2 Repulsion Type Moving Iron Instruments

Figure 18 shows a repulsion type moving iron instrument. In this instrument, the spindle is rigidly attached to a damping vane, to a pointer and to a piece of iron, "A". A winding (surrounding this arrangement) is stationary and carries current proportional to the quantity to be measured. Another piece of iron "B" is rigidly connected with the stationary part of the instrument, i.e. the winding.

The current carried by stationary winding develops magnetic flux. Both the iron pieces "A" and "B" get magnetized in the similar way. Since they develop same magnetic polarities, they repel each other. Since piece B is fixed, A moves away from B. This action is similar in AC as well as DC currents. The movement of A is proportional to the pole strengths of A and B, and hence is proportional to the quantity to be measured. The spindle moves and the pointer indicates the reading on the scale.

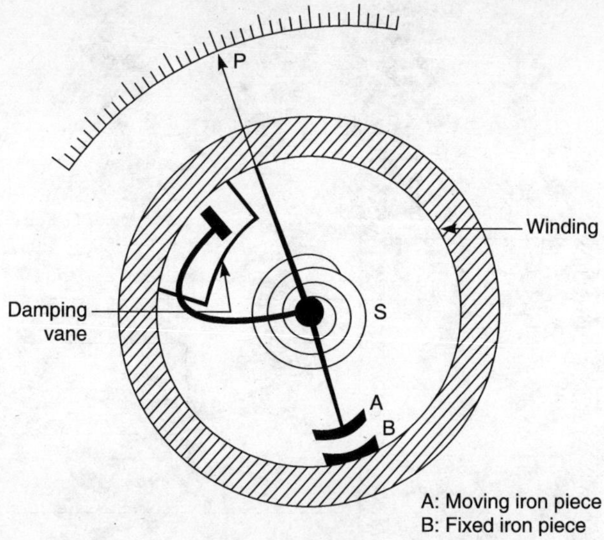

Figure 18 Constructional details of repulsion type moving iron instrument.

3.10 Single-Phase Dynamometer Type Wattmeter

In this instrument, there are two low-resistance current coils (AA′ and BB′), which are fixed at their positions. Figures 19(a) and (b) show the cross-sections of these coils. A high-resistance moving coil called the potential or pressure coil a, b, c, d is placed between the coils AA′ and BB′, such that it may cut the magnetic field created by the two coils. The spindle, carrying spring S and pointer P on it, is rigidly connected with the coil a, b, c, d.

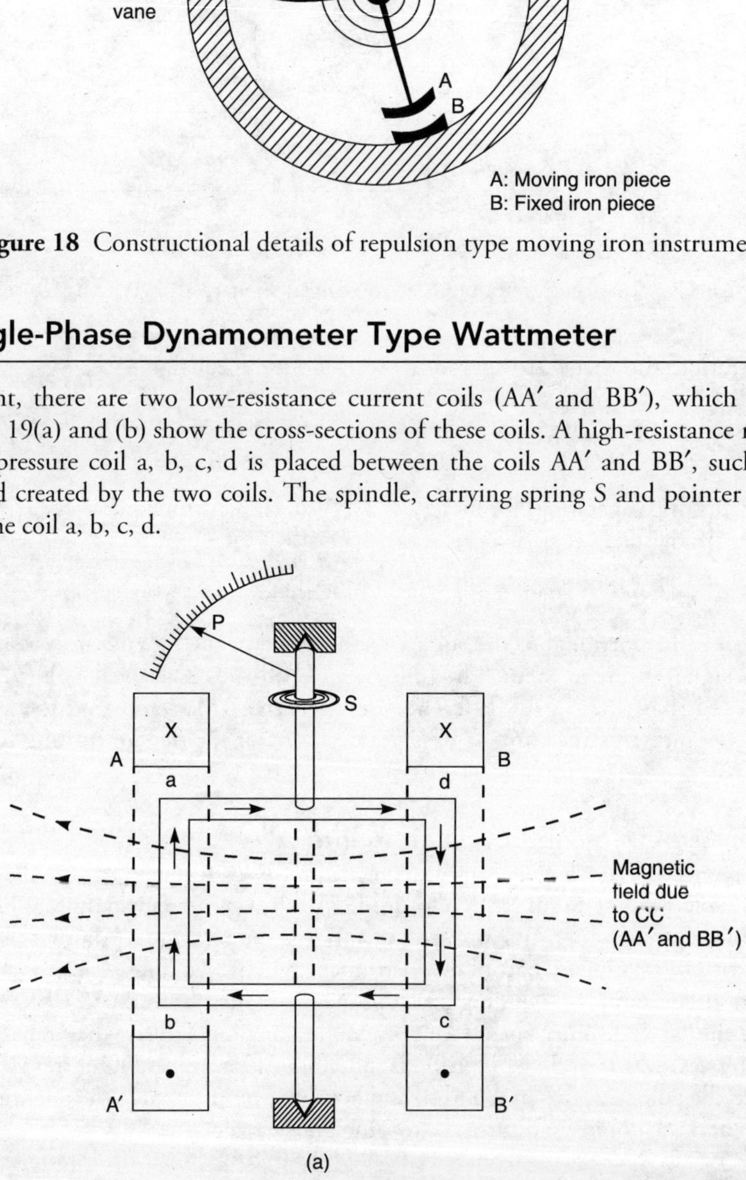

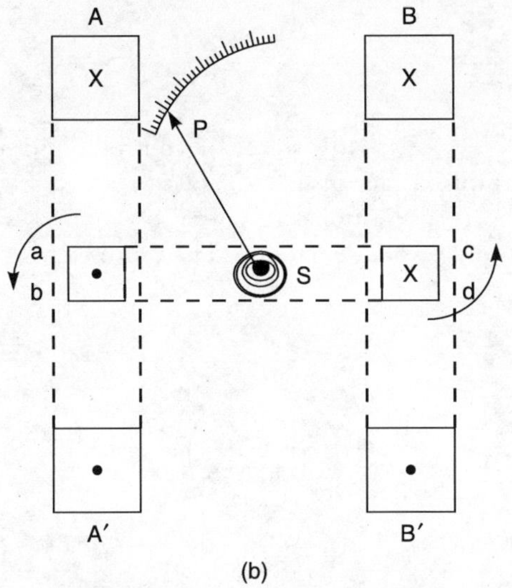

Figure 19 (a) Arrangements of the current coils AA′, BB′; (b) top view of the same.

The electrical connections of the coils are shown in Figure 20. The two current coils (CC) are connected in series with each other. The sense of winding in them is such that they produce magnetic field in the same direction (Fleming's right-hand rule), as shown in Figure 19(a). The pressure coil (PC) shown by coil a, b, c, d, is connected in parallel, as shown in Figure 20. Figure 19(a) explains that in the same magnetic field, the coil side "ab" carries current in one direction, while the coil side "cd" carries it in opposite direction. Applying Fleming's left-hand rule, it can be seen in Figure 19(a) that "ab" experiences a force in forward direction (out of the plane of paper) while "cd" experiences the same amount of force in backward direction (into the plane of paper). Thus, the coil a, b, c, d turns in anti-clockwise direction [when seen from the top as shown in Figure 19(b)] proportional to the cumulative effects of the currents in CC and PC. We know that

$$\text{Force} \propto \phi i, \text{ where } \phi \propto I_{CC} \text{ and } i = I_{PC}$$

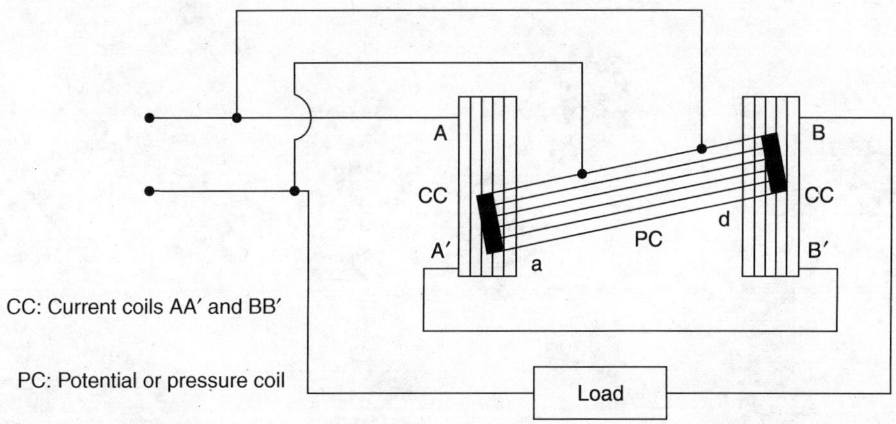

Figure 20 Working principle of single-phase dynamometer type wattmeter.

Also,

$$T_D\,(\text{deflecting torque}) = \text{Force} \times r \qquad [\text{where } r = \text{radius of coil (which is constant)}]$$

Therefore, $T_D \propto \phi i$. That is, $T_D \propto I_{CC} I_{PC}$. Now since T_C (controlling torque) $\propto \theta$, the pointer stops at a reading, where $T_D = T_C$. Thus at this instant $\theta \propto I_{CC} I_{PC}$.

In other words, the deflection is proportional to power, because I_{CC} is proportional to current in the circuit while I_{PC} is proportional to the voltage across it. The wattmeter is connected in a circuit, as shown in Figure 21.

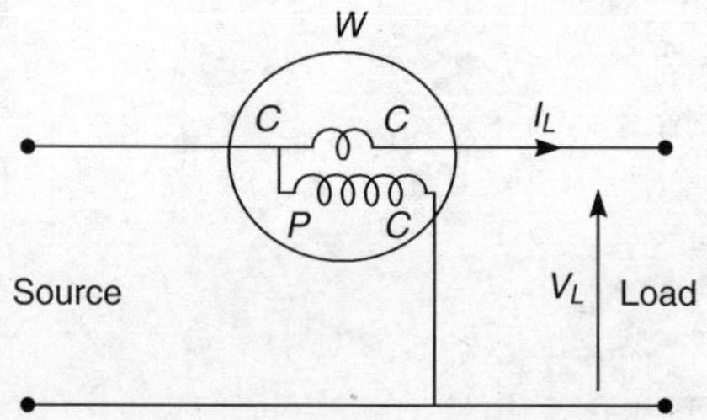

Figure 21 Wattmeter connection diagram.

3.11 Use of Shunts and Multipliers

Shunts and multipliers are used along with the ammeters and voltmeters, respectively, in order to extend their ranges. Let the internal resistance of the meter be represented by R_m. This R_m decides the maximum amount of current I_m allowed to flow through the meter. The value of R_m is such that I^2R_m does not exceed the thermal limit of that meter. If an ammeter is to be used to measure a load current, I_L, which is greater than its rated current, I_m, then a small shunt resistance, R_{sh}, is placed in parallel with the ammeter, as shown in Figure 22.

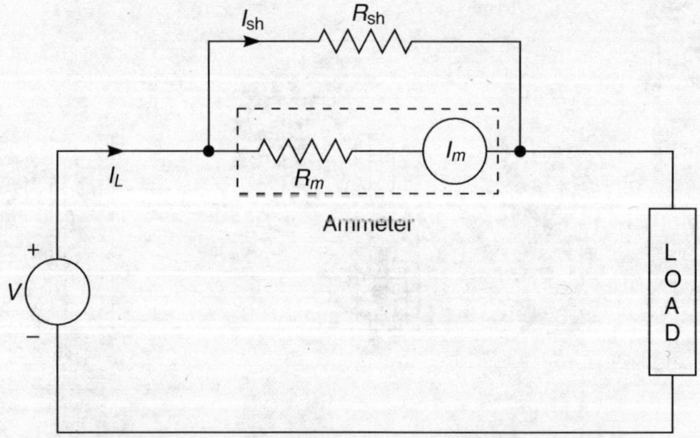

Figure 22 Ammeter with a shunt resistance.

In this case,

$$I_m R_m = I_{sh} R_{sh} = (I_L - I_m) R_{sh} \quad \text{or} \quad R_{sh} = \frac{I_m}{I_L - I_m} R_m$$

I_m and R_m are fixed (and given quantities) for an ammeter. Thus in order to measure current equal to I_L, an external resistance R_{sh} is required to be placed in shunt with the ammeter.

Similarly, if it is desired to extend the range of a voltmeter, a resistance R_{se} is placed in series with it. Again the constraint is this that the current through the voltmeter should not exceed the rated voltmeter current I_m for safe operation of the voltmeter. From Figure 23, it can be seen that

$$\text{Total resistance, } R_t = R_m + R_{se}$$

Therefore,

$$I_m = \frac{V}{R_t} = \frac{V}{R_m + R_{se}}$$

But

$$I_m = \frac{V_m}{R_m}$$

Hence

$$\frac{V_m}{R_m} = \frac{V}{R_m + R_{se}}$$

Hence

$$R_{se} = \frac{V - V_m}{V_m} R_m = \frac{VR_m}{V_m} - R_m = \frac{V}{I_m} - R_m$$

Again, the values of V_m and R_m, pertaining to the given voltmeter, are fixed (given quantities). Hence in order to measure a voltage V greater than V_m a resistance equal to R_{se} is required to be placed in series with the voltmeter.

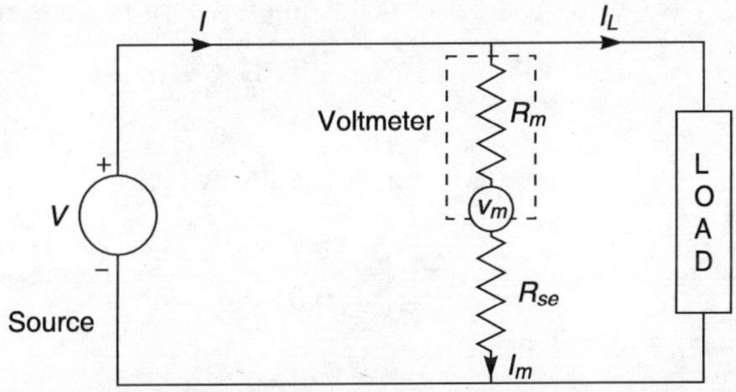

Figure 23 Voltmeter with a series resistance, R_{se}.

Solved Examples

Example 1

A moving-coil milli-ammeter, having a resistance of 8 Ω, gives full-scale deflection when a current of 5 mA is passed through it. Explain how this instrument can be used for measurement of: (a) current up to 2 A and (b) voltage up to 8 V. **(UPTU 2009–10)**

Solution

(a) For current up to 2 A:

$$R_{sh} = \frac{I_m}{I - I_m} R_m = \frac{5\times 10^{-3}}{2 - 5\times 10^{-3}} \times 8 = 0.0201\ \Omega \text{ (in shunt with the meter)}$$

(b) For voltage up to 8 V:

$$R_{se} = \frac{V}{I_m} - R_m = \frac{8}{5\times 10^{-3}} - 8 = 1592\ \Omega \text{ (in series with the meter)}$$

Example 2

A moving coil ammeter has a full-scale deflection of 50 μA and a coil resistance of 100 Ω. What will be the value of the shunt resistance required for the instrument to be converted to read a full-scale reading of 1A? **(UPTU 2008–09)**

Solution

The required shunt resistance is

$$R_{sh} = \frac{I_m}{I - I_m} R_m = \frac{50\times 10^{-6}}{1 - 50\times 10^{-6}} \times 100 = 0.005\ \Omega$$

Example 3

A moving coil instrument has coil of 200 turns. The width and depth of the coil are 20 mm and 30 mm, respectively. Find the deflecting torque in the instrument when the air-gap flux density is 0.12 Wb/m^2 and the current measured is 1 A. Also, find the angle of deflection if the spring constant is 4.8×10^{-4} Nm/degree.

Solution

Here

$$\begin{aligned}\text{Deflecting torque} &= (2r)(BilN) \text{ Nm}\\ &= (20 \times 10^{-3})(0.12 \times 1 \times 30 \times 10^{-3} \times 200)\\ &= 0.01440 \text{ Nm}\end{aligned}$$

On the verge of operation, $T_D = T_C$. Hence

$$\text{Deflection} = \frac{T_D}{\text{Spring constant}} = \frac{0.01440}{0.00048} = 30^\circ$$

Example 4

A three-phase balanced star-connected load has an impedance of (3 + *j*4) Ω per phase. The line voltage is 400 V, 50 Hz. Find the line currents, real power, reactive power and apparent power on a three-phase basis. Repeat the question, if the same load is connected in delta configuration.

Solution

Star connection: The line voltage being given as 400 V, the phase voltage will be

$$V_{\text{ph}} = \frac{V_L}{\sqrt{3}} = \frac{400}{\sqrt{3}} = 231 \text{ volts/phase}$$

And with per phase impedance = $(3 + j4)\ \Omega$, the phase current according to Ohm's law will be

$$I_{\text{ph}} = \frac{V_{\text{ph}}}{Z} = \frac{231\angle 0^\circ}{3+j4} = \frac{231\angle 0^\circ}{5\angle 53.13^\circ} = 46.19\angle -53.13^\circ$$

Since the line and phase currents are same in star, this will also be the line current, that is

$$I_L = 46.19\angle -53.13^\circ$$

Now

$$\text{Three-phase real power } = \sqrt{3}\, V_L I_L \cos\phi$$

Here $\phi = -53.13^\circ$. So $\cos\phi = 0.6$. Putting the values, we get

$$P_R = \sqrt{3} \times 400 \times 46.19 \times 0.6 = 19200 \text{ W}$$

Three-phase reactive power,

$$Q = \sqrt{3} V_L I_L \sin\phi = \sqrt{3} \times 400 \times 46.19 \times 0.8 = 25{,}601.1 \text{ VAR}$$

and

$$\text{Three-phase apparent power } = \sqrt{3}\, V_L I_L = \sqrt{3} \times 400 \times 46.19 = 32001.37 \text{ VA}$$

Powers could also have been found as

$$P_R = 3I^2R = 3 \times 46.19^2 \times 3 = 19201 \text{ W}$$

$$Q = 3I^2 \times X_L = 3 \times 46.19^2 \times 4 = 25602 \text{ VAR}$$

$$S \text{ (apparent power)} = 3 \times V_{\text{ph}} I_{\text{ph}} = 32001.4 \text{ VA}$$

Delta connection: Here line voltage = phase voltage. $V_{\text{ph}} = 400$ V, and hence

$$I_{\text{ph}} = \frac{V_{\text{ph}}}{Z} = \frac{400\angle 0}{3+j4} = \frac{400\angle 0}{5\angle 53.13} = 80\angle -53.13$$

and

$$I_L = \sqrt{3} I_{\text{ph}} = \sqrt{3} \times 80\angle -53.13 = 138.56\angle -53.13$$

As before,

$$P_R = \sqrt{3}V_L I_L \cos\phi = \sqrt{3}\times 400\times 138.56\times 0.6 = 57598.3\ \text{W}$$

$$Q = \sqrt{3}V_L I_L \sin\phi = \sqrt{3}\times 400\times 138.56\times 0.8 = 76797.74\ \text{VAR}$$

$$S = \sqrt{3}V_L I_L = 95997.2\ \text{VA}$$

As can be noticed, "With line voltage remaining fixed, the three powers (active, reactive and apparent) become three-fold on just connecting the load from star to delta."

Example 5

Three identical coils, each having a resistance of 8 Ω and inductance of 6 Ω, are connected in star across a three-phase, 50 Hz, 400 V supply. The power (real) input is measured by the two-wattmeter method. Find the readings of the two wattmeters.

Solution

Per phase impedance is $Z_{\text{ph}} = 8 + j6$ (given). For the star-connected load

$$\text{Voltage per phase} = \frac{400}{\sqrt{3}} = 231\ \text{V}$$

So,

$$\text{Phase (and hence line) current} = \frac{231\angle 0^\circ}{8 + j6} = \frac{231\angle 0^\circ}{10\angle 36.87} = 23.1\angle -36.87$$

The expressions for the two wattmeter readings are

$$W_1 = V_L I_L \cos(30 + \phi) \quad \text{and} \quad W_2 = V_L I_L \cos(30 - \phi)$$

Putting the values of $V_L = 400$, $I_L = 23.1$ and $\phi = 36.87°$, we get

$$W_1 = 400\times 23.1\cos(30 + 36.87) = 3629.64\ \text{W}$$

$$W_2 = 400\times 23.1\cos(30 - 36.87) = 9173.65\ \text{W}$$

Now

$$\text{3-phase real power} = \sqrt{3}\times V_L I_L \cos\phi = \sqrt{3}\times 400\times 23.1\cos(-36.87) = 12803\ \text{W}$$

Also

$$W_1 + W_2 = 12803\ \text{W}$$

Example 6

A balanced three-phase, star-connected load of 120 kW takes a leading current of 85 A when connected across a three-phase, 1100 V, 50 Hz supply. Obtain the values of resistance, capacitance and impedance of the load per phase and also the power factor of the load.

Solution

Given the three phase real power = 120 kW:

$$\sqrt{3}V_L I_L \cos\phi = 120{,}000$$

With $V_L = 1100$ V, $I_L = 85$ A, we have

$$\sqrt{3} \times 1100 \times 85 \times \cos\phi = 120{,}000$$

Solving, we get

$$\cos\phi = 0.74 \text{ leading} \quad \text{or} \quad \phi = \cos^{-1} 0.74 = 42.27^\circ$$

We have

$$Z = \frac{V_{\text{ph}}}{I_{\text{ph}}} = \frac{1100/\sqrt{3}}{85} = 7.47\ \Omega \ \ [\text{as } I_{\text{ph}} = I_L \text{ for a star-connected load}]$$

So, complete expression for Z becomes

$$Z = 7.47\angle -42.27$$

The negative sign is taken because it is given that the load is capacitive. So,

$$Z = 5.530 - j5.02\ \Omega$$

Hence

$$R = \text{Re}\{Z\} = 5.530\ \Omega \quad \text{and} \quad X_C = \text{Im}\{Z\} = 5.02\ \Omega$$

That is,

$$\frac{1}{\omega C} = 5.02 \quad \text{or} \quad C = \frac{1}{314 \times 5.02} = 634\ \mu\text{F}$$

Example 7

A balanced three-phase star-connected load, having a resistor of 10 Ω in each phase, is fed through a balanced three-phase, 400 V supply with the phase sequence being RYB. Find the line current and the three-phase active power. Also, find these readings with phase sequence being RBY.

Solution

With phase sequence being RYB, we have

$$V_{RY} = 400\angle 0^\circ,\ V_{YB} = 400\angle -120^\circ,\ V_{BR} = 400\angle 120^\circ$$

So, the three-phase currents will be

$$I_R = \frac{V_{RN}\angle 0^\circ}{10} = \frac{400\angle 0^\circ}{\sqrt{3}\times 10} = \frac{40}{\sqrt{3}}\angle 0^\circ$$

$$I_Y = \frac{V_{YN}}{10}\angle -120^\circ = \frac{400}{\sqrt{3}\times 10}\angle -120^\circ = \frac{40}{\sqrt{3}}\angle -120^\circ$$

$$I_B = \frac{V_{BN}\angle 120°}{10} = \frac{400\angle 120°}{\sqrt{3}\times 10} = \frac{40}{\sqrt{3}}\angle 120°$$

$$|I_L| = |I_{ph}| = \frac{40}{\sqrt{3}} = 23.09\text{ A}$$

The sum of the three phase currents (and also line current, since the load is star connected) is

$$I_R + I_Y + I_B = 0$$

This is because the load is balanced.

Three-phase power $= \sqrt{3}V_L I_L \cos\phi$ with $\cos\phi = 1$. We have

$$P_{real} = \sqrt{3}\times 400\times 23.09\times 1 = 15.997\text{ kW} = 16\text{ kW}$$

There will be absolutely no difference in the individual values on changing the phase sequence to RBY.

The only change will be in their respective phasors. I_R will be equal to $23.09\angle 0°$, $I_Y = 23.09\angle 120°$, $I_B = 23.09\angle -120°$.

Example 8

The ohmic values of the resistance and reactance of an impedance are equal. Three such impedances are connected in star across a three-phase supply. The three-phase power was measured by two-wattmeter method. The wattmeter giving the lower (of the two wattmeters) reading indicated 400 W. Find the per phase reactive power in the circuit.

Solution

We have $Z = R + jx$. Then

$$\text{Angle of } Z = \tan^{-1}\frac{X}{R} \text{ with } R = X, \text{ Angle of } Z = \tan^{-1}1 = 45°$$

Now

$$\phi = \tan^{-1}\left(\sqrt{3}\frac{W_2 - W_1}{W_2 + W_1}\right)$$

So

$$\tan\phi - 1 - \frac{\sqrt{3}W_2 - \sqrt{3}W_1}{W_2 + W_1}$$

or

$$2.732W_1 = 0.732W_2$$

or

$$W_2 = 3.732W_1$$

So, W_1 is lower whose value is 400 W. Thus, $W_1 = 400$ W, $W_2 = 1492.8$ W. Now,

$$\text{Per phase reactive power } = \frac{\sqrt{3}(W_2 - W_1)}{3}\text{VAR} = \frac{\sqrt{3}(1492.8 - 400)}{3} = 631\text{ VAR}$$

Example 9

A balanced three-phase star-connected load of $(8 + j6)\,\Omega$ per phase is connected to a three-phase, 400 V supply. Find the line current, power factor, three-phase real power and three-phase volt-ampere.

(UPTU 2002–03)

Solution

The circuit of the problem will look like as in Figure 24. We have

$$V_{\text{ph}} = \frac{V_{\text{Line}}}{\sqrt{3}} = \frac{400}{\sqrt{3}} = 231\,\text{V} = Z\angle\theta = 8 + j6 = 10\angle 36.87$$

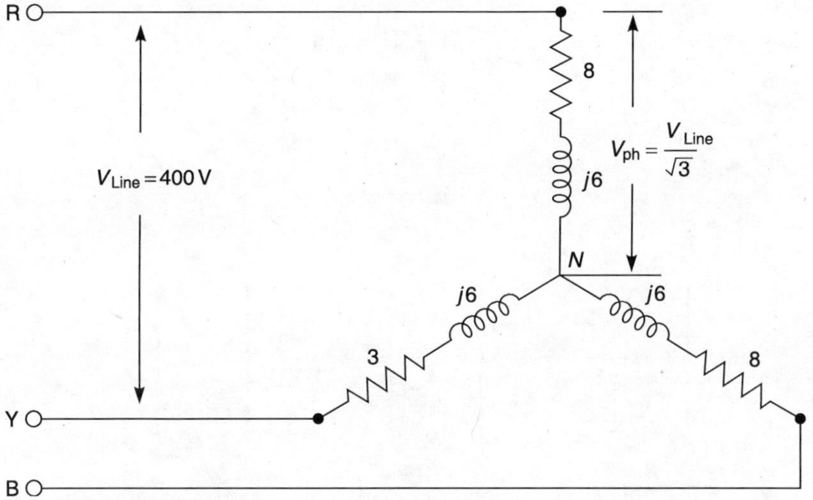

Figure 24 Figure for Example 9.

1. Applying Ohm's law in any phase, we get:

$$\text{Phase current, } I_{\text{ph}} = \frac{V_{\text{ph}}}{Z\angle\theta} = \frac{231\angle 0^\circ}{10\angle 36.87} = 23.1\angle -36.87$$

In star the phase and line currents are same. So the line current:

$$I_L = I_{\text{ph}} = 23.1\angle -36.87^\circ$$

The other two line currents will also have the same magnitude and a progressively lagging phase angle of 120°.

2. Power factor: The angle between phase voltage and phase current is 36.87° with current lagging. So the power factor is

$$\text{pf} = \cos 36.87 = 0.8 \text{ lagging}$$

3. The three-phase real power is given by

$$P = \sqrt{3}V_L I_L \cos\phi = \sqrt{3} \times 400 \times 23.1 \times 0.8 = 12803.32\,\text{W}$$

4. Volt-ampere drawn by the load will be

$$S = \sqrt{3}V_L I_L = \sqrt{3} \times 400 \times 23.1 = 16004.15\,\text{W}$$

Example 10

A three-phase balanced load connected across a three-phase 400 V AC supply draws a line current of 10 A. Two wattmeters are used to measure the input power. The ratio of the two-wattmeter readings is 1:2. Find the readings of the two wattmeters.

Solution

The three-phase real power is given by the expression

$$P = \sqrt{3}V_L I_L \cos\phi$$

Here $V_L = 400$ V, $I_L = 10$ A $\cos\phi = ?$. Also, the load angle (in terms of the two wattmeter readings, W_1 and W_2) is

$$\phi = \tan^{-1}\left[\frac{\sqrt{3}(W_2 - W_1)}{W_2 + W_1}\right]$$

Here we know that

$$\frac{W_1}{W_2} = \frac{1}{2}$$

So, $W_2 = 2W_1$. Putting the values, we get

$$\phi = \tan^{-1}\left[\frac{\sqrt{3}\,W_2\left(1 - \dfrac{W_1}{W_2}\right)}{W_2\left(1 + \dfrac{W_1}{W_2}\right)}\right] = \tan^{-1}\left[\frac{\sqrt{3}\left(1 - \dfrac{1}{2}\right)}{1 + (1/2)}\right] = \tan^{-1}\left(\frac{1}{\sqrt{3}}\right) = 30^\circ$$

So,

$$P = \sqrt{3}\,V_L I_L \cos\phi = \sqrt{3} \times 400 \times 10 \times \cos 30^\circ = 6000 \text{ W}$$

Now, $W_2 + W_1 = 6000$. With $W_2 = 2W_1$, we have $3W_1 = 6000$. This gives $W_1 = 2000$ W and $W_2 = 4000$ W.

Example 11

In a certain measurement of three-phase power by two-wattmeter method, one of the wattmeters gave a reading of 20 kW and the other 5 kW (after the connections of the pressure coil were reversed). Calculate (a) total real, reactive and apparent powers; (b) power factor of the load.

Solution

(a) The total real power will be $W_1 + W_2$ (with $W_1 = 20$ kW and $W_2 = -5$ kW. So,

$$W_1 + W_2 = P = 20 + (-5) = 15 \text{ kW}.$$

The total reactive power is

$$Q = \sqrt{3}(W_1 - W_2)$$

Putting the values, we get

$$Q = \sqrt{3}[20 - (-5)] = \sqrt{3} \times 25 = 43.3 \text{ kVAR}$$

The total apparent power will be

$$S = \sqrt{P^2 + Q^2} = \sqrt{15^2 + 43.3^3} = 45.82 \text{ kVA}$$

(b) Angle of the load is given by

$$\phi = \tan^{-1}\left(\sqrt{3}\frac{W_1 - W_2}{W_1 + W_2}\right)$$

Putting the values of W_1 and W_2, we get

$$\phi = \tan^{-1}\left[\sqrt{3}\frac{20-(-5)}{20-5}\right] = 70.89°$$

Power factor = cos ϕ = cos 70.89 = 0.327. The mere knowledge of W_1 and W_2 does not tell whether the load is inductive or capacitive.

Example 12

Prove that if the two wattmeters, in the measurement of three-phase power by two-wattmeter method, show equal readings, the power factor of the load is unity. What will be the value of the power factor, when the wattmeters show opposite and equal readings?

Solution

If the two wattmeter readings are W_1 and W_2, the angle of the load in terms of these readings is

$$\phi = \tan^{-1}\left(\sqrt{3}\frac{W_2 - W_1}{W_2 + W_1}\right)$$

Here the choice of taking ($W_2 - W_1$) or ($W_1 - W_2$) depends upon which one among W_1 and W_2 is bigger. So, if $W_1 = W_2$, the value of $\phi = 0$. Thus the power factor, that is $\cos\phi = \cos 0 = 1$. Also, if $W_1 = -W_2$, the denominator of the ϕ expression will become zero, thus making $\phi = \tan^{-1}\infty$. So, $\phi = 90°$. Here, the power factor $\cos\phi = \cos 90° = 0$. Thus, here the load must be purely reactive.

Example 13

A balanced three-phase delta connected load of 150 kW takes a leading current of 80 A, when connected across a three-phase, 2000 V, 50 Hz supply. Find the circuit constants of the load per phase.

Solution

We know that the three-phase real power is given by $P = \sqrt{3}\, V_L I_L \cos\phi$. Here P = 150,000 W, V_L = 2000 V, I_L = 80 A. Putting these values, we get

$$150{,}000 = \sqrt{3} \times 2000 \times 80 \times \cos\phi \Rightarrow \cos\phi = 0.54 \Rightarrow \sin\phi = 0.84$$

Now impedance per phase is

$$Z_{\text{ph}} = \frac{V_{\text{ph}}}{I_{\text{ph}}}$$

In delta:

$$V_{ph} = V_L \quad \text{and} \quad I_{ph} = \frac{I_L}{\sqrt{3}}$$

So

$$V_{ph} = 2000 \text{ V and } I_{ph} = \frac{80}{\sqrt{3}} = 46.19 \text{ A}$$

So

$$Z_{ph} = \frac{2000}{46.19} = 43.3 \ \Omega$$

Hence the resistance per phase is

$$R = Z_{ph} \cos\phi = 43.3 \times 0.54 = 23.38 \ \Omega$$

The reactance (capacitive) per phase is

$$X_c = Z_{ph} \sin\phi = 43.3 \times 0.84 = 36.37 \ \Omega$$

So,

$$\frac{1}{\omega C} = 36.37; \qquad C = \frac{1}{\omega \times 36.37} = \frac{1}{314 \times 36.37} = 87.56 \ \mu\text{F}$$

Thus,

$$R = 23.38 \ \Omega \quad \text{and} \quad C = 87.56 \ \mu\text{F}$$

Example 14

Two ammeters–one with a current scale of 10 A, resistance of 0.01 Ω, and the other with a current scale of 15 A, resistance of 0.005 Ω–are connected in parallel. What can be the maximum current carried by this parallel combination, so that none of the ammeter reading goes out of scale? **(UPTU 2003–04)**

Solution

The circuit diagram is shown in Figure 25.

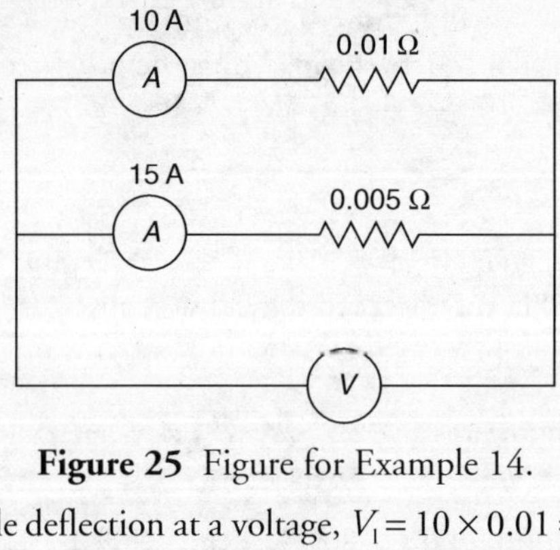

Figure 25 Figure for Example 14.

The first ammeter gives full-scale deflection at a voltage, $V_1 = 10 \times 0.01 = 0.1$ V, and second ammeter gives full-scale deflection at a voltage, $V_2 = 0.005 \times 15 = 0.075$ V. Hence the voltage applied to the parallel circuit should be 0.075 V, because if the voltage increases then the reading of second ammeter will go out of the scale.

Now, current through first ammeter, $I_1 = 0.075/0.01 = 7.5$ A, the current through the second ammeter, $I_2 = 0.075/0.005 = 15$ A. Hence, the total maximum current, $I = I_1 + I_2 = 7.5 + 15 = 22.5$ A.

Example 15

A moving coil instrument has a resistance of 2 Ω, and it reads up to 250 V when a resistance of 5000 Ω is connected in series with it. Find the current range of the instrument, when it is used as an ammeter with the coil connected across a shunt resistance of 2 mΩ. **(UPTU 2000–01)**

Solution

Given: moving coil resistance, $R_m = 2\ \Omega$; series resistance $R_{sc} = 5000\ \Omega$. Now, they are connected in series, as shown in Figure 26.

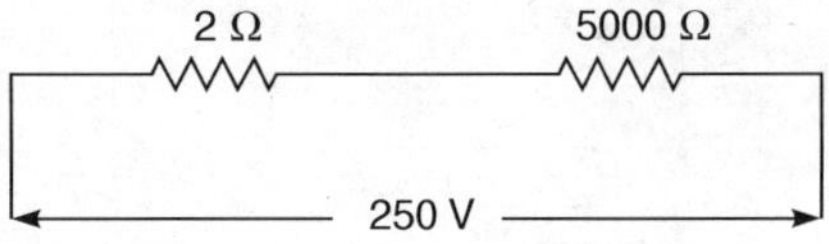

Figure 26 Figure for Example 15.

The current flowing in the circuit is

$$I_{fd} = \frac{250}{2+5000} = 0.049\ \text{A}$$

So, voltage across the coil at full-scale deflection is

$$V_C = I_{fd} \times 2 = 0.04998 \times 2 = 0.09996\ \text{V}$$

The diagram for a shunt resistance of 2 mΩ when connected in parallel with the coil is shown in Figure 27.

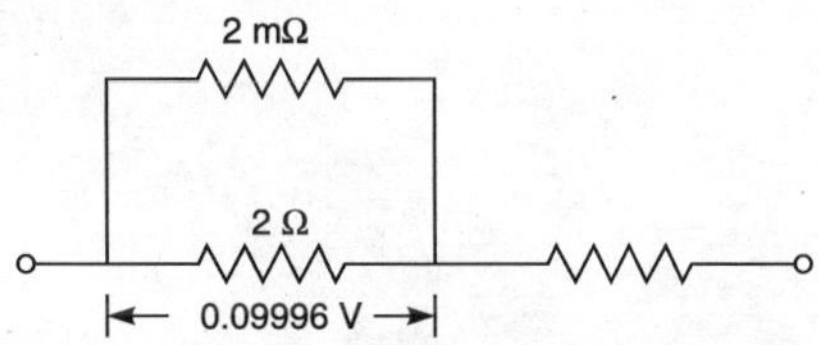

Figure 27 Shunt resistance for Example 15.

Therefore

$$I_{sh} = \frac{V_c}{R_{sh}} = \frac{0.09996}{2\times 10^{-3}} = 49.98\ \text{A}$$

So, the range of the instrument will be:

$$I = I_{fd} + I_{sh} = 0.04998 + 49.98 = 50\ \text{A}$$

Example 16

A moving iron instrument, having internal resistance of 50 Ω, indicates full-scale deflection with a current of 10 mA. How can it be made to work as: (a) a voltmeter to read 100 V on full scale, (b) an ammeter of 1 A on full scale? **(UPTU 2001–02)**

Solution

Given: internal resistance, $R_m = 50\ \Omega$; full-scale deflection current, $I_m = 10$ mA; maximum voltage drop across the instrument,

$$V_m = 10 \times 10^{-3} \times 50 = 0.5\ \text{V}$$

(a) As a voltmeter: A high resistance will be connected in series to read 100 V, as shown in Figure 28. Now,

$$V = I_m R_m + I_m R = V_m + RI_m$$

or

$$R = \frac{V - V_m}{I_m} = \frac{100 - 0.5}{10 \times 10^{-3}} = 9950\ \Omega$$

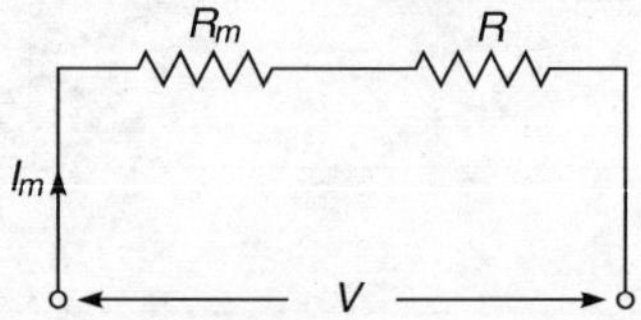

Figure 28 Figure for Example 16(a).

(b) As an ammeter: A low resistance will be connected in parallel, as shown in Figure 29.

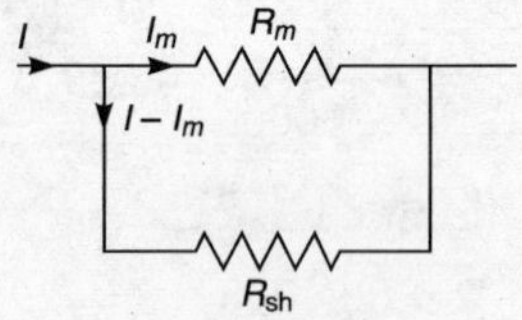

Figure 29 Figure for Example 16(b).

Now, $V_m = V_{sh}$. Hence

$$I_m R_m = (I - I_m) R_{sh}$$

or

$$R_{sh} = \frac{I_m}{I - I_m} \times R_m = \frac{10 \times 10^{-3} \times 50}{1 - 0.01} = 0.505\ \Omega$$

Example 17

A moving iron millimeter gives full-scale deflection with 15 mA and has a resistance of 5 Ω. Calculate the resistance to be connected: (a) in parallel to enable the instrument to read up to 1 A, (b) in series to enable it to read up to 10 V. **(UPTU 2001–02)**

Solution

Given: $I_{fd} = 15$ mA; $R_m = 5\ \Omega$.

(a) Let a shunt resistance R_{sh} be connected in parallel with the coil, as shown in Figure 30. Now, the voltage across the coil is

$$V = I_{fd} \times R_m = 15 \times 10^{-3} \times 5 = 75 \text{ mV}$$

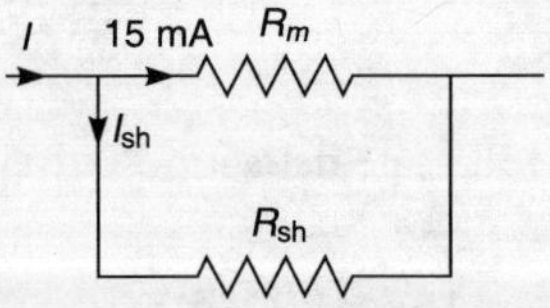

Figure 30 Figure for Example 17(a).

Now, $I_{sh} = I - 15 \times 10^{-3} = 0.985$ A

Hence $$R_{sh} = \frac{V}{I_{sh}} = \frac{75 \times 10^{-3}}{985 \times 10^{-3}} = 0.076\ \Omega$$

(b) Let the series resistance R be connected in series with coil, as shown in Figure 31. Now voltage across R is

$$V_R = V - I R_m = 10 - (15 \times 10^{-3} \times 5) = 9.925 \text{ V}$$

or $$R = \frac{V_R}{I} = \frac{9.925}{15 \times 10^{-3}} = 661.667\ \Omega$$

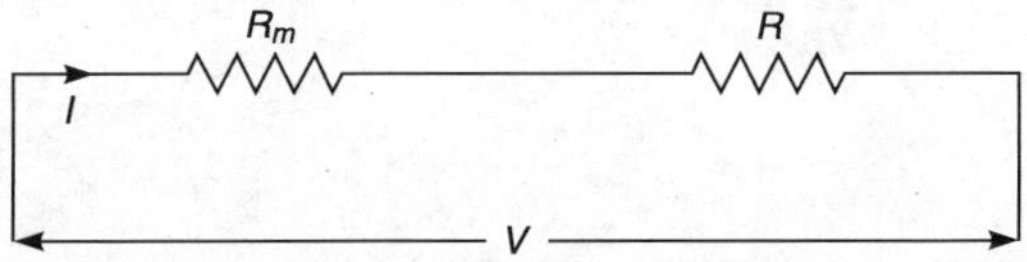

Figure 31 Figure for Example 17(b).

Summary

This chapter mainly highlights the three-phase balanced electrical supply and load system. It also presents the description of some commonly used measuring instruments used for the measurement of voltage and current. The three-phase load having equal impedances per phase is called a *balanced load.* There is a definite sequence followed while giving the three-phase supply, which is called *phase sequence*, usually represented by RYB. In a star-connected three-phase (balanced or unbalanced) load,

$$V_L = \sqrt{3} V_{ph} \quad \text{and} \quad I_L = I_{ph}$$

In a delta-connected three-phase (balanced or unbalanced) load:

$$V_L = V_{ph} \quad \text{and} \quad I_L = \sqrt{3} I_{ph}$$

Two-wattmeter method is popular for measurement of power of three-phase (balanced or unbalanced) load. This is given by $W_1 + W_2 = \sqrt{3}\, V_L I_L \cos\theta$. If the difference of the two-wattmeter readings is multiplied by $\sqrt{3}$, the answer gives the total three-phase reactive power.

The various instruments used to measure voltage, current and power are categorized into mainly four types: permanent magnet moving coil (useful for measuring the average values of voltages and currents), moving iron (useful for measuring the RMS values of voltages and currents) and electrodynamometer type (useful for measuring voltages currents and mainly power).

The extension of the range of ammeter is done by connecting a shunt (parallel resistor) of very low ohmic value to the ammeter, whereas that of a voltmeter is done by connecting a series resistance of high ohmic value to the voltmeter.

Short Answer Questions

1. What are the functions of current coil and voltage coil in a wattmeter?
2. In the three-phase power measurement by two-wattmeter method, if the numerical readings of the two wattmeters are equal, what can be concluded about the nature of the load?
3. If the aim is to provide more voltage to a load, which connection ought to be made among star and delta for a fixed line voltage?

4. From the readings of the two wattmeters in the measurement of three-phase power by two-wattmeter method, what among the following can be conclusively found: Three-phase real power, three-phase reactive power, three-phase apparent power, phase angle of the load, power factor of the load?
5. If a pure sine wave is fed to the input of a PMMC instrument, what will be its reading?
6. What is the reading displayed by a MI instrument in Q5?
7. Which torque is not present in an integrating type instrument?
8. In the measurement of three phase power by two wattmeter method, the real and reactive powers were found to be numerically equal. What can be said about the phase angle of the load?
9. In the measurement of three phase power by two wattmeter method, what actually happens when one of the wattmeters starts showing negative reading?
10. State whether a delta connected three-phase supply can be fed to a household or not? Give reasons.

Review Questions

1. Derive the two relationships between the line and phase voltages and currents for star-and delta-connected balanced loads. Does this relationship change for unbalanced loads?
2. Prove that the sum of two-wattmeter readings, in the measurement of three-phase power by two-wattmeter method, gives the total three-phase real power.
3. Derive the expression of the power factor in terms of the two wattmeter readings in Problem 2.
4. What do you understand by phase sequence? Explain its significance in the three-phase power system.
5. What are the basic torques acting on an indicating instrument? Briefly discuss them.
6. What do you understand by attraction-type and repulsion-type instruments? Explain.
7. Explain with sketch the principle of operation and working of a PMMC type instrument. Explain how is it unsuitable for measuring the AC quantities?
8. Explain with sketch the principle of operation and working of an attraction type MI instrument. Explain how is it suitable for measuring the AC quantities?
9. Differentiate between moving-coil and moving-iron instrument.
10. Bring out the advantages and disadvantages of *MI* and *MC* instrument.

Problems

1. A moving coil instrument requires 20 mA and 100 mV for a full-scale deflection. What is the value of the shunt resistance required to make the instrument read up to 10 A?
2. A moving coil instrument has the coil resistance of 50 Ω and gives maximum deflection at 5 mA. Find the external resistance to be connected in series to make it suitable to read up to 30 V.
3. A moving coil instrument, having a resistance of 50 Ω, has a full-scale deflection of 1 mA. Calculate the shunt resistance to convert the instrument into an ammeter of 2 A range. **(UPTU 2006–07)**
4. Three similar coils connected in delta across a 400 V, 50 Hz, three-phase AC supply take a line current of 17.32 A at a lagging power factor of 0.8. Find: (a) per phase current, (b) the coil parameters and (c) the three powers drawn by each coil.
5. Three similar coils, each of impedance 10 Ω and impedance angle of 36.87, are connected

across a 400 V, three-phase, 50 Hz supply. Find the line current, power input and also the reactive and apparent powers, if the star connection exists.

6. The power input to a three-phase balanced delta connected *R-L* load is measured by the two-wattmeter method, whose readings were found to be $W_1 = 3$ kW, $W_2 = 1$ kW with its pressure coil reversed. Find the real, reactive and apparent power per phase. If the line current is 10 A, find the values of *R* and *L*. Given frequency = 50 Hz.
7. In the measurement of three-phase power by the two wattmeter method, it was found that the reading of one wattmeter is thrice the other. Find the power factor. Can it be concluded whether the power factor is leading or lagging comment?
8. A balanced star connected load is supplied from a symmetrical three phase, 400 V (line-to-line) supply. The current in each phase is 50 A and lags 30° behind the phase voltage. What is the phase impedance?
(UPTU June 2009)
9. A moving coil instrument gives a full-scale deflection with 20 mA. The resistance of the coil is 4 Ω. What is the value of the series resistance needed for the instrument to read up to 30 V?
(UPTU December 2009)
10. A moving coil ammeter has a full-scale deflection with 50 μA and a coil resistance of 100 Ω. What will be the value of shunt resistance required for the instrument to be converted to read a full-scale reading of 1 A?
(UPTU December 2008)

Answers

Short Answer Questions

1. The current coil measures the current through and the voltage coil measures the voltage across the load.
2. In case the two wattmeter readings are equal, the load will be purely resistive. In case one of the wattmeter reading is equal to the other but its current or voltage coil connections have been reversed, then the load will be purely inductive or capacitive.
3. Since the phase voltage is equal to the line voltage in delta connection (and is equal to 57.7% of the line voltage in star connection), hence for providing more phase voltage to the load, delta connection is preferred.
4. All can be found except whether the power factor is leading or lagging.
5. It will show zero reading because PMMC shows the average value of the input signal and the average value of a pure sine wave is zero.
6. Since an MI instrument shows the RMS value of the input waveform hence its pointer will show a value equal to $Vm/\sqrt{2}$ where V_m is the peak value of the input sine waveform.
7. Controlling torque is usually not present in an integrating type instrument because here the readings have to be integrated, that is, cumulated, so that there is no role of restoring the pointer back to its zero position at the removal of the input.
8. Since the sum of the two wattmeter readings W_1 and W_2 is equal to the difference $W_1 - W_2$ multiplied by $\sqrt{3}$, this condition can be put in the expression of phase angle to give an answer of 45°.
9. The phase angle between the voltage of the voltage coil and the current of the current coil becomes 90°. Hence the wattmeter starts showing zero real power.
10. Since most of the household equipment are designed for a single-phase supply, a delta-connected supply cannot be given there. This is because there is no neutral terminal in a delta connection and hence the reduced (phase) voltage will not be available in delta connection. Thus a 230 V RMS device will have to face 440 V if given a phase supply of delta which may damage it.

Problems

1. 0.01 Ω
2. R_{se} = 5950 Ω
3. 0.025 Ω
4. **(a)** 10 A,
   (b) $R = 32\ \Omega$, $L = 102$ mH
   (c) $P_{(real)} = 3200$ W, $P_{(reactive)} = 2400$ VAR, $P_{(apparent)} = 4000$ VA
5. **(a)** Line current = 23.1 A.
   (b) Real power = 12.8 kW.
   (c) Reactive power = 9.6 kVAR (lagging).
   (d) Apparent power = 16 kVA.
6. Real power = 667 W, Reactive power = 2309.4 VARs, Apparent power = 2403.8 VAs, $R = 6.65\ \Omega$, $L = 23.06$ mH
7. $\cos \phi = 0.76$; No, because the nature of the load needs to be predefined.
8. $4.618\angle 30° = 4 + j2.309\ \Omega$
9. 1496 Ω
10. $5 \times 10^{-3}\ \Omega$

4 Power System, Magnetic Circuit and Single-Phase Transformers

LEARNING OBJECTIVES

After a careful study of the chapter, you would be able to understand:

- The need for earthing in electrical equipment.
- Definition of magnetic circuit and its performance.
- A typical use of the magnetic circuit, that is single-phase transformer.
- Dot convention.
- Principle of operation of a single-phase transformer.
- Equivalent circuit of a single-phase transformer.
- Efficiency calculations of a single-phase transformer.
- Introduction to autotransformer.

4.1 Need for Earthing

The power is fed through electrical connections from SUPPLY to the EQUIPMENT. Every supply system has a switchboard which contains switch and SOCKET. Each equipment has a PLUG which, when inserted into the socket, connects the equipment to the electric supply.

As shown in Figure 1 three wires are used for single-phase system: One is (RED colored) LIVE wire, the second is (BLACK colored) NEUTRAL wire; and the third is (thicker and GREEN colored) EARTH wire. The electricity flows from the supply through live wire and returns through neutral wire.

Ideally the outer body of any equipment must be electrically insulated so that any person touching the equipment should not get electric shock. The outer body is always internally connected to the earth wire of plug. The earth wire of the supply system is always connected through a low resistance path to earthing plate buried under ground.

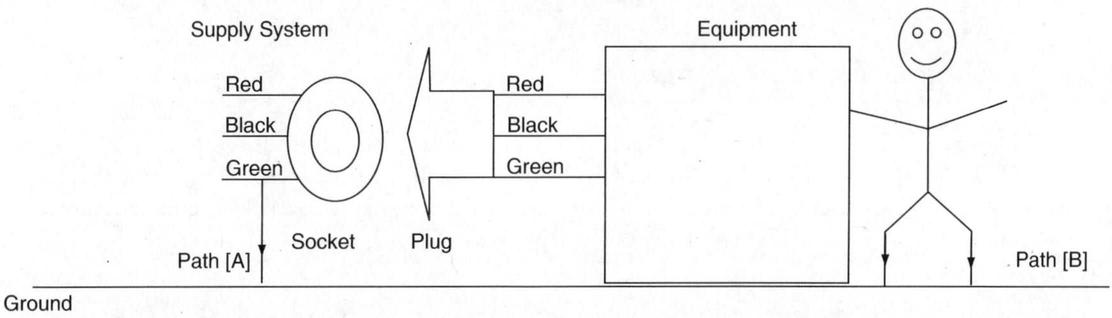

Figure 1 Arrangement of a household earthing unit.

Let us say the earthing is not proper. That is, the equipment does not have earth connection or the earth wire of the supply system is not properly grounded, then there is a possibility of the operating person getting an electric shock. Suppose, by chance, the live part of the equipment touches its outer body, there would be two possibilities. One is that the earthing of equipment and the supply are intact (path [A] is intact), so the leakage current would flow to ground through earth wire and the person will not get electric shock. In second case, where earthing of either the equipment or the supply system is not proper, the person touching the outer body of the equipment will get the electric shock (as the leakage current will now flow through path [B] instead of path [A]).

Thus, earthing of equipment and supply system is of utmost importance to prevent human being from getting electric shock while operating electrical equipment. Improper supply system earthing (earthing of neutral) may also lead to insulation breakdowns causing fire hazards.

Further in order to provide a low-resistance path to the earth wire, the size of the earth pin (in plug) and the corresponding hole (in the socket) are kept bigger, both in length and the area as shown below in Figure 2.

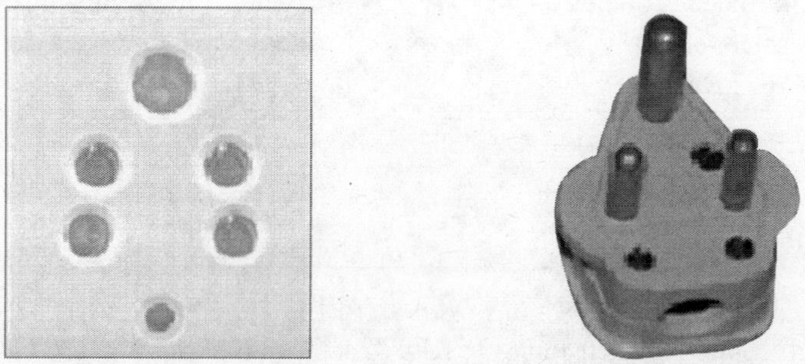

(a) (a)

Figure 2 (a) Socket and (b) a three-pin plug top with earth pin thicker and longer.

The current flowing through the body of a person may be fatal (kill the person). There is a safe value of current in amp (rms) which a human body can tolerate. This is given by $I = 0.165/\sqrt{t}$ for $t < 3$ s. For $t > 3$ s, $I = 9$ mA. There are some codes and standards defining the earthing methods. The important codes are Indian Electrical Rules, National Electrical Code (India), IS:3043 (Code of Practice for Earthing), etc.

4.2 Magnetic Circuit Concepts

Any coil, having N number of turns and carrying I A of current through it, develops magnetomotive force (MMF) of $NI = \Gamma$ (say) ampere-turns. This force results in development of flux of ϕ Wb in the magnetic circuit. The flux has a tendency to close its path through the path of the least reluctance. In the arrangement shown in Figure 3 (as an example), the flux and MMF follow a law similar to Ohm's law:

$$\text{MMF} = \text{Flux} \times \text{Reluctance}$$

that is

$$\Gamma = \phi \Re$$

where $\Re$, the reluctance in the path of flux, is given by

$$\Re = \frac{l}{\mu_0 \mu_r A}$$

Here l is the length of magnetic path and A is the cross-sectional area perpendicular to the magnetic path. μ_0 and μ_r are the absolute permeability and relative permeability, respectively. From this, it is clear that there exists an analogy between electrical circuits and magnetic circuits. Figure 3 shows the torroidal ring made of a magnetic material.

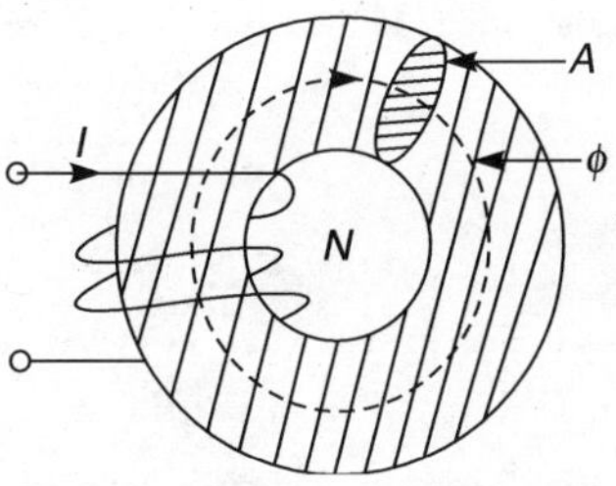

Figure 3 Torroidal ring made of magnetic material.

N number of turns of a coil surrounds the ring. This coil carries a current of I A. Depending upon the reluctance offered by the magnetic circuit, a flux ϕ gets developed in the ring. Here "A" is the cross-sectional area of the torroidal ring and "l" is equal to $2\pi r$, where "r" is the mean radius of the torroidal ring. Table 1 shows the analogy between an electrical circuit and a magnetic circuit.

Table 1 Analogy between electrical circuit and magnetic circuit concepts

Magnetic circuit	*Electrical circuit*
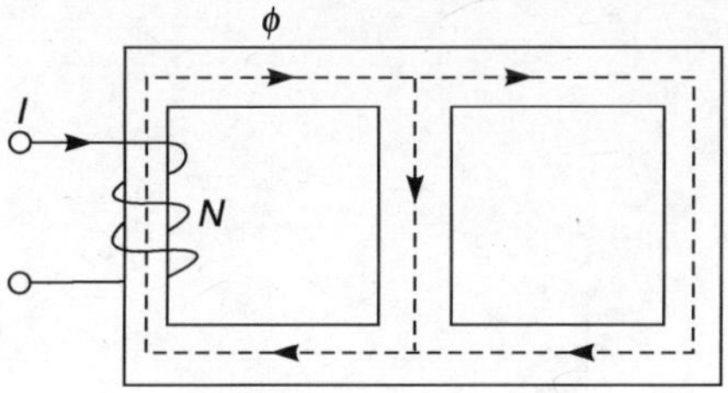 A magnetic circuit	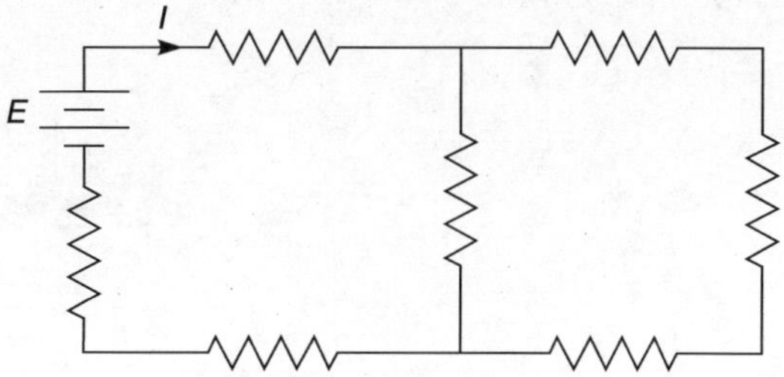 An electrical circuit
1. MMF (Γ ampere-turns) (Magnetomotive force)	**1.** EMF (E volts) (Electromotive force)
2. Flux (ϕ weber)	**2.** Current (I ampere)
3. Reluctance ($\Re$ ampere-turn/weber)	**3.** Resistance (R ohm)

(*continued*)

Table 1 Continued

Magnetic circuit	*Electrical circuit*
4. If μ = permeability of magnetic material, l = length of magnetic path and A = cross-sectional area of the path, then $\mathfrak{R} = l/(\mu A)$	**4.** If ρ = resistivity of conductor material, l = length of conductor and A = cross-sectional area of the conductor, then $R = \rho l/A$
5. Permeability (μ henry/meter)	**5.** Conductivity (σ) $\left(\text{mho/meter as } \frac{1}{\text{ohm}} = \text{mho}\right)$
6. Reluctivity = $1/\mu$ meters/henry	**6.** Resistivity, $\rho = 1/\sigma$ (ohm meter)
7. Flux density (**B** weber/m^2) ($\mathbf{B} = \phi/A$)	**7.** Current density **J** ampere/m^2 ($\mathbf{J} = I/A$)
8. Magnetic field intensity (or magnetizing force), **H** $\mathrm{H} = NI/l = \Gamma/l$ (ampere-turn/m)	**8.** Electric field intensity, **E** $\mathbf{E} = EI/l$ (volt/m)
9. The MMF is $\Gamma = \mathfrak{R}\phi$	**9.** The EMF is $E = RI$
10. $\mathbf{B} = \mu\mathbf{H}$	**10.** $\mathbf{J} = \sigma\mathbf{E}$
11. For series circuit: $\mathfrak{R}_{\text{total}} = \mathfrak{R}_1 + \mathfrak{R}_2 + \mathfrak{R}_2 + \cdots$ (similar to KVL)	**11.** For series circuit: $R_{\text{total}} = R_1 + R_2 + \cdots$ (follows KVL)
12. For parallel circuit: $\mathfrak{R}_{\text{total}} = \left(\frac{1}{\mathfrak{R}_1} + \frac{1}{\mathfrak{R}_2} + \frac{1}{\mathfrak{R}_3} + \cdots\right)^{-1}$ (similar to KCL)	**12.** For parallel circuit: $R_{\text{total}} = \left(\frac{1}{R_1} + \frac{1}{R_2} + \cdots\right)^{-1}$ (follows KCL)

4.2.1 B–H Curve

Let a core of magnetic material have relative permeability of μ_r, length l and cross-section area A. Let it be magnetized by a coil, having N number of turns and carrying I ampere current. A flux of ϕ Wb gets developed, as shown in Figure 4.

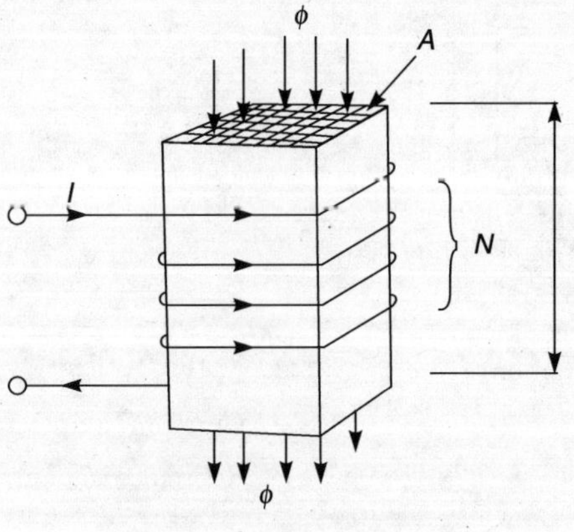

Figure 4 Magnetic core.

The flux density **B** is given by

$$\mathbf{B} = \frac{\phi}{A}\ \text{Wb/m}^2$$

and the magnetizing force **H** is given by

$$\mathbf{H} = \frac{NI}{l}\ \text{AT/m}$$

Here **B** and **H** are related by

$$\mathbf{B} = \mu\mathbf{H}$$

where $\mu = \mu_o\ \mu_r$; μ_o = absolute permeability (permeability of free space) = $4\pi \times 10^{-7}$ H/m, μ_r = relative permeability of the magnetic material of the core. A change in magnetizing force **H** results in corresponding change in flux density **B**. This plot is known as **B–H** *curve* for the given magnetic material (see Figure 5).

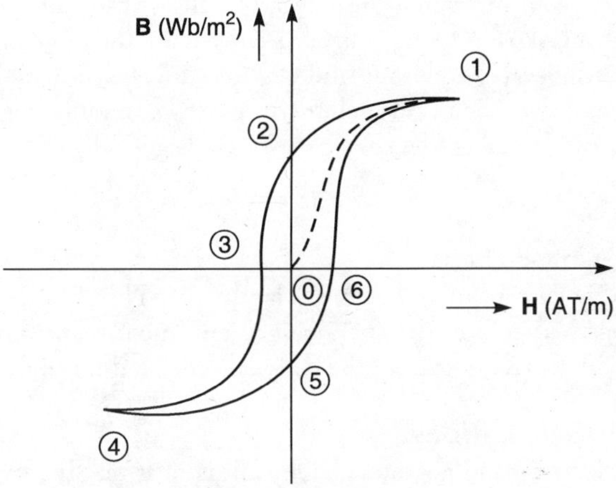

Figure 5 **B–H** curve.

When a magnetic core is magnetized for the first time, the curve begins from origin O. With increase in current through the coil, **H** increases, and hence the flux increases to increase **B** over the path (0–1), as shown in Figure 5. Beyond certain **H**, the value of **B** becomes nearly constant (i.e. a very slight change occurring in it on making a big change in **H**). This is called *the magnetic saturation of the core*. When the current (and hence **H**) is decreased and made equal to 0, the curve does not trace back the (1–0) path. Instead, it follows a new path (1–2). At point 2, corresponding to the zero value of **H**, the flux density **B** does not become zero. This flux density (0–2) is called the *residual magnetism*.

When the current is increased in reverse direction, the curve follows the path (2–3). Observe that **B** becomes zero when negative magnetizing force (0–3) is applied. When **H** is further increased in negative direction, the core gets saturated in reverse direction. Thus, it is seen that for repeated magnetization and demagnetization of the core, the curve follows the path

1–2–3–4–5–6–1–2⋯

The value of **H**, required to bring **B** equal to zero, is called the *coercive force*, represented by (0-3) and (0–6) in the positive and the negative directions, respectively, as shown in Figure 6. The size and shape of **B–H** curve varies from material to material, but the basic nature of the curves remains the same.

4.2.2 Hysteresis and Eddy Current Losses

Hysteresis loss

The magnetization and demagnetization of a magnetic core, due to AC excitation, do not follow a linear and similar path. Hence, in every cycle, some energy gets wasted, which is represented by the area of the **B–H** curve. It is, therefore, preferred to use material having thinner **B–H** curve. The power loss per unit volume of the core due to hysteresis, P_h, is given by

$$P_h \propto fB^{1.6}\,V\ \text{W/m}^3$$

where V is the volume of iron and f is the frequency of supply current in Hz (which is same for the flux).

Eddy-current loss

In the above-mentioned process of magnetization and demagnetization of the magnetic core, the magnetic dipoles in it reverse their orientation 50 times in a second (for 50 Hz supply). Magnetic core is also a good conductor of electricity. In random movement of the dipoles inside the core material, small loops of eddy currents get induced (inside and all over the core) in accordance with the Lenz's law. These tiny eddy-current loops randomly and temporarily keep developing and eloping, in various possible planes in the core. These currents lead to small i^2r losses due to them. All such tiny i^2r losses, combined together, result in a substantial power loss. This power loss, P_e, also called the *eddy-current loss*, is given by

$$P_e \propto B^2t^2f^2V\ \text{W/m}^3$$

where t is the thickness of the laminations.

In order to reduce the eddy-current loss, the core is laminated. Thin insulated sheets (called lamina) of core material are stacked together to form the core. This arrangement reduces the possible numbers of the planes for development of eddy currents. Hence the eddy currents and the eddy current loss get reduced.

4.2.3 Magnetic Circuit Calculations

From the analogy between electrical and magnetic circuits, it is clear that the two circuits behave similarly. Table 2 details out the fact that the magnetic circuit also follows Ohm's law, Kirchhoff's voltage law and Kirchhoff's current law.

Table 2 Analogy between electrical circuit and magnetic circuit calculation

Electrical circuit	*Magnetic circuit*
1. For the electrical circuit shown below 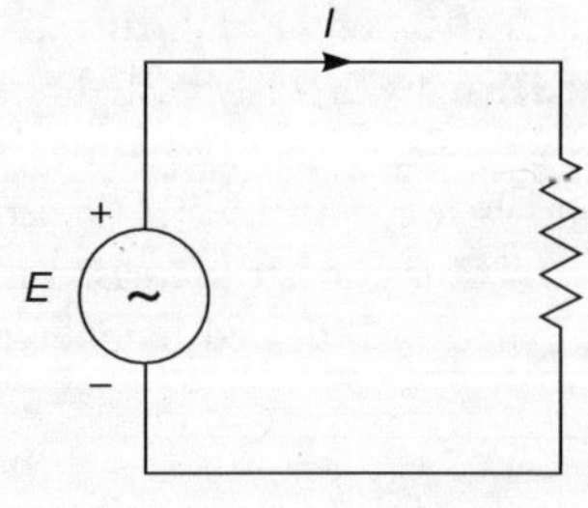 Ohm's law: $\mathbf{E} = RI$, where, $R = \dfrac{\rho l}{A}$	**1.** For the magnetic circuit shown below 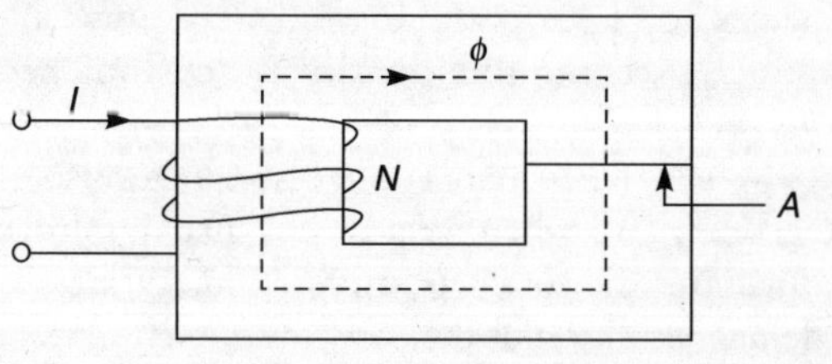 Similar to Ohm's law: $\Gamma = \Re\phi$, where $\Gamma = NI,\ \Re = \dfrac{l}{\mu_0\mu_r A}$

(*continued*)

Table 2 Continued

Electrical circuit	*Magnetic circuit*
2. For a *series circuit*, as shown below 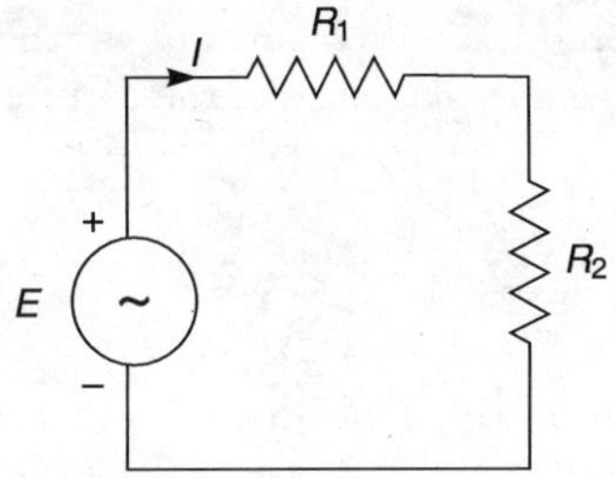 By using KVL: $\mathbf{E} = R_T I$, where $R_T = R_1 + R_2$	**2.** For a *series circuit*, as shown below 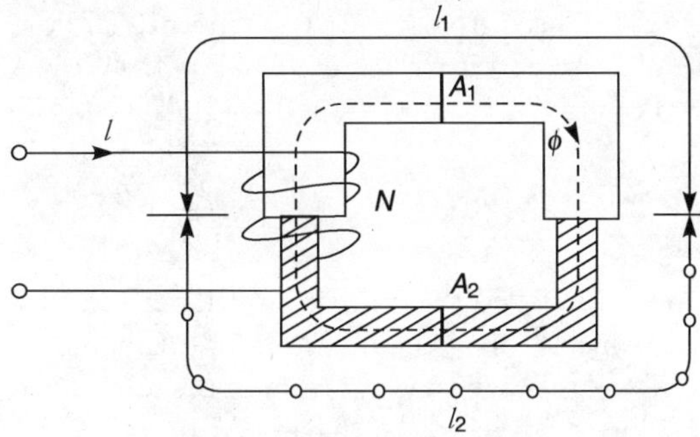 Similar to KVL: $\Gamma = \Re_T \phi$, where $\Re_T = \Re_1 + \Re_2$ where $\Re_1 = \dfrac{l_1}{\mu_0 \mu_{r1} A_1}$ and $\Re_2 = \dfrac{l_2}{\mu_0 \mu_{r2} A_2}$ μ differs if the magnetic materials are different, l differs as the path lengths and A differs as the cross-sectional areas are different.
3. For a *parallel circuit* shown below 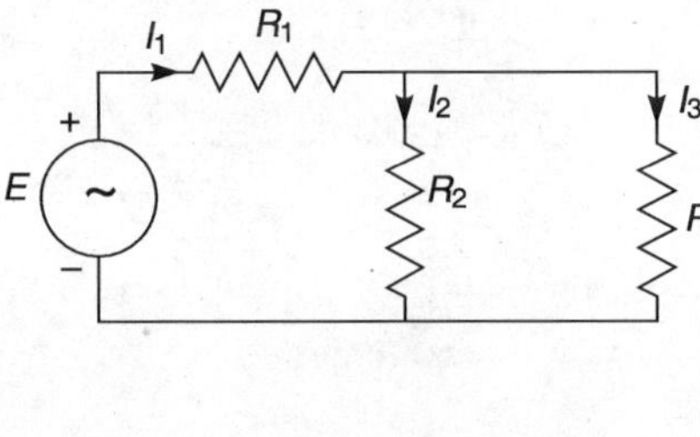 $R_1 = \dfrac{\rho_1 l_1}{A}$, $R_2 = \dfrac{\rho_2 l_2}{A}$, $R_3 = \dfrac{\rho_3 l_3}{A}$ Thus, the total resistance is $R_T = R_1 + R_2 \parallel R_3$ or $R_T = R_1 + \dfrac{R_2 R_3}{R_2 + R_3}$ KVL gives $\mathbf{E} = R_T I_1$. Applying KCL for current $I_1 = I_2 + I_3$	**3.** For a *parallel circuit* shown below 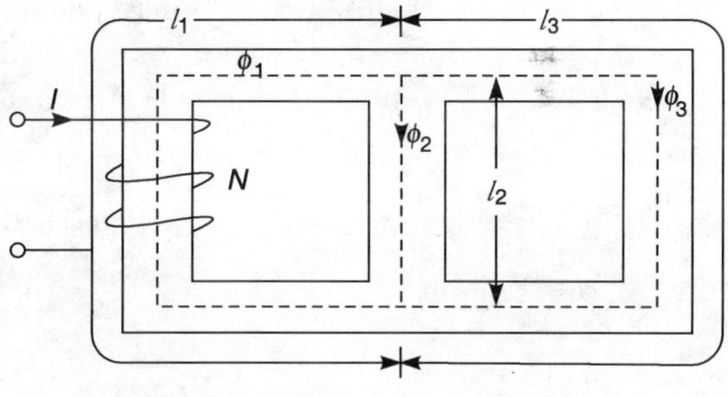 $\Re_1 = \dfrac{l_1}{\mu_0 \mu_r A}$, $\Re_2 = \dfrac{l_2}{\mu_0 \mu_r A}$, $\Re_3 = \dfrac{l_3}{\mu_0 \mu_r A}$ Thus, the total reluctance is $\Re_T = \Re_1 + \Re_2 \parallel \Re_3$ or $\Re_T = \Re_1 + \dfrac{\Re_2 \Re_3}{\Re_2 + \Re_3}$ Similarly $\Gamma = \Re_T \phi_1$. Applying KCL for flux $\phi_1 = \phi_2 + \phi_3$

4.2.4 Mutual Coupling

If there are two coils lying nearby, and if an alternating current flows through one of them, a portion of the flux thus produced, links with the second coil. The flux being alternating (due to the current which has produced it being AC) causes the induction of an emf in the second coil. This phenomenon is called *mutual magnetic coupling,* and the circuit containing the two coils is called *coupled circuit.*

As can be seen from Figure 6, a current $i_1(t)$ flowing in coil 1 produces a flux, a part of which links with the coil 2. An emf is induced in coil 2, which is proportional to the rate of change of the current $i_1(t)$ with regard to time (according to the Faraday's law). This implies

$$v_2(t) \propto \frac{di_1(t)}{dt} \quad \text{or} \quad v_2(t) = M_{21}\frac{di_1(t)}{dt}$$

where M_{21} is called the mutual inductance in coil 2 due to current in coil 1. As there is no current in the coil 2 (which is open circuited), no mutually induced voltage will exist in coil 1 circuit. From Figures 6(a) and (b), we see that only a portion of the total flux generated by the flow of current in the coil 1 (having N_1 turns) links with the coil 2 (having N_2 turns).

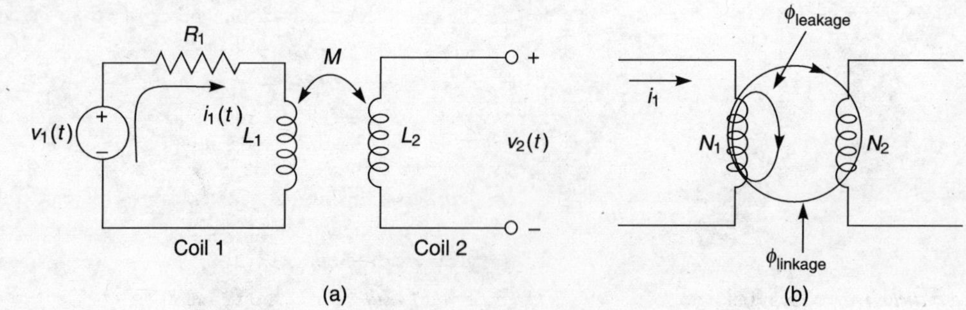

Figure 6 (a) Coupled circuit; (b) linkage and leakage flux.

The ratio of linkage flux to total flux is called the *coupling coefficient* (usually denoted by the letter k). Thus

$$k = \frac{\text{Linkage flux}}{\text{Linkage flux} + \text{Leakage flux}} = \frac{\phi_{\text{linkage}}}{\phi_{\text{total}}} = \frac{\phi_{\text{linkage}}}{\phi_{\text{linkage}} + \phi_{\text{leakage}}}$$

Now since $\psi = N\phi = MI$,

$$v_2(t) = M_{21}\frac{di_1(t)}{dt} = N_2\frac{d(\phi_{\text{linkage}})}{dt}$$

This gives

$$M_{21} = N_2\frac{d(\phi_{\text{linkage}})}{di_1(t)}$$

Had there been a current flowing only in coil 2 (say i_2) also similar $v_1(t)$ would have got induced in coil 1. In that case, the value of mutual induction M_{12} would have been the mutual inductance in coil 1 due to current in coil 2, and the equation then would have been

$$v_1(t) = M_{12}\frac{di_2(t)}{dt} = N_1\frac{d(\phi_{\text{linkage}})}{dt}$$

Thus

$$M_{12} = N_1 \frac{d(\phi_{\text{linkage}})}{di_2}$$

Now, if $M_{21} = M_{12} = M$, then

$$M_{21} \times M_{12} = M^2 = N_2 \frac{d(\phi_{\text{linkage}})}{di_1} \times N_1 \frac{d(\phi_{\text{linkage}})}{di_2} = N_2 \frac{d(k\phi_{\text{total}})}{di_1} \times N_1 \frac{d(k\phi_{\text{total}})}{di_2}$$

as

$$\phi_{\text{total}} = \phi_{\text{linkage}} + \phi_{\text{leakage}} \quad \text{and} \quad k = \frac{\phi_{\text{linkage}}}{\phi_{\text{total}}}$$

as already stated before. So, we have

$$M^2 = N_2 \frac{d(k\phi_{\text{total}})}{di_1} \times N_1 \frac{d(k\phi_{\text{total}})}{di_2} = k^2 N_1 \frac{d(\phi_{\text{total}})}{di_1} \times N_2 \frac{d(\phi_{\text{total}})}{di_2} = k^2 \times L_1 \times L_2$$

$$\left(\text{because } L = N \frac{d\phi}{d_2}\right)$$

Thus we get the relationship between mutual and self-inductances as

$$M = k\sqrt{L_1 \times L_2}$$

The coefficient of coupling "k" is equal to 1 if there is no leakage flux (i.e. if all the flux produced by either of the two coils links with the other) and the coupling is said to be tight. If the value of k is less than 1 the coupling is said to be loose.

Dot convention

Dot convention helps in deciding the polarity of the induced emfs due to mutual induction in the coupled circuits. According to it, dots are placed at one (of the two) end of each of the two coils, which are mutually coupled. Figure 7 shows the circuit for the explanation of the dot convention. If the current (say, i_1) is entering through the dot (at a) in coil 1, there are two conventions for denoting the second winding, as shown in Figures 7(a) and (b). The dotted end of the coil 2 will have the same polarity as that of the dotted end of the coil 1.

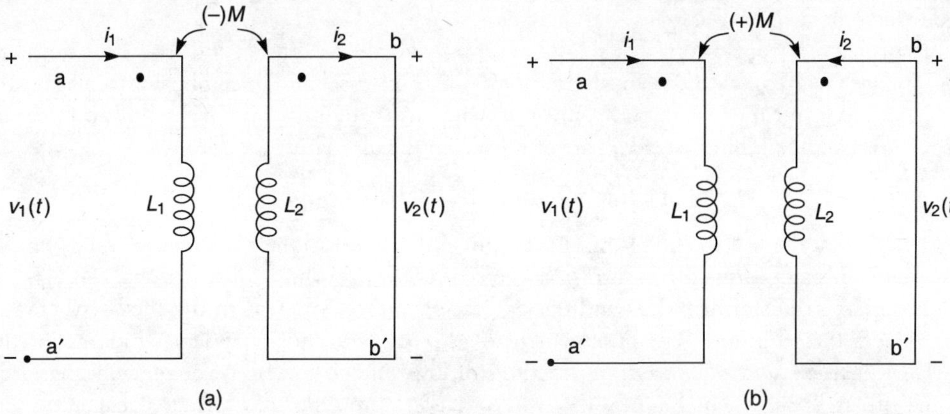

Figure 7 Understanding the dot convention.

The discussion can be interpreted in a different way: In Figure 7(a), the dot in the second winding is placed at b where the current (i_2) is leaving from, while in Figure 7(b), the dot is placed at b where the current (i_2) is entering the coil. It shows that if both the currents i_1 and i_2 are simultaneously entering the dotted terminals, the sign of the M term (in the KVL equation) is *same* as that of the L term. And, if the current i_1 is entering at and i_2, is leaving from the dotted terminals, the sign of the M term (in the KVL equation) is opposite to that of the L terms. Thus the KVL equation for Figure 7(a) would be

$$v_1(t) = L_1 \frac{di_1(t)}{dt} - M \frac{di_2(t)}{dt}$$

whereas that for Figure 7(b) will be

$$v_1(t) = L_1 \frac{di_1(t)}{dt} + M \frac{di_2(t)}{dt}$$

4.3 Single-Phase Transformer

It is an electrical device to convert the voltage and current to different levels without affecting the power and the frequency. In transmission lines, at the sending (generation) ends, voltages are raised to extremely high levels in order to ensure that the corresponding currents for a given power to be transmitted are low, and hence the power (I^2R) losses and the size of conductors in transmission are also low. Similarly, at the receiving (distribution) end, the voltage levels need to be stepped down so as to be made usable for the consumers. At both these places, step-up/step-down transformers are, respectively, used.

4.3.1 Construction of Single-Phase Transformer

A single-phase transformer consists of two windings, *high-voltage winding* and *low-voltage winding*. The winding in which the input is applied is called the *primary winding*, and the one across which the electrical load is connected is called the *secondary winding*. Thus, depending upon whether the high voltage (or the low voltage) is applied at the primary winding and the low voltage (or high voltage) is obtained at the secondary winding, the transformer is said to be operated as a *step-down* (or a *step-up*) transformer, respectively. Since nothing in a transformer moves, it is a *static device*. The two windings are wound on a high-permeability cold rolled grain oriented (CRGO) steel core in order to ensure an effective mutual coupling between them.

The basic principle of operation of a single-phase transformer is mutual induction phenomenon. An alternating voltage in a coil induces an alternating voltage in a second (neighboring) coil according to Faraday's law of electromagnetic induction. But it so happens that all the flux of the first coil does not link with that of the second coil, but instead a part of it leaks in the air. Thus, we can say

$$\text{Total flux} = \text{Linkage flux} + \text{Leakage flux}$$

In order to ensure that the leakage flux is minimum, the physical position of the two windings should be as close as possible. This fact forms the basis of the two types of transformer constructions: *core type* and *shell type*. In the core-type transformer, the windings embrace the core, whereas in the shell-type transformer, the core embraces the windings. The arrangement of the primary and secondary windings in shell-type transformer is called *sandwiched winding*, as fractions of low-voltage and high-voltage windings are placed one after the other, as shown in Figure 8(b), while that in core-type transformer is called a concentric winding, as shown in Figure 8(a).

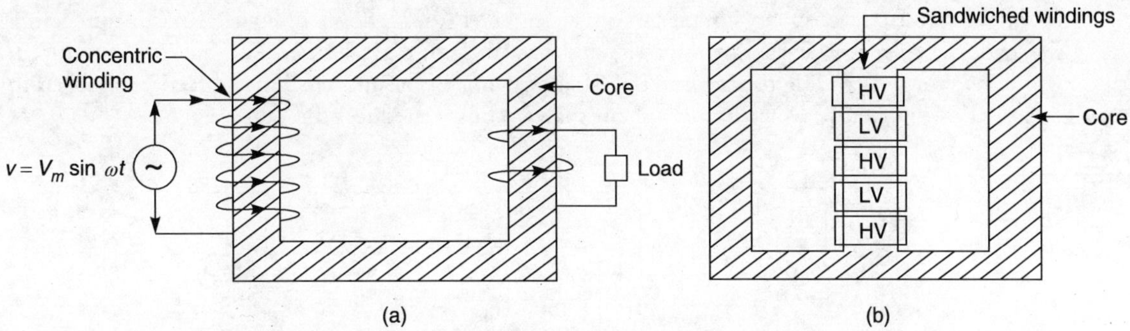

Figure 8 (a) Core-type transformer; (b) shell-type transformer.

4.3.2 Principle of Operation of a Single-Phase Transformer

We start with the operation of a transformer at no-load.

1. As shown in Figure 9, when a rated voltage is applied to the primary winding of a single-phase transformer with the secondary winding kept open (a condition called *no-load condition*), a small current flows in the primary winding, which is called the *no-load current*, I_0 (say).

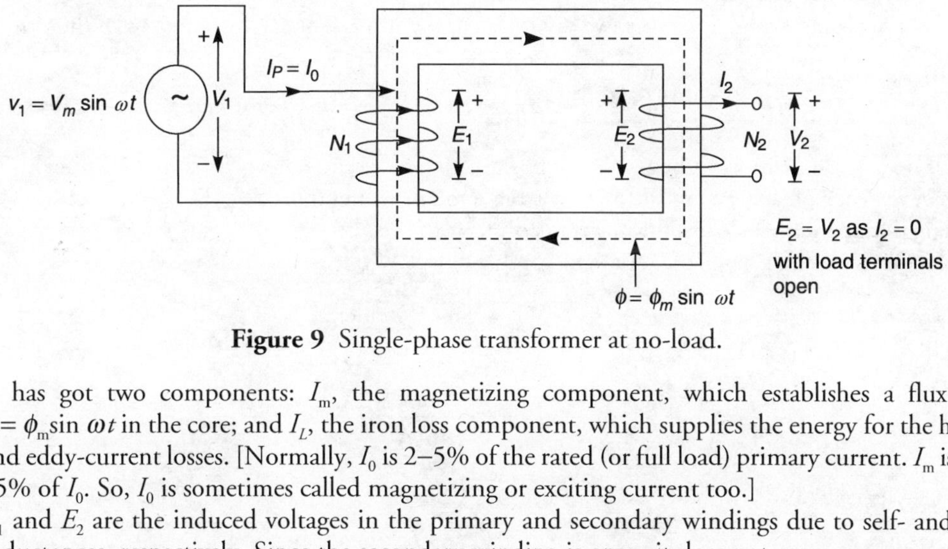

Figure 9 Single-phase transformer at no-load.

2. I_0 has got two components: I_m, the magnetizing component, which establishes a flux pattern $\phi = \phi_m \sin \omega t$ in the core; and I_L, the iron loss component, which supplies the energy for the hysteresis and eddy-current losses. [Normally, I_0 is 2–5% of the rated (or full load) primary current. I_m is around 85% of I_0. So, I_0 is sometimes called magnetizing or exciting current too.]
3. E_1 and E_2 are the induced voltages in the primary and secondary windings due to self- and mutual inductances, respectively. Since the secondary winding is open, it does not carry any current. So the load voltage, $V_2 = E_2$ under no-load condition. Also,

$$E_1 = \frac{d\phi}{dt} = \frac{d\phi_m}{dt} \sin \omega t = \phi_m \omega \cos \omega t \tag{4.1}$$

4. The primary winding resistance is r_1 (say), so the circuit reason for the flow of the primary current I_0 is

$$I_P = I_0 = \frac{V_1 - E_1}{r_1} \tag{4.2}$$

Had $r_1 = 0$, the induced emf would have been exactly equal and opposite to V_1.

5. Now let us close the secondary circuit by connecting the load, as shown in Figure 10. On connecting the load, the current I_2 starts flowing to it, as shown in Figure 10. Due to I_2, a flux ϕ_2 is produced, which (as shown according to ampere's right-hand rule) is in the anti-clockwise direction, as it has to oppose the original flux ϕ (according to Lenz's law). Thus, ϕ_2 reduces the value of ϕ.

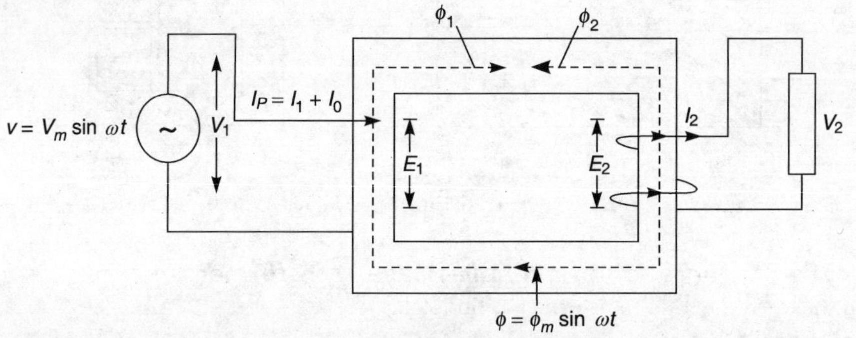

Figure 10 Single-phase transformer on load.

6. The reduction of ϕ results in the reduction of E_1 [by Eq. (4.1)], and hence the numerator of Eq. (4.2) increases, causing an increase in the magnitude of the primary current I_P.
7. Instead of saying that I_0 increases, we say that a new component (say) I_1 emerges in the primary current, which thus becomes

$$I_P = I_0 + I_1$$

8. This new component I_1 of the primary current (I_P) gives rise to a new component of the flux ϕ_1 (say). Now, I_1 keeps on increasing till this ϕ_1 is big enough to neutralize ϕ_2. This finally ensures that the original flux ϕ remains constant in a transformer.
9. Now Ohm's law for magnetic circuit states that

$$\text{MMF} = \text{Flux} \times \text{Reluctance}$$

that is

$$\text{NI} = \phi \times \frac{l}{\mu_0 \mu_r \text{A}}$$

Thus, we have for the primary side

$$N_1 I_1 = \phi_1 \times \frac{l}{\mu_0 \mu_r A}$$

and for the secondary side

$$N_2 I_2 = \phi_2 \times \frac{l}{\mu_0 \mu_r A}$$

The reluctance (of the core) is constant. Since the two windings link with the same core flux (i.e. $\phi_2 = \phi_1$), the left-hand side of the above two equations should be equal. Hence,

$$N_2 I_2 = N_1 I_1$$

4.3.3 EMF Equation

We know that with the sinusoidal voltage source connected at primary, the flux established in the core of the transformer is sinusoidal, as shown in Figure 11. The average value of the emf induced can be found by considering one-fourth of a cycle for which the time taken is 1/(4*f*), as shown in the figure. Thus,

$$e \text{ (per turn)} = \frac{d\phi}{dt} = \frac{\phi_m - 0}{1/(4f) - 0} = 4f\phi_m$$

For N turns, the average value of induced emf would be $E = 4f\phi_m N$ volts. For finding the RMS value, we need to multiply the expression by the form factor, that is 1.11. So, $E_{\text{RMS}} = 4.44f\phi_m N$ volts. For primary side, it is $E_1 = 4.44f\phi_m N_1$, and for the secondary side, it will be $E_2 = 4.44f\phi_m N_2$ volts. The ratio of these two induced emfs yields

$$\frac{E_1}{E_2} = \frac{N_1}{N_2}$$

This ratio is called the *turn ratio*.

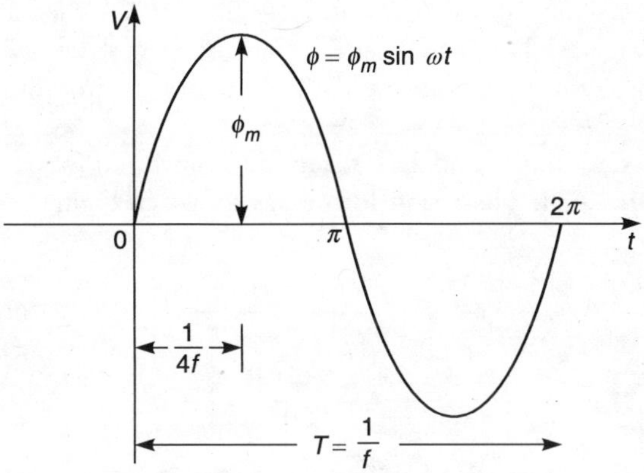

Figure 11 Deriving the emf equation of a single-phase transformer.

4.3.4 Equivalent Circuit of a Single-Phase Transformer

The physical behavior of a single-phase transformer is represented by its electrical equivalent circuit. This can be explained as follows:

1. As explained earlier, the primary and secondary winding fluxes (due to the currents I_1 and I_2) exactly neutralize each other. The flux in the core is only due to I_0, the no-load current. This is true for an ideal case, where there is no leakage flux. But in an actual single-phase transformer, there is always some leakage flux both in the primary and the secondary windings.
2. Due to leakage in ϕ_1, E_1 becomes less than what it should have been. Also due to leakage in ϕ_2, the value of V_2 obtained is less than what it should have been without leakage. These reductions in voltages (due to flux leakages) are represented by putting two fictitious reactances, x_1 and x_2. The corresponding winding resistances r_1 and r_2 of primary and secondary windings are placed in series with x_1 and x_2, respectively. The reductions in E_1 and V_2 are thus taken into account as the voltage drops $I_p x_1$ and $I_2 x_2$, respectively. The equivalent circuit of transformer is shown in Figure 12.

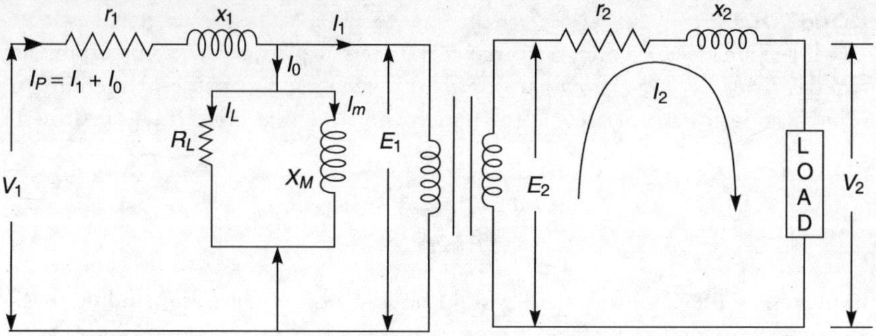

Figure 12 Exact equivalent circuit of a single-phase transformer.

3. As hysteresis and eddy currents are complex phenomena, they cannot be represented naturally. So, another fictitious quantity, a resistance R_L, is put in the equivalent circuit where the current I_L (also known as the *working component of the no-load current*) flows. $I_L^2 R_L$ represents the total hysteresis and eddy current losses, and is said to be the total iron loss.
4. The magnetization of the core (which is done by the current I_m) is represented by making the current I_m flow in an inductive reactance X_M, which is also a fictitious quantity.
5. Thus, we see that the resistances of the two windings, r_1 and r_2, are the only *actually existing quantities.* The remaining four, that is x_1, x_2, R_L and X_M, *do not actually exist.* They are put in the equivalent circuit to represent the actual physical phenomena like leakage of flux, iron loses and magnetization of the iron core.
6. If the shunt branch (containing R_L and X_M) is shifted to the left of r_1 and x_1, we get the "approximate equivalent circuit" of the single-phase transformer, as shown in Figure 13.

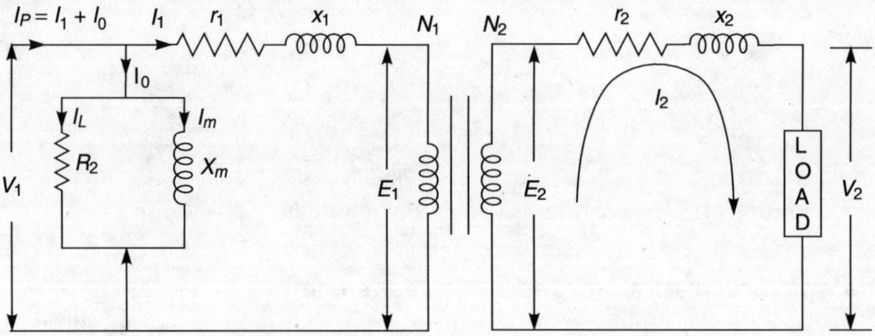

Figure 13 Approximate equivalent circuit of a single-phase transformer.

Here since I_0 is 2% to 5% of the I_P (rated current), there happens to be *no significant loss of accuracy*. For example, if I_0 = 3% of I_P (say), we will have a voltage drop of 0.97 $I_P r_1$ and 0.97 $I_P x_1$ instead of $I_P r_1$ and $I_p x_1$ voltage drops before. But, the advantage of this circuit is that the iron loss can be calculated without finding the value of I_L. The iron loss will be V_1^2/R_L, as the voltage across R_L is now V_1. As the transformer analysis can be done easily at any side (high voltage or low voltage side), it becomes necessary to place all the parameters of both the windings to that side only. For this the parameters of the other side are referred to the side under consideration. The following text describes the referral of the quantities from one side to another.

In Figure 13, the voltage across r_2 is $I_2 r_2$. If this voltage has to be taken (referred) to the primary side, we just have to find the volt/turn and multiply it by the number of turns in the primary side, that is $I_2 r_2/N_2$ is to be multiplied by N_1. This can be safely done as the voltage per turn of the primary is same as that of the secondary in a transformer. So,

$$I_2 r_2 \times \frac{N_1}{N_2}$$

is the value of the voltage $I_2 r_2$, when it is taken from the secondary side to the primary side. But in the primary side, the current flowing in the series branch parameters (r_1 and x_1) is I_1. So, we try to represent

$$I_2 r_2 \times \frac{N_1}{N_2}$$

in terms of I_1 by making use of the relation

$$I_1 N_1 = I_2 N_2 \quad \text{or} \quad I_2 = I_1 \frac{N_1}{N_2}$$

Substituting for I_2, we get

$$I_2 r_2 \frac{N_1}{N_2} = I_1 r_2 \left(\frac{N_1}{N_2}\right)^2$$

Thus the right-hand side expression becomes the value of the voltage drop across r_2, when it is shifted from the secondary side to the primary side. It can be said that r_2 becomes $r_2(N_1/N_2)^2$ when referred from the secondary side to the primary side. In other words, for referral of resistances and reactances, the square of turn ratio is used. Since

$$\frac{Z_1}{Z_2} = \left(\frac{V_1}{I_1} \div \frac{V_2}{I_2}\right) = \left(\frac{V_1}{V_2} \times \frac{I_2}{I_1}\right) = \left(\frac{N_1}{N_2}\right)^2$$

Hence

$$\left(\frac{Z_1}{Z_2}\right) = \left(\frac{N_2}{N_1}\right)^2$$

Now since r_1 already exists in the primary side and r_2 is referred from secondary to primary side, so the total equivalent series resistance in the primary will be equal to r_{1e} (suffix "1" for primary and "e" for equivalent).

$$r_{1e} = r_1 + r_2\left(\frac{N_1}{N_2}\right)^2$$

Similarly,

$$x_{1e} = x_1 + x_2\left(\frac{N_1}{N_2}\right)^2, \quad x_{2e} = x_2 + x_1\left(\frac{N_2}{N_1}\right)^2, \quad r_{2e} = r_2 + r_1\left(\frac{N_2}{N_1}\right)^2$$

$$R_L' = R_L\left(\frac{N_2}{N_1}\right)^2, \quad X_M' = X_M\left(\frac{N_2}{N_1}\right)^2$$

If r_1, r_2, x_1 and x_2 are not given, and only x_{1e} and r_{1e} are given, then x_{2e} and r_{2e} can be found as

$$x_{2e} = x_{1e}\left(\frac{N_2}{N_1}\right)^2 \quad \text{and} \quad r_{2e} = r_{1e}\left(\frac{N_2}{N_1}\right)^2$$

4.3.5 Power Losses

A transformer comprises two major parts: One is the *magnetic core* (made of iron) and the other is *winding* (made of copper). Magnetic losses take place in the iron part (called *iron losses*) and I^2R losses take place in copper windings (called *copper losses*). As already stated, the sum of the hysteresis and eddy-current losses (both occurring in the iron core) is called the *iron loss*, *core loss* or *fixed loss*. Recalling the equations for hysteresis and eddy current losses

$$P_h = K_h B^{1.6} f \quad \text{and} \quad P_e = K_e B^2 f^2 t^2$$

We see that since the frequency and flux density of a single-phase transformer normally remain fixed, for a constant voltage and constant frequency supply, these losses are usually fixed, and hence the name *fixed* or *constant losses*.

The *real* (or *true* or *active*) *power losses* that occur in the resistance of the primary and secondary windings (i.e. in r_1 and r_2) are called *copper losses*. So,

$$\text{Total copper loss} = I_1^2 r_1 + I_2^2 r_2 = I_1^2 r_{1e} = I_2^2 r_{2e}$$

Since the load is a variable quantity (I_2) is variable, and as a result I_1 is also variable so, the copper loss is also termed as the *variable loss*.

4.3.6 Efficiency

Efficiency is defined as the ratio of output to input:

$$\text{Efficiency } (\eta) = \frac{\text{Output power}}{\text{Input power}}$$

In a transformer, the output power (in watts) is always less than the input power (in watts), because of the fixed and variable losses. So, we can write

$$\eta = \frac{\text{Output power}}{\text{Output power } + \text{ Losses}}$$

and according to Figure 12 with the angle of the load impedance being θ_2

$$\eta = \frac{V_2 I_2 \cos\theta_2}{V_2 I_2 \cos\theta_2 + P_i + I_2^2 r_{2e}} \tag{4.3}$$

where $I_2^2 r_{2e}$ represents the total copper loss of both sides referred to secondary (i.e. the load) side.

There are two variables in Eq. (4.3): θ_2 and I_2. Keeping I_2 constant, if Eq. (4.3) is differentiated with regard to θ_2 and the answer equated to zero (to find the condition of $\eta_{\max}$), we get $\theta_2 = 0$ or $\cos\theta_2 = 1$ as the answer. Thus for a constant load magnitude (i.e. constant I_2), the efficiency is maximum for a resistive load.

Next keeping θ_2 constant, if we differentiate Eq. (4.3) with regard to I_2 and equate the answer to zero, we get

$$\frac{d\eta}{dI_2} = \frac{(V_2\cos\theta_2)(V_2 I_2 \cos\theta_2 + P_i + I_2^2 r_{2e}) - (V_2 I_2 \cos\theta_2)(V_2\cos\theta_2 + 2I_2 r_{2e})}{(V_2 I_2 \cos\theta_2 + P_i + I_2^2 r_{2e})^2} = 0$$

or

$$V_2 I_2 \cos\theta_2 + P_i + I_2^2 r_{2e} = I_2(V_2\cos\theta_2 + 2I_2 r_{2e})$$

or

$$P_i = I_2^2 r_{2e} \tag{4.4}$$

Thus, the condition for the occurrence of maximum efficiency (for a fixed load angle) is that "The fixed losses and the variable losses should be equal."

Since I_2 is variable in the above derivation, it can be concluded that "It is definitely not necessary that the value of I_2 featuring in Eq. (4.4) is the rated (or full load) current value of the transformer." However, since the maximum efficiency of the transformer is occurring for this value of I_2, let us give it a special name, say $I_{2\eta M}$. So Eq. (4.4) becomes

$$I_{2\eta M}^2 r_{2e} = P_i$$

Hence

$$I_{2\eta M} = \sqrt{\frac{P_i}{r_{2e}}}$$

Dividing both sides the rated full load current by (I_{2fl}) we get

$$\frac{I_{2\eta M}}{I_{2fl}} = \sqrt{\frac{P_i}{I_{2fl}^2 r_{2e}}} \tag{4.5}$$

The denominator of the RHS of Eq. (4.5) is a fixed quantity, that is "the copper loss at the rated load." Now with P_i already fixed, we can say that the RHS is a fixed quality for a given transformer. Let, the RHS = K. Now, LHS = RHS. So,

$$\frac{I_{2\eta M}}{I_{2fl}} = K$$

This means that the maximum efficiency of the transformer occurs at a fraction K of the full load, which can be said as the K% of the full load if the LHS is multiplied by 100. That is, if the rated current is $I_{2fl} = 10$ A (say) and $I_{2\eta M} = 7$ A (say), the value of K is 0.7, which means that the maximum efficiency of the transformer will occur at 70% of the full load. [Actually, since in a transformer, the voltage is usually fixed (rated), the words "load" and "current" are interchangeably used. Thus increase in load means increase in current.] Thus if the $P_i = 49$ W (say) and rated copper loss is 100 W (say), the maximum efficiency will occur at $\sqrt{49/100} = 0.7$, that is, 70% of the full load. Two sentences need special mention at this point in time:

1. The copper (Cu) losses are dependent on the square of the load (current).
2. Cu losses increase with increase in load and decrease with decrease in load.

Thus if the Cu loss is 100 W (say) at the full load (i.e. 100% load), the Cu loss at half full load will

1. vary as $(1/2)^2$,
2. be less than 100 W.

Thus Cu loss at 1/2 full load will be $100 \times (1/2)^2 = 25$ W.

Similarly, if the Cu loss at 80% full load is 900 W (say), the Cu loss

1. at 90% full load will be $(90/80)^2 \times 900 = 1140$ W.
2. at 70% full load will be $(70/80)^2 \times 900 = 690$ W.

4.3.7 Introduction to Autotransformer

An *autotransformer* is a single-winding transformer in which one of its sides uses all the turns of the winding, while the other side is formed by tapping a part from the same winding, as shown in Figure 14(a). Either of the two sides can function as a primary (or the secondary) depending upon connecting the supply to it. If the supply is connected to the side having smaller number of turns, it acts as a step-up transformer and vice versa.

In an autotransformer, the electrical isolation between the primary and secondary windings is no more existent, as there is a direct conduction between the two windings. Thus, the power is transferred from primary to secondary winding, both through induction and conduction. This is the reason why the VA rating of a two winding transformers increases when it is redesigned and used as an autotransformer. Also, the efficiency of an autotransformer is more than a corresponding two-winding transformer, because the losses remain the same wheras the VA increases.

In order to understand the above discussion, consider a case where the two windings [of a normal transformer shown in Figure 14(b)] are physically connected together to form an autotransformer, represented by Figure 14(c). This autotransformer is a step-up transformer with input voltage being V_1 volts and the output voltage being $(V_1 + V_2)$ volts.

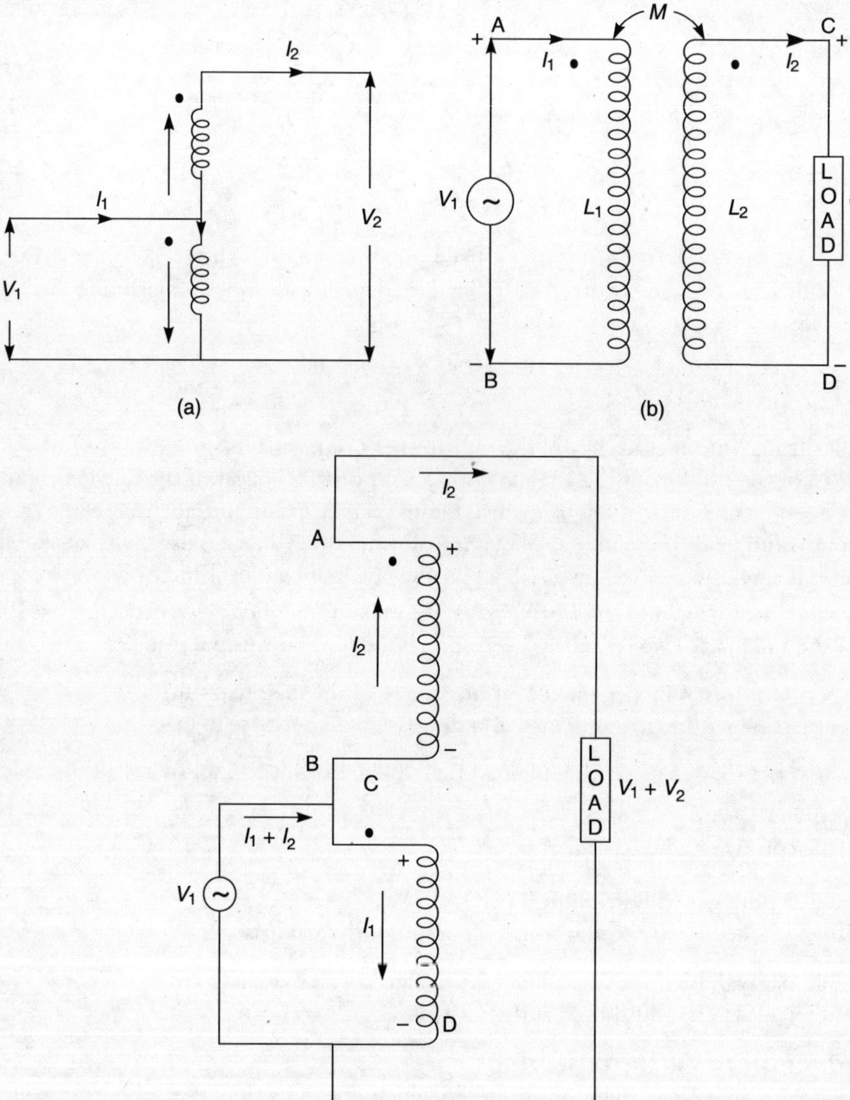

Figure 14 (a) Autotransformer winding arrangement; (b) two-winding transformer; (c) two windings of transformer shown in Figure 14(b) connected to form an autotransformer.

The input power in the autotransformer shown in Figure 14(c) is given by

$$V_1 I_1$$

while the output power is given by

$$(V_1 + V_2)I_2 = V_1 I_2 + V_2 I_2$$

Here, the term $V_1 I_2$ is the power transferred due to conduction and the term $V_2 I_2$ is transferred due to induction.

Solved Examples

Example 1

A cast-steel electromagnet has an air gap of 2 mm and a metallic path of length 30 cm. Find the number of ampere-turns required to produce a flux density of 0.8 Wb/m² (8000 lines/cm²) in the gap. Neglect leakage and fringing. Assume the relative permeability of cast steel is equal to 300.

Solution

The fringing effect is neglected. Hence the cross-sectional area (A) of the core and that of the air gap would be the same. Then by definition

$$\Re = \frac{l}{\mu_0 \mu_r A}$$

where

$$\text{Reluctance of air gap, } \Re_g = \frac{lg}{\mu_0 \mu_{rg} Ag} = \frac{lg}{\mu_0 Ag} \quad (\text{as } \mu_{rg} = 1)$$

and

$$\text{Reluctance of metal part, } \Re_m = \frac{l_m}{\mu_0 \mu_m A_m} = A_m$$

Let $A_g = A_m = A$. Let the total reluctance of magnetic path = $\Re_T$. Since the metal part and the air gap are in series,

$$\Re_T = \Re_g + \Re_m \tag{1}$$

We know that ampere-turn (AT) = $\Re_T\, \phi$, where $\phi = \mathbf{B} \times A = (0.8) \times A$. Therefore,

$$\text{AT} = R_T\,(0.8\text{ A}) \tag{2}$$

Substituting Eq. (1) into Eq. (2), we get

$$\begin{aligned}\text{AT} &= \left(\frac{l_g}{\mu_0 A} + \frac{l_m}{\mu_0 \mu_{rm} \text{A}}\right)(0.8\text{ A})\\ &= \frac{0.8}{\mu_0}\left(0.002 + \frac{0.30}{300}\right)\\ &= \frac{0.8}{4\pi \times 10^{-7}}\ \frac{0.9}{300}\\ &= 1910.8 \text{ ampere-turns}\end{aligned}$$

Example 2

An iron ring of mean length equal to 100 cm has an air gap of 2 mm and a winding of 500 turns (see Figure 15). Find the flux density in the coil, if 1A current flows through the winding and the relative permeability of iron is 900.

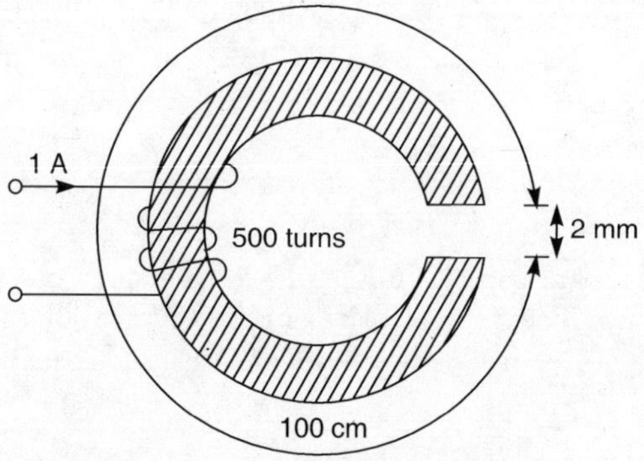

Figure 15 Iron ring (for Example 2).

Solution

When length is given in such problems, it is assumed that it is the mean length of the path of flux. So,

$$\phi = \frac{AT}{\Re_T} = \frac{1 \times 500}{\dfrac{l_g}{\mu_0 \mu_g A} + \dfrac{l_m}{\mu_0 \mu_m A}} = \frac{500 \times 4\pi \times 10^{-7} A}{(0.2/1) + (100/900)}$$

or

$$\mathrm{B} = \frac{\phi}{A} = \frac{500 \times 4\pi \times 10^{-7}}{0.2 + (1/9)} = 20.185 \text{ Wb/cm}^2$$

Example 3

In the magnetic circuit, as shown in Figure 16, the coil has 10 turns carrying a current of 2 A DC. Find the flux densities in each material, the inductance of coil and the energy stored in the circuit. Neglect the leakage and fringing. The following data are given:

Material	*Length (cm)*	*Area (cm²)*	μr
A	20	3.225	300
B	20	3.225	500
C	50	6.45	1000
Air gap (*g*)	0.1	6.45	1

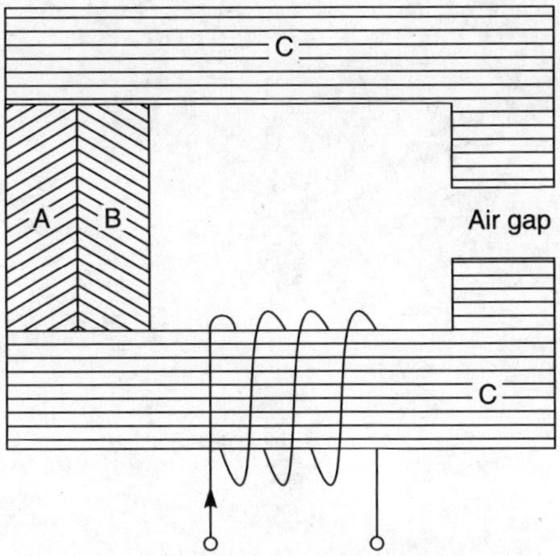

Figure 16 Magnetic circuit.

Solution

The equivalent circuit is shown in Figure 17. From the figure, it is seen that

$$\mathfrak{R}_T = 2\mathfrak{R}_C + \frac{\mathfrak{R}_A \mathfrak{R}_B}{\mathfrak{R}_A + \mathfrak{R}_B} + \mathfrak{R}_g$$

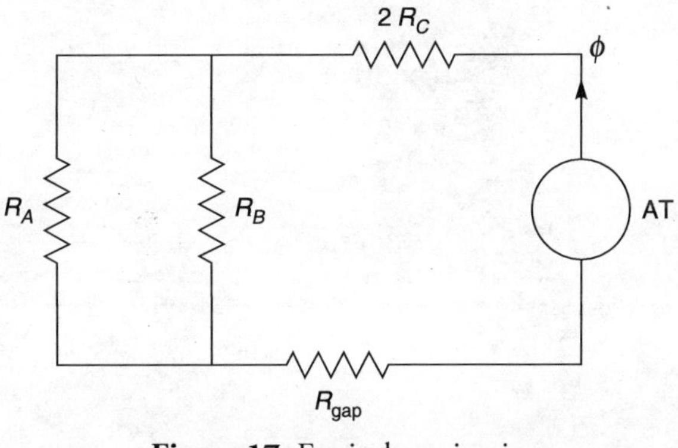

Figure 17 Equivalent circuit.

As parts A and B are in parallel and appear in series with part 2C and the air gap, hence, we get

$$\mathfrak{R}_A = \frac{l}{\mu_0 \mu_r A} = \frac{20\times10^{-2}}{4\pi\times10^{-7}\times300\times3.225\times10^{-4}} = 1.645\times10^6$$

$$\mathfrak{R}_B = \frac{l}{\mu_0 \mu_r A} = \frac{20\times10^{-2}}{4\pi\times10^{-7}\times500\times3.225\times10^{-4}} = 0.987\times10^6$$

$$\mathfrak{R}_C = \frac{l}{\mu_0 \mu_r A} = \frac{50\times10^{-2}}{4\pi\times10^{-7}\times1000\times6.45\times10^{-4}} = 0.617\times10^6$$

$$\mathfrak{R}_g = \frac{l}{\mu_0 A} = \frac{0.1\times10^{-2}}{4\pi\times10^{-7}\times6.45\times10^{-4}} = 1.234\times10^6$$

Therefore,

$$\mathfrak{R}_T = \left(2\times 0.617 + \frac{1.645\times 0.987}{1.645+0.987} + 1.234\right)\times 10^6 = 3.08\times 10^6$$

Thus

$$\phi = \frac{\text{AT}}{\mathfrak{R}_T} = \frac{1000\times 2}{3.08\times 10^6} = 6.49\times 10^{-6}\ \text{Wb}$$

Hence

$$\phi_B = \frac{\mathfrak{R}_A}{\mathfrak{R}_A + \mathfrak{R}_B}\phi \quad \text{and} \quad \phi_A = \frac{\mathfrak{R}_B}{\mathfrak{R}_A + \mathfrak{R}_B}\phi$$

Then

$$\phi_B = 4.056\times 10^{-6}\ \text{Wb} \quad \text{and} \quad \phi_A = 2.434\times 10^{-6}\ \text{Wb}$$

In parallel path, $\phi_A\mathfrak{R}_A = \phi_B\mathfrak{R}_B$. Also, $\phi_A = B_A A_A$ and $\phi_B = B_B A_B$. Therefore,

$$B_A = \frac{2.434\times 10^{-6}}{3.225\times 10^{-4}} = 7.547\times 10^{-3}\ \text{Wb/m}^2; \qquad B_B = \frac{4.056\times 10^{-6}}{3.225\times 10^{-4}} = 0.012\ \text{Wb/m}^2$$

Inductance of coil:

$$L = \frac{N\phi}{I} = \frac{10\times 6.49\times 10^{-6}}{2} = 32.45\ \mu\text{H}$$

Energy stored in field:

$$W = \frac{1}{2}LI^2 = \frac{1}{2}\times 32.45\times 10^{-6}\times 2^2 = 0.065\times 10^{-3}\ \text{J}$$

Example 4

A coil of 2000 turns is wound uniformly over a hollow wooden ring having a mean circumference of 600 cm and a uniform cross-sectional area of 50 cm^2. If the current through the coil is 10 A, calculate

(a) Magnetic field strength.
(b) Flux density.
(c) Total flux.

Solution

Here

(a) $\mathbf{H} = 10\times 2000/0.6 = 0.033\times 10^6$ AT/m
(b) $\mathbf{B} = \mu\mathbf{H} = 4\pi\times 10^{-7}\times 0.033\times 10^6 = 0.04146$ Wb/m^2
(c) $\phi = \mathbf{B}\times A = 0.041\times 50\times 10^{-4} = 0.2073$ mWb

Example 5

The magnetic circuit shown in Figure 18 has a cross-sectional area of 15 cm^2 throughout. The relative permeability of the magnetic material used is 2000. Find the flux density in the air gap. Neglect the magnetic leakage and fringing.

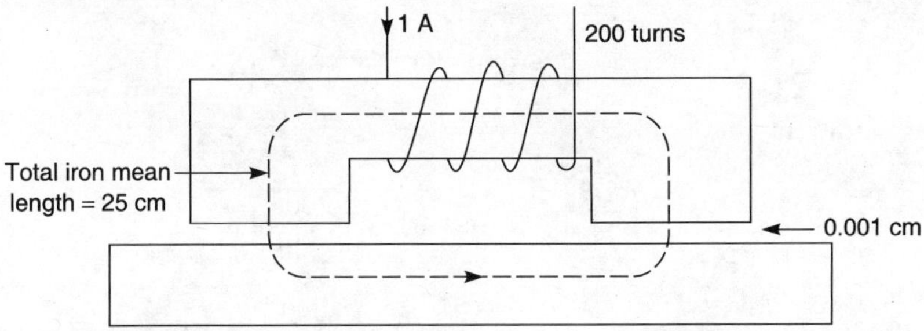

Figure 18 Magnetic circuit (for Example 5).

Solution

Here

$$\Re = \frac{l_{\text{core}}}{\mu_{\text{core}}\mu_0 A_{\text{core}}} + 2\frac{l_{\text{gap}}}{\mu_0 A_{\text{gap}}}$$

$$= \frac{25\times10^{-2}}{2000\times4\pi\times10^{-7}\times15\times10^{-4}} + 2\frac{0.001\times10^{-2}}{4\pi\times10^{-7}\times15\times10^{-4}} \quad \text{[25 cm is the iron mean length]}$$

$$= 0.077\times10^6$$

Therefore,

$$\phi = \frac{AT}{\Re} = \frac{200\times1}{0.077\times10^6} = 2.6\,\text{mWb}$$

and

$$B_{\text{airgap}} = \frac{\phi}{A} = \frac{2.6\times10^{-3}}{15\times10^{-4}} = 1.73\ \text{Wb/m}^2$$

Example 6

A 20 kVA, 2000 V/200 V, single-phase, 50 Hz transformer has a primary resistance of 1.5 Ω and a reactance of 2 Ω. The secondary resistance and reactance are 0.015 Ω and 0.02 Ω, respectively. The no-load current of transformer is 1 A at 0.2 power factor. Determine

(a) Equivalent resistance, reactance and impedance referred to primary
(b) Supply current
(c) Total copper loss

Solution

The turn ratio is given as

$$\frac{N_1}{N_2} = \frac{2000}{200} = 10$$

We also have: $r_1 = 1.5\ \Omega$, $X_1 = 2\ \Omega$. Also, $r_2 = 0.015\ \Omega$, $X_2 = 0.02\ \Omega$.

(a) The equivalent resistance is

$$r_{1e} = r_1 + r_2\left(\frac{N_1}{N_2}\right)^2 = 1.5 + 0.015 \times \left(\frac{2000}{200}\right)^2 = 3\ \Omega$$

$$x_{1e} = x_1 + x_2\left(\frac{N_1}{N_2}\right)^2 = 2 + 0.02 \times \left(\frac{2000}{200}\right)^2 = 4\ \Omega$$

Impedance referred to the primary side will be

$$\text{Impedance} = \sqrt{r_{le}^2 + x_{le}^2} = \sqrt{3^2 + 4^2} = 5\ \Omega$$

Full load current at primary will be

$$I_1 = \frac{\text{kVA}}{V_1} = \frac{20 \times 10^3}{2000} = 10\ \text{A}$$

(b) Taking the load power factor to be unity, we get the supply current as

$$I_s = I_1 + I_0 = 10\angle 0° + 1\angle 78.46° = 10.246\angle 5.48°$$

(c) Tottal copper loss

$$P_{\text{Cu}} = 10^2 \times 3 = 300\ \text{W}$$

Example 7

The maximum efficiency of a 10 kVA, 2000/500 V, 50 Hz, single-phase transformer is 98%, and it occurs at 80% full load, at 0.9 pf lagging. Find the efficiency at

(a) Full load 0.75 pf lagging
(b) 70% full load 0.92 pf lagging

Solution

The rating of a transformer (or any machine for that matter) is its output at full load or rated conditions. Although the transformer rating is done in volt-amperes, the efficiency calculations are always done in watts. Here the output (at full load or 100% load) 10 kVA and 10,000 × 0.9 = 9000 W (with 0.9 pf of the load). The output at 80% full load will be 9000 × 0.8 = 7200 W. It is given that the efficiency at these loadings is 98%. Hence,

$$0.98 = \frac{7200}{\text{Input}} \quad \text{or} \quad \text{Input} = \frac{7200}{0.98} = 7347 \text{ W}$$

Therefore,

$$\text{Losses} = \text{Input} - \text{Output} = 7347 - 7200 = 147 \text{ W}$$

Since the efficiency (98%) is given to be the maximum one, the two losses will be equal. Therefore,

$$P_i = \frac{147}{2} = 73.5 \text{ W}$$

Hence,

$$P_{\text{Cu}}\text{(at 80\% full load)} = 73.5 \text{ W}$$

$$P_{\text{Cu}}\text{(at 100\% full load) will be } \left(\frac{100}{80}\right)^2 \times 73.5 = 115 \text{ W}$$

$$P_{\text{Cu}}\text{ (at 70\% full load) will be } \left(\frac{70}{80}\right)^2 \times 73.5 = 56.27 \text{ W}$$

(a) Now

$$\eta \text{ at full load (0.75 pf lagging)} = \frac{\text{Output (at full load)}}{\text{Output (at full load)} + P_i + P_{\text{Cu}}\text{(at full load)}}$$

Hence,

$$\text{Output at full load, 0.75 pf lagging} = 10{,}000 \times 1 \times 0.75 = 7500 \text{ W}$$

Here P_i is constant = 73.5 W, P_{Cu} (at full load) = 115 W (as calculated above). Therefore,

$$\eta \text{ at full load (0.75 pf lagging)} = \frac{7500}{7500 + 73.5 + 115} = 97.54\%$$

(b) We also have

$$\eta \text{ at 70\% full load, 0.92 pf lagging} = \frac{\text{Output (at 70\% full load)}}{\text{Output (at 70\% full laod)} + P_i + P_{\text{Cu}}\text{ (at 70\% full laod)}}$$

and

$$\text{Output at 70\% full load, 0.92 pf lagging} = 10{,}000 \times 0.7 \times 0.92 = 6440 \text{ W}$$

Here $P_i = 73.5$ W, P_{Cu} (at 70% fl) = 56.27 W (as calculated above). Hence,

$$\eta \text{ (at 70\% fl)} = \frac{6440}{6440 + 73.5 + 56.27} = 98.02\%$$

Note that voltage and current are not used for the efficiency calculations unless required specifically.

Example 8

A 10 kVA, single-phase transformer has 200 turns in one winding and 100 turns in the other winding. The frequency of the supply is 50 Hz. If the induced emf in the primary side (when the transformer is used as a step-down transformer) is 250 V, find the value of maximum flux density if the core area is 64×10^{-4} m^2.

Solution

The induced emf $\mathbf{E} = 4.44\phi_m fN$, where ϕ_m = maximum value of the flux in the core, f = input supply frequency and N = number of turns. Here $\mathbf{E}$ = 250 V, ϕ_m = ?, f = 50 Hz and N = 200 turns. Hence,

$$\phi_m = \frac{250}{4.44 \times 50 \times 200} = 5.6 \text{ mWb}$$

Then the maximum flux density is

$$\mathbf{B}_{\max} = \frac{5.6 \times 10^{-3}}{64 \times 10^{-4}} = 0.875 \text{ Wb/m}^2$$

Example 9

A voltage of $V = 200 \sin 314t$ is applied to the primary of a 2 kVA, 50 Hz single-phase transformer, while the secondary is kept open. The resultant primary current is $i_0 = 4 \sin (314t - 30)$. Find the shunt resistance R_L (corresponding to iron loss), shunt reactance (corresponding to core magnetization) and the iron loss.

Solution

The equivalent circuit under the no-load condition mentioned in the question is shown in Figure 19.

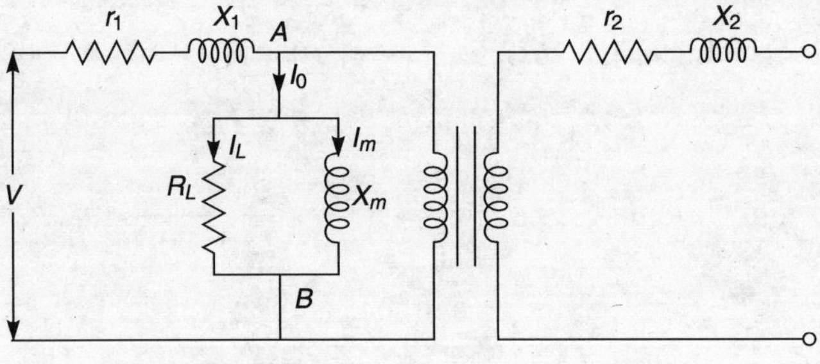

Figure 19 Equivalent circuit (for Example 9).

Neglecting the voltage drops in r_1 and X_1, due to the flow of the current is through them, we can say that the applied voltage, $v \equiv V_{AB}$. Thus,

$$Z_{AB} = \frac{V_{AB}}{i_0} = \frac{v}{i_0} = \frac{(200/\sqrt{2})\angle 0^\circ}{(4/\sqrt{2})\angle -30^\circ} = 50\angle 30^\circ$$

Hence Current in R_L, $i_L = i_o \cos\ 30^\circ = 4\cos 30^\circ = \dfrac{4 \times \sqrt{3}}{2} = 2\sqrt{3}$ A

and Current in X_m, $i_m = i_0 \sin 30^\circ = \dfrac{4 \times 1}{2} = 2$ A

Thus

$$R_L = \frac{v}{i_L} = \frac{200\angle 0°}{2\sqrt{3}} = 57.73\ \Omega$$

and

$$X_m = \frac{v}{i_m} = \frac{200}{2} = 100\ \Omega$$

Iron loss is the power consumed in R_L. Therefore,

$$P_i = i_L^2 R_L = \frac{(2\sqrt{3})^2}{(\sqrt{2})^2} \times 57.73 = 346.4\ \text{W}$$

and

$$I_L\,(\text{RMS}) = \frac{2\sqrt{3}}{\sqrt{2}}; \quad (2\sqrt{3}\ \text{is the peak value of}\ i_L) = 2.45\ \text{A}$$

Thus, $R_L = 57.73\ \Omega$, $X_m = 100\ \Omega$, P (iron loss) = 346.4 W.

Example 10

Derive the expression of the voltage at the load terminals (i.e. V_2) in a single-phase transformer with all its leakage parameters and primary voltage referred to the secondary side.

Solution

The equivalent circuit of a single-phase transformer referred to the secondary side is shown in Figure 20.

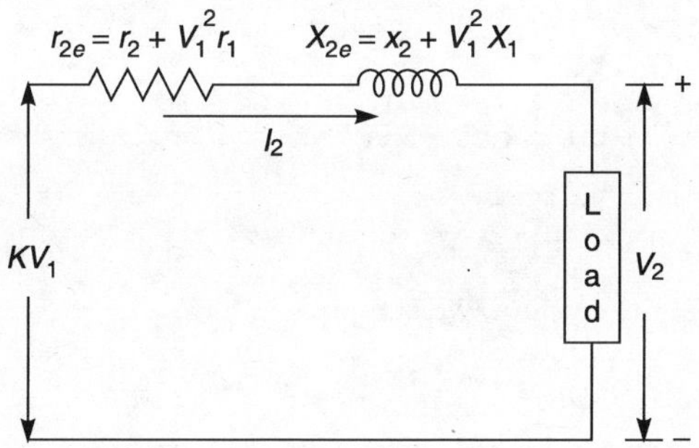

Figure 20 Equivalent circuit (for Example 10).

Here

$$\frac{N_2}{N_1} = K$$

The shunt branch parameters have been neglected. The phasor diagram corresponding to the above equivalent circuit is as shown in Figure 21.

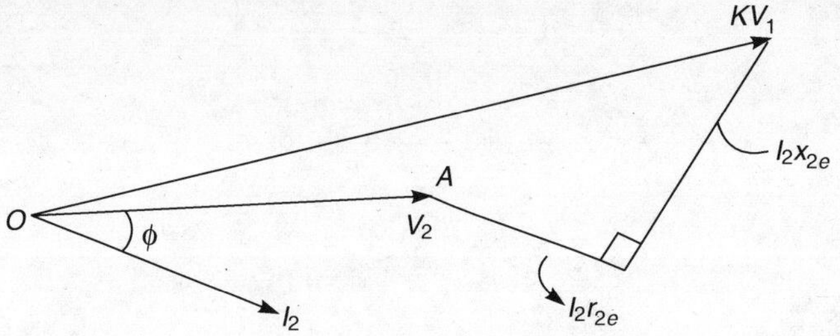

Figure 21 Phasor diagram corresponding to Figure 20.

This phasor diagram is for lagging load (as the current I_2 is shown lagging V_2 by the angle ϕ). Now Figure 21 can be redrawn as shown in Figure 22.

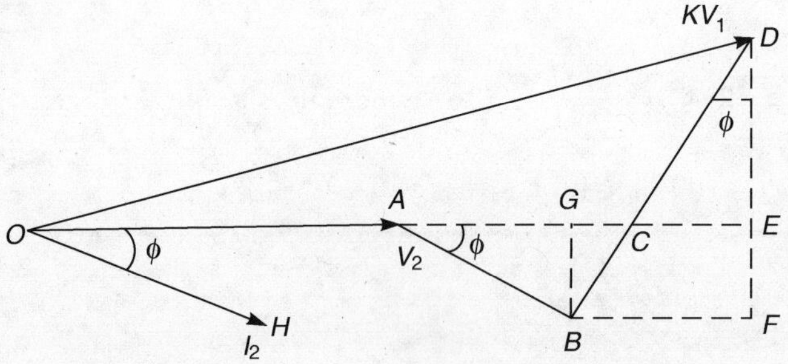

Figure 22 Redrawn phasor diagram for Figure 21.

Hence

$$\angle AOH = \angle GAB = \angle CDE = \phi$$

In the right-angled triangle OED

$$OD^2 = OE^2 + ED^2$$

Now,

$$OE = OA + AG + GE = OA + AG + BF = V_2 + I_2 r_{2e} \cos\phi + I_2 X_{2e} \sin\phi$$

and

$$ED = DF - EF = DF - GB = I_2 X_{2e} \cos\phi - I_2 r_{2e} \sin\phi$$

Also, OD = KV. So, on putting the values, we get

$$(KV_1)^2 - (V_2 + I_2 r_{2e} \cos\phi + I_2 X_{2e} \sin\phi)^2 (I_2 X_{2e} \cos\phi - I_2 r_{2e} \sin\phi)^2$$

Now the second (bracketed) term on the RHS of the above equation is very less in magnitude and can be safely neglected in front of the first term. So,

$$V_2 + I_2 r_{2e} \cos\phi + I_2 X_{2e} \sin\phi = KV_1$$

or

$$V_2 = KV_1 - I_2 r_{2e} \cos\phi - I_2 X_{2e} \sin\phi$$

The above is the derived expression.

Example 11

A 25 kVA, 2200/220 V, 50 Hz, single-phase transformer has the following parameters $r_1 = 1.2\ \Omega$, $X_1 = 6.0\ \Omega$, $R_1 = 1000\ \Omega$, $X_m = 300\ \Omega$; $r_2 = 0.03\ \Omega$, $X_2 = 0.06\ \Omega$ (low-voltage side). Draw the equivalent circuit referred to low-voltage side for a 0.8 pf lagging load supplied at a 220 V terminal voltage.

Solution

We have the transformer equivalent as shown in Figure 23. Here from the equivalent circuit, we see that the load voltage, $V_2 = 220$ V. Assuming its angle to be zero, the angle of the load current (i.e. I_2) will be $\cos^{-1} 0.8 = 36.87$ lagging. Now with

$$I_2 = \frac{25{,}000}{220} = 113.64\text{A}$$

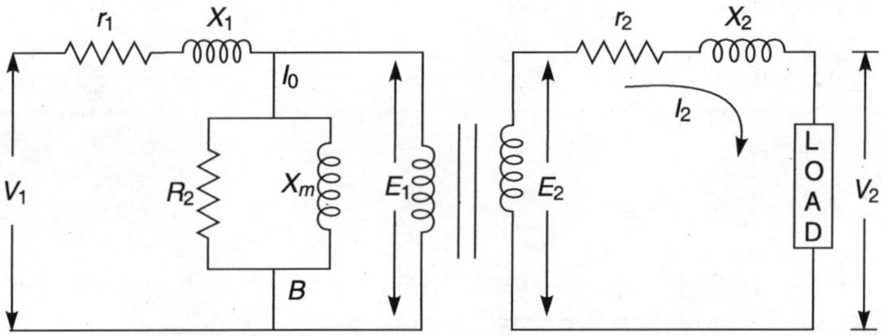

Figure 23 Equivalent circuit for Example 11.

we have complete phasor expression of I_2, as

$$I_2 = 113.64\angle{-36.87°}, \qquad V_2 = 220\angle 0°$$

Now, in the low voltage (i.e. secondary side), we can get the induced emf, E_2, by adding the voltage drops $I_2 r_2$ and $I_2 X_2 V_2$. Thus with $r_2 = 0.03\Omega$ and $X_2 = j0.06$, we have

$$\begin{aligned} E_2 &= 220\angle 0° + 113.64\angle{-36.87°} \times 0.03\angle 0° + 113.64\angle{-36.87°} \times 0.06\angle 90° \\ &= 226.83 + j3.41 \\ &= 226.85\angle 0.86° \end{aligned}$$

Now we refer to r_1, X_1, R_L and X_m towards the secondary (i.e. low-voltage) side to get their values as shown in Figure 24.

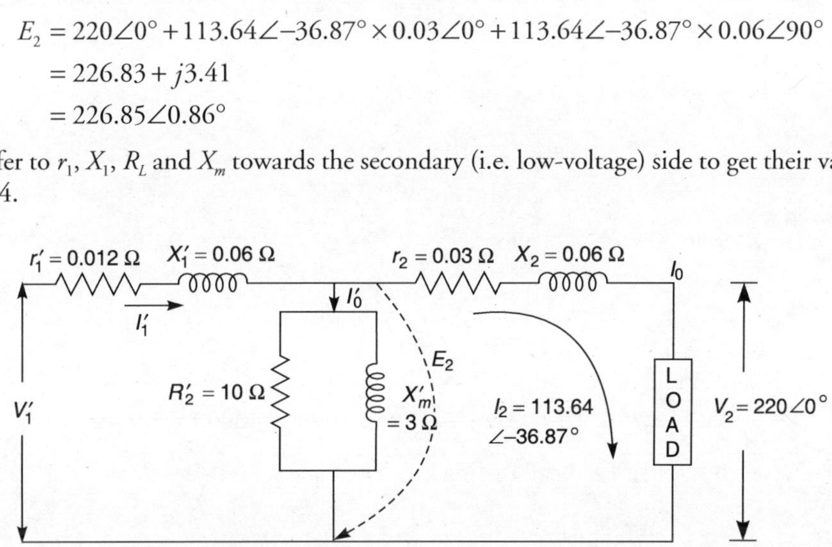

Figure 24 Values of variables in Figure 23.

We need to find V_1'. For that we need to calculate, $I_1' = I_2 + I_0'$. Now I_0' (no-load current referred to the secondary) is

$$I_0' = \frac{E_2}{R_L'} + \frac{E_2}{jX_m'} = \frac{22685\angle 0.86°}{10} + \frac{226.85\angle 0.86°}{3\angle 90°} = 23.814 - j75.26 = 78.94\angle -74.44$$

I_2 is already known as

$$I_2 = 113.64(0.8 - j0.6) = 90.912 - j68.18$$

and

$$I_1' = I_0' + I_2 = 114.73 - j143.45 = 183.68\angle -51.35$$

And finally

$$V_1 = E_2 + I_1'(0.012 + j0.06) = 226.85\angle 0.86° + 183.68\angle -51.35° \times 0.061\ \angle 78.69°$$

Thus $V_1' = 237$ V will complete the equivalent circuit.

Example 12

A 10 kVA, 2000/1000 V single-phase transformer is having an efficiency of 95% at full load 0.8 power factor lagging and also at half full load 0.7 power factor lagging. Find

(a) Full load copper losses,
(b) Iron loss,
(c) Maximum efficiency at unity power factor and the percentage of full load at which it occurs.

Solution

We have efficiency

$$\eta = \frac{\text{Output}}{\text{Input}}$$

Here, η is 95% at full load 0.8 pf lagging. Therefore,

$$0.95 = \frac{10{,}000 \times 1 \times 0.8}{\text{Input}}$$

All η calculations are done in watts. Therefore,

$$\text{Input} = \frac{8000}{0.95} = 8421 \text{ W}$$

and

$$\text{Losses} = P_i + P_{\text{cufl}} = \text{Input} - \text{Output} = 8421 - 8000 = 421 \text{ W}$$

Also, given that η is 95% at half full load, 0.7 power factor lagging. So,

$$0.95 = \frac{10{,}000 \times 0.5 \times 0.7}{\text{Input}} \quad \text{or} \quad \text{Input} = \frac{3500}{0.95} = 3684.2\text{ W}$$

Then

$$\text{Losses (at half full load)} = P_i + P_{\text{cufl}} = 3684.2 - 3500 = 184.2 \text{ W}$$

Now $$P_{\text{cuhl}} = \frac{P_{\text{cufl}}}{4}$$

Cu loss depends upon the square of the load, so at half full load, it reduces to one-fourth. Thus, we have two equations

$$P_i + P_{\text{cufl}} = 421; \qquad P_i + \frac{P_{\text{cufl}}}{4} = 184.2$$

Solving them, we get $P_{\text{cufl}} = 315.74$ W and $P_i = 105.27$ W. Hence, we get

(a) $P_{\text{cufl}} = 315.74$ W
(b) $P_i = 105.27$ W
(c) The percentage of full load at which the maximum efficiency occurs is given by the factor

$$k = \sqrt{\frac{P_i}{P_{\text{cufl}}}}$$

Putting the values here, we get

$$k = \sqrt{\frac{105.27}{315.74}} = 0.577 \text{ or } 57.7\%$$

and

$$\eta_{\max}(\text{at upf}) = \frac{\text{Output at 57.7\% full load, upf}}{\text{Input at 57.7\% full, upf}}$$

$$= \frac{0.577 \times 10{,}000 \times 1}{0.577 \times 10{,}000 \times 1 + 105.27 + 105.27}$$

$$= 96.47\%$$

At $\eta_{\max}$, the Cu loss will be equal to iron loss, that is 105.27 W.

Example 13

A 20 kVA, 2000/100 V, single-phase transformer has $r_1 = 1.5\ \Omega$, $r_2 = 0.005\ \Omega$. The leakage reactances are, $x_1 = 2.8\ \Omega$ and $x_2 = 0.008\ \Omega$. Find

(a) Equivalent resistance referred to primary (r_{1e})
(b) Equivalent resistance referred to secondary (r_{2e})
(c) Equivalent leakage reactance referred to primary and secondary (x_{1e} and x_{2e})
(d) Equivalent impedance referred to primary and secondary (Z_{1e} and Z_{2e})
(e) Total copper loss

Solution

The turns ratio given is

$$\frac{2000}{100} = 20 = \frac{N_1}{N_2}$$

Higher values of resistance and leakage reactance are r_1 and x_1. Hence these are the high voltage side values. Lower values, that is r_2 and x_2 are the low voltage side values.

(a) Equivalent resistance referred to primary is

$$r_{1e} = r_1 + r_2\left(\frac{N_1}{N_2}\right)^2 = 1.5 + 0.005 \times (20)^2 = 3.5\ \Omega$$

(b) Equivalent resistance referred to secondary is

$$r_{2e} = r_2 + r_1\left(\frac{N_2}{N_1}\right)^2 = 0.005 + 1.5 \times \left(\frac{1}{20}\right)^2 = 0.00875\ \Omega$$

(c) Equivalent leakage reactance referred to primary is

$$x_{1e} = x_1 + x_2\left(\frac{N_1}{N_2}\right)^2 = 2.8 + 0.008 \times 20^2 = 6\ \Omega$$

Equivalent leakage reactance referred to secondary is

$$x_{2e} = x_2 + x_1\left(\frac{N_2}{N_1}\right)^2 = 0.008 + 2.8\left(\frac{1}{20}\right)^2 = 0.015\ \Omega$$

(d) Equivalent impedance referred to primary is

$$Z_{1e} = r_{1e} + jx_{1e} = (3.5 + j6)\ \Omega = 6.94\ \Omega$$

Equivalent impedance referred to secondary is

$$Z_{2e} = r_{2e} + jx_{2e} = (0.00875 + j0.015)\ \Omega = 0.017\ \Omega$$

We also have

$$Z_{2e} = Z_{1e} \times \left(\frac{N_2}{N_1}\right)^2 = 6.94 \times \frac{1}{400} = 0.017\ \Omega$$

(e) Total copper losses are given by $I_1^2 r_{1e}$ or $I_2^2 r_{2e}$. Here,

$$I_1 = \text{High voltage side current} = \frac{20{,}000}{2000} = 10\ \text{A}$$

Therefore,

$$\text{Total Cu loss} = I_1^2 r_{1e} = 10^2 \times 3.5 = 350\ \text{W} \quad \text{(obtained at primary side)}$$

Also since,

$$I_2 = \text{Low voltage side current} = \frac{20{,}000}{100} = 200\ \text{A}$$

we get

$$\text{Total Cu loss} = I_2^2 r_{2e} = 200^2 \times 0.00875 = 350\ \text{W} \quad \text{(obtained at secondary side)}$$

It is clear from above that it is immaterial whether the total copper losses are calculated from primary side or from secondary side. The total copper losses come out to be the same.

Example 14

A 5 kVA, single-phase 50 Hz, 1000/500 V transformer has the following equivalent circuit parameter values: $r_1 = 0.1\ \Omega$, $x_1 = 0.3\ \Omega$; $r_2 = 0.004\ \Omega$, $x_2 = 0.012\ \Omega$. Find the equivalent resistance and reactance referred to the low voltage side. Also, find the full-load resistance drop in the low-voltage side (taken as secondary here).

Solution

We know that

$$r_{2e} = r_2 + r_1\left(\frac{N_2}{N_1}\right)^2$$

Here,

$$\frac{N_1}{N_2} = \frac{1000}{500} = 2 \quad \text{or} \quad \frac{N_2}{N_1} = \frac{1}{2}$$

Therefore,

$$r_{2e} = 0.004 + 0.1 \times \frac{1}{2^2} = 0.029\ \Omega$$

Similarly,

$$x_{2e} = x_2 + x_1\left(\frac{N_2}{N_1}\right)^2 = 0.012 + 0.3 \times \frac{1}{2^2} = 0.087\ \Omega$$

Full-load secondary current is

$$I_2 = \frac{5{,}000}{500} = 10\ \text{A}$$

Full-load resistance drop is

$$I_2 r_{2e} = 10 \times 0.029 = 0.29\ \text{V}$$

Example 15

A 1 kVA, 200/400 turns, single-phase transformer is fed with a 220 V, 50 Hz supply at the low-voltage end. Assuming ideal conditions, calculate the maximum flux density, if the area of cross section of the core is 2 cm^2.

Solution

Since the ideal conditions have been given, the resistances of the windings and the leakage reactances would be zero. Hence the 220 V applied at the input would also be the E_1, that is the induced emf in the primary side. Since the low-voltage side (i.e. the side having 200 turns) is the primary (given), we have

$$E_1 = 4.44 f \phi_m N_1 \text{ volts} \Rightarrow 220 = 4.44 \times 50 \times \phi_m \times 200$$

Hence $\phi_m = 0.00495$ Wb with area, $A = 5 \times 10^{-4}$ m^2. Therefore,

$$\text{Maximum flux density, } B_m = \frac{\phi_m}{\text{Area}} = \frac{0.00495}{5 \times 10^{-4}} = 9.9\ \text{Wb/m}^2$$

Example 16

A 50 kVA, 22/11 kV, single-phase transformer working at 50 Hz has a full-load efficiency of 95% at the unity power factor. The efficiency at 80% full load is 94% at 0.8 pf. Find the

(a) Iron loss.
(b) Copper loss at full and half load.
(c) Maximum efficiency at unity pf.

Solution

Given full load efficiency, $\eta_{\text{fl}} = 95\%$. Therefore,

$$0.95 = \frac{50{,}000 \times 1 \times 1}{\text{Input}}$$

where 1×1 is for full-load unity power factor. So,

$$\text{Input} = \frac{50{,}000}{0.95} = 52632 \text{ W}$$

Thus,

$$\text{Losses} = 52632 - 50{,}000 = 2632 \text{ W}$$

Thus, we have

$$P_i + P_{\text{cufl}} = 2632 \qquad (1)$$

Also,

$$\eta_{(\text{at }80\%\text{ full load})} = 0.94 = \frac{50{,}000 \times 0.8 \times 0.8}{\text{Input}} \quad (\text{as pf} = 0.8)$$

Then

$$\text{Input} = \frac{32{,}000}{0.94} = 34042.55 \text{ W}$$

Thus,

$$\text{Losses} = 34042.55 - 32{,}000 = 2042.55$$

Now, copper loss at 80% full load = $0.64P_{\text{cufl}}$. Therefore,

$$P_i + 0.64P_{\text{cufl}} = 2042.55 \qquad (2)$$

Solving Eqs. (1) and (2), we get

$$P_{\text{cufl}} = 1637.35 \text{ W} \quad \text{and} \quad P_i = 994.65 \text{ W}$$

Hence,

$$\text{Copper loss at full load} = 1637.35$$

$$\text{Copper loss at half full load} = \frac{1637.35}{4} = 409.34 \text{ W}$$

For finding the maximum efficiency, we need to know the percentage of full load at which it occurs. Thus, the fraction of full load at which η_m occurs is

$$k = \sqrt{\frac{P_i}{P_{\text{cufl}}}} = \sqrt{\frac{994.65}{1637.35}} = 0.78$$

Thus, the η_m occurs at 78% of the full load. So,

$$\eta_m = \frac{\text{Output at 78\% full load}}{\text{Input at 78\% full load}} = \frac{0.78 \times 50{,}000 \times 1}{0.78 \times 50{,}000 \times 1 + 994.65 + 994.65} = 95.15\%$$

Example 17

A rectangular magnetic core shown in Figure 25 has a square cross-section of area 16 cm². An air gap of 2 mm is cut across one of its limbs. Find the existing current needed in a coil having 1000 turns wound on the core to create an air gap flux 4 mWb. The relative permeability of the core is 2000.

(UPTU 2001–02)

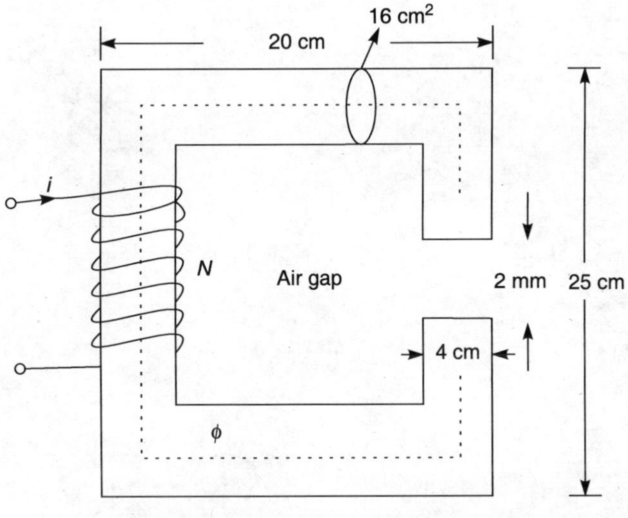

Figure 25 Figure for Example 17.

Solution

Given flux, $\phi = 4 \times 10^{-3}$ Wb; $\mu_r = 2000$; $N = 1000$; $A = 16\ \text{cm}^2 = 16 \times 10^{-4}$; limb thickness $= 4 \times 10^{-2}$ m. Mean length of flux path for magnetic core is

$$L_c = 2 \times (20 - 4) + 2(25 - 4) - 0.2 = 73.8\ \text{cm} = 73.8 \times 10^{-2}\ \text{m}$$

We have $\mu_0 = 4\pi \times 10^{-7}$. Then

$$\text{Core reluctance, } \mathfrak{R}_c = \frac{L_c}{\mu_0 \mu_r A} = \frac{73.8 \times 10^{-2}}{2000 \times 4\pi \times 10^{-7} \times 16 \times 10^{-4}} = 183.618 \times 10^{3}\ \text{AT/Wb}$$

and

$$\text{Air gap reluctance, } \mathfrak{R}_y = \frac{L_y}{\mu_0 A} = \frac{2 \times 10^{-3}}{4 \times 10^{-7} \times \pi \times 16 \times 10^{-4}} = 994.77 \times 10^{-3}\ \text{AT/Wb}$$

As we know that mmf (AT) = flux × total reluctance, so

$$i = \frac{\phi(\mathfrak{R}_E + \mathfrak{R}_Y)}{N} = \frac{4 \times 10^{-3}(183.618 \times 10^{3} + 994.77 \times 10^{3})}{1000} = 4.713\ \text{A}$$

Example 18

An iron ring of mean length 50 cm and relative permeability 300 has an air gap of 1 mm (see Figure 26). If the ring is provided with a winding of 200 turns and a current of 1 A is allowed to flow through, find the flux density across the air gap. **(UPTU 2000–01)**

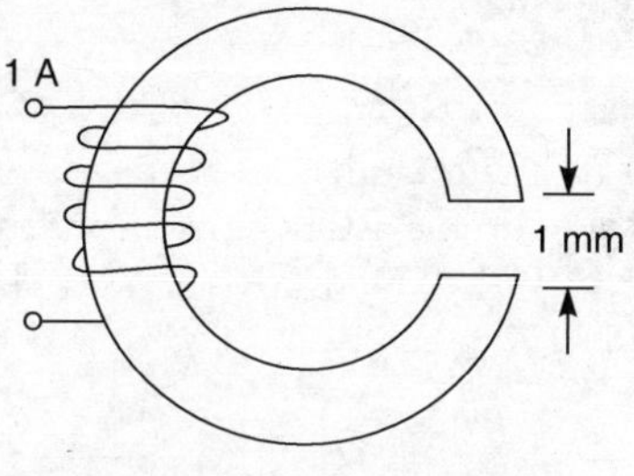

Figure 26 Figure for Example 18.

Solution

Given mean length, $L = 50 \times 10^{-2}$ m; $N = 200$; $I = 1$ A; $\mu_r = 300$; $L_y = 1 \times 10^{-3}$ m. Then

$$\text{Reluctance of the core, } \Re_c = \frac{L_c}{\mu_0 \mu A}; \text{ Reluctance of air gap, } \Re_y = \frac{L_y}{\mu_0 A}$$

Hence

$$\text{mmf} = \phi\,(\Re_y + \Re_e)$$

or

$$NI = BA\left(\frac{L_y}{\mu_0 A} + \frac{L_c}{\mu_0 \mu A}\right) \quad (\text{as } \phi = BA)$$

or

$$200 \times 1 = B\left(\frac{1 \times 10^{-3}}{4\pi \times 10^{-7}} + \frac{50 \times 10^{-2}}{4\pi \times 10^{-7} \times 300}\right)$$

or

$$B = \frac{200 \times 4\pi \times 10^{-7}}{1 \times 10^{-3} + 1.670 \times 10^{-3}} = 94.1 \text{ mWb}$$

Example 19

Two coils having 100 and 50 turns, respectively, are wound on a core with $\mu = 4000\ \mu_0$. Effective core length is 60 cm and core area 9 cm^2. Find the mutual inductance between the coils.

(UPTU 2004–05)

Solution

The mutual inductance is

$$M = \frac{N_1 N_2 \mu_0 \mu_v A}{l}$$

Given $N_1 = 100$; $N_2 = 50$; $\mu = 4000\mu_0$ (as $\mu = \mu_0\mu_r$); $A = 9 \times 10^{-4}$ m^2; $L = 60 \times 10^{-2}$ m. Then

$$M = \frac{100 \times 50 \times 4000 \times 4\pi \times 10^{-7} \times 9 \times 10^{-4}}{60 \times 10^{-2}} = 0.038 \text{ H}$$

Example 20

A coil of insulated wire of 500 turns and of resistance 4 Ω is closely wound on an iron ring (see Figure 27). The ring has a mean diameter 0.25 m and a uniform cross-sectional area of 700 mm^2. Calculate the total flux in the ring when a DC supply of 6 V is applied to the ends of the winding. Assume a relative permeability of 550. **(UPTU 2001–02)**

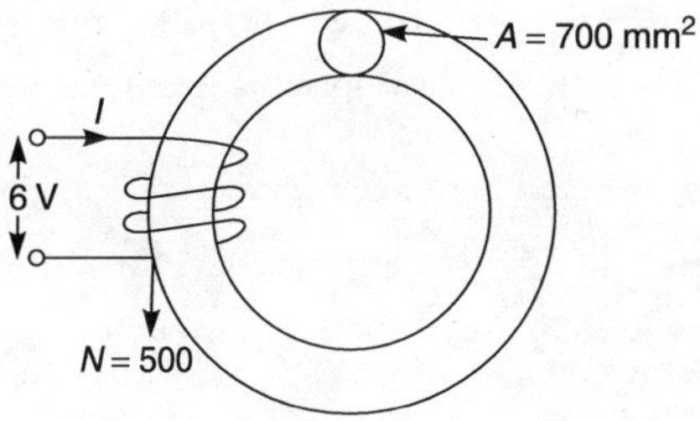

Figure 27 Figure for Example 20.

Solution

Given $N = 500$; $d_{\text{mean}} = 0.25$ m; $\mu_r = 550$; $A = 700$ mm^2. Then

$$L = \pi d_{\text{mean}} = 3.14 \times 0.25 = 0.785 \text{ m}$$

and

$$\Re = \frac{L}{\mu_0 \mu_r A} = \frac{0.785}{4\pi \times 10^{-7} \times 550 \times 700 \times 10^{-6}} = 1.62338 \times 10^6 \text{ AT/Wb}$$

We also have

$$\text{mmf} = NI = \frac{NV}{R}$$

Then

$$\phi = \frac{\text{mmf}}{\Re} = \frac{500 \times 6 \times 10^{-6}}{4 \times 1.62338} = 462 \ \mu\text{Wb}$$

Example 21

A coil of 1000 turns is wound on a laminated core of steel having a cross-section of 5cm^2 (see Figure 28). The core has an air gap of 2 mm cut at right angle. What value of current is required to have an air gap flux density of 0.5 T? Permeability of steel may be taken as infinity. Determine the coil inductance.

(UPTU 2003–04)

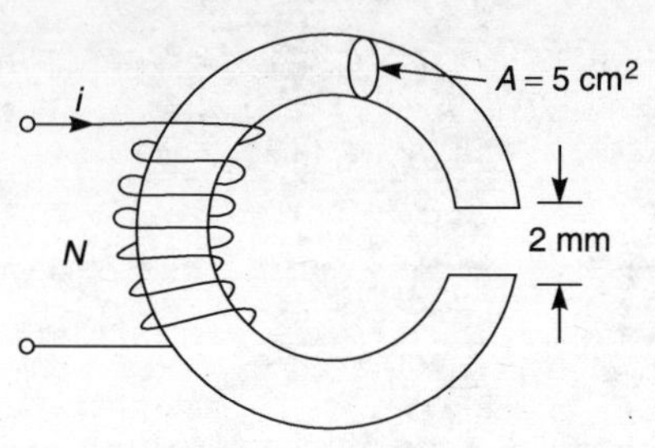

Figure 28 Figure for Example 21.

Solution

Given B = 0.5 T; $A = 5 \times 10^{-4}$ m²; $L_y = 2 \times 10^{-3}$ m; $N = 1000$. We know that $\phi = BA$. Therefore,

$$\phi = 0.5 \times 5 \times 10^{-4} = 2.5 \times 10^{-4} \text{ Wb}$$

In the given problem, μ_r is ∞ for iron core. So no mmf is required to produce a flux in the core. It is required only for the air gap. Hence

$$\Re_y = \frac{l_y}{\mu_0 A} = \frac{2 \times 10^{-3}}{4\pi \times 10^{-7} \times 5 \times 10^{-4}} = 3.1847 \times 10^6 \text{ AT/Wb}$$

We know that

$$\phi = \frac{\text{mmf}}{\Re} = \frac{NI}{\Re}$$

Therefore,

$$I = \frac{\phi \Re}{N} = \frac{2.5 \times 10^{-4} \times 3.1847 \times 10^6}{1000} = 0.7961 \text{ A}$$

Now,

$$L = \frac{N^2}{\Re} = \frac{(1000)^2}{3.1847 \times 10^6} = 0.3140 \text{ H}$$

Summary

The present chapter introduces the electrical power system, bringing about the roles of its different components. The general layout of a typical electrical power system in India is also presented in the chapter. The properties and performance of magnetic circuits are described. A magnetic circuit bears a perfect analogy with an electrical circuit, where in the flux, magnetomotive force and reluctance in a magnetic circuit are analogous to the current, electromotive force and resistance in an electrical circuit. The magnetic circuits, like electrical circuits, follow the Ohm's law (for linear region of **B–H** curve) and Kirchhoff's laws. The hysteresis and the eddy-current losses take place in magnetic circuit, which increase with the increase in frequency of the supply.

Single-phase transformers work on constant-flux density, constant frequency and constant power (VA). The leakage flux is responsible for reduction in induced emfs on both the primary and secondary windings. This reduction, in equivalent circuit, corresponds to the voltage drops across the fictitious series inductances on respective sides. The iron losses in the magnetic circuit and the amount of emfs induced across the windings of a transformer depend on the supply voltage, which are represented by the fictitious resistance and inductance, respectively, in the equivalent circuit.

In a transformer, the iron losses remain constant for constant-supply voltage. The copper (I^2R) losses vary with the square of the load current. The efficiency of a transformer is maximum at a load for which the corresponding copper loss becomes equal to the iron loss. An autotransformer is a single-winding transformer. Its VA and efficiency are higher than a two-winding transformer.

Short Answer Questions

1. Why is earthing so important?
2. Why are the iron losses also called fixed losses?
3. If the copper loss and iron loss for a resistive load of 5 ohms are 100 W and 200 W, respectively, what will be their values when the load is purely inductive of an ohmic value 5 ohms?
4. If the thickness of the laminations is reduced to half, what effect will it have on the iron and copper losses for the same load?
5. If the copper loss of a single-phase transformer is 100 W and the iron loss is 200 W, what will be the percentage of full load at which the maximum efficiency occurs?
6. What are the values of leakage reactances of a single-phase transformer if the coupling between primary and secondary winding fluxes is perfect?
7. What is the necessity of dot convention?
8. What is the value of reluctance if the length of the magnetic path is doubled?
9. What is the difference between electric and magnetic fluxes?
10. What is the difference between permeability and permittivity?

Review Questions

1. What is a **B–H** curve? Explain the hysteresis and eddy current loss. How are they minimized **(UPTU 2008–2009)**
2. Explain the following:
   (a) Magnetic flux and its properties
   (b) Flux density
   (c) Fleming's right-hand rule
   (d) Fleming's left-hand rule
   (e) Lenz's law **(UPTU 2009–10)**
3. Draw and explain the hysteresis loop. What is it significance?
4. What is the basic principle of operation of a single-phase transformer?
5. Derive the expression for the following for a single-phase transformer:
   (a) EMF
   (b) Efficiency
   (c) Condition of occurrence of the maximum efficiency
6. Develop the equivalent circuit of a single-phase transformer on no-load and on-loaded conditions. **(UPTU 2008–09)**
7. Describe the analogies that can be made between electric and magnetic circuit regarding the following items: (a) Driving force, (b) field intensity, (c) impedance drops, (d) equivalent circuits. **(UPTU 2008–09)**
8. Explain the single-phase autotransformer and give its application. **(UPTU 2008–09)**

Problems

1. A single-phase transformer has 350 primary and 1050 secondary turns. The net cross-sectional area of the core is 80 cm^2. If the primary winding is connected across a 600 V, 50 Hz, single-phase supply, calculate: (a) The peak flux density in the core and (b) Voltage induced in the secondary winding. Neglect leakage impedances in the transformer.
2. A 30 kVA, 2000 V/200 V, single-phase, 50 Hz transformer has a primary resistance of 3.5 Ω and reactance of 4.5 Ω. The secondary resistance and reactance values are 0.015 Ω and 0.02 Ω, respectively. Find: (a) equivalent resistance, reactance and impedance referred to primary side and (b) total copper loss in the transformer.
3. A 100 kVA transformer has an efficiency of 95% at full load and also at 80% full load, both at unity power factor. Find its maximum efficiency and the percentage load at which it will occur.
4. The efficiency of a 400 kVA, single-phase transformer is 98.77% at full load 0.75 pf lagging and 99.15% at 75% full load, unity power factor. Find iron losses and copper losses at half and full loads.
5. 40 kVA, 50 Hz, single-phase transformer has rated Cu loss and iron loss as 800 W and 400 W, respectively. Find the percentage of full load at which the maximum efficiency occurs. Also, find the value of that maximum efficiency at unity power factor.
6. A cast-steel electromagnet, having relative permeability of 300, has an iron path of length 50 cm and air gap of length of 1 mm (see Figure P4.1). Find the number of ampere-turns required to produce a flux density of 0.0942 Wb/m^2 in the air gap. Neglect the leakage and fringing.

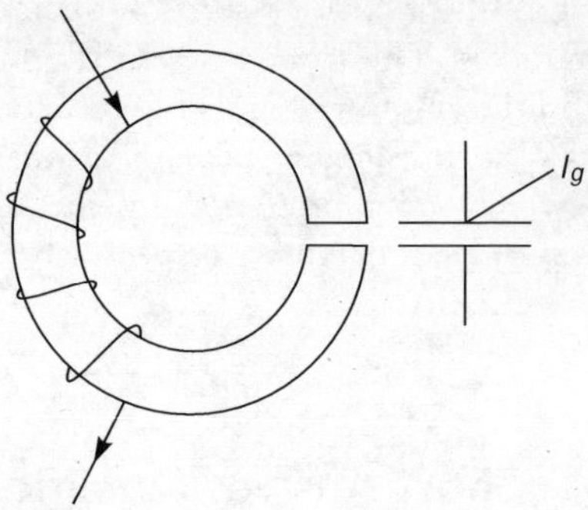

Figure P4.1 Figure for Problem 6.

7. An iron circuit with a small air gap has a cut in it. A 5000-turn coil carries a current I = 30 mA, which sets up a flux within the iron and across the air gap. If the iron cross-section is 0.6×10^{-4} m^2, the mean length of flux path in iron is 0.15 m, μ_r = 800 in iron and air gap length is 0.75 mm, calculate the air gap flux density. It may be assumed that the flux lines flow straight across the air gap. That is, the air gap cross-section is also 0.6×10^{-4} m^2.
8. A cast-steel magnetic core of permeability 600 has 800-turn coil placed at its central limb. The cross-sectional areas of the central and side limbs are 10 cm^2 and 8 cm^2, respectively. Calculate the current required to be driven through the coil to set up a magnetic field of 1.4 mWb in the central limb. The lengths of magnetic path in the central and side limbs are 15 cm and 35 cm, respectively. Neglect the fringing and leakage of flux.
9. A mild steel ring, having a cross-sectional area of 50 mm^2 and a mean circumference of 500 mm, has a coil of 900 turns wound uniformly around it. Calculate: (a) the reluctance of the ring, (b) the current required to produce a flux of 700 μWb in the ring. (Given that μ_r is about 400.)

Answers

Short Answer Questions

1. In the event of a fault in an electrical equipment, the current may follow an unusual, alternate path through the body of the equipment. If a person standing nearby happens to touch the equipment, that current may pass through his body down to the earth. While the current passes through the body of the person, it may damage some important body parts including the heart which may lead to a fatal accident. On the other hand, if the equipment is properly earthed, the fault current will find its way through the earth wire whose resistance is definitely lower than that of a human body.
2. The iron losses are made up of the hysteresis loss and the eddy current loss. Both these losses depend upon the frequency and the applied voltage both of which are usually kept fixed in a transformer operation. Hence, the iron losses are termed as fixed or constant losses. But the moment either or both of the frequency and voltage change, these losses will no longer remain fixed.
3. The losses in a transformer do not depend upon the power factor of the load (hence the transformer rating is always done in voltamperes). Hence for the same ohmic value of the load, the two losses will remain unchanged.
4. The eddy current loss depends upon the square of the thickness of the laminations, hence if the thickness is halved, the eddy current loss becomes one-fourth which reduces the overall iron loss. The copper loss remains unchanged.
5. The percentage of full load at which maximum efficiency occurs is given by multiplying by 100, the square root of the ratio of iron loss to rated copper loss. In this case it comes out to be 70.7%.
6. In the case when the coupling between primary and secondary winding fluxes is perfect, there will be no leakage flux and hence the values of leakage reactances will be zero.
7. Whenever there are two coils placed near each other and alternating current flows through any one of them, a magnetic field is produced in the vicinity which links (interacts) with the other coil. This linking induces an emf in the other coil and if its circuit is closed, a current flows through the other (second) coil. This current in turn gives rise to another magnetic field whose flux should oppose the flux due to the original field (according to the Lenz's law). The direction of the flux produced due to the current flowing in the second coil (i.e. the induced current) depends upon the manner in which the second coil has been wound. Thus every time the second coil induced emf flux direction is desired, the manner of winding the second coil needs to be known and hence shown. This makes the whole drawing big and cumbersome to handle. The dot convention has been developed precisely to counter this problem. Here instead of showing the winding and its direction, only two symbolic inductances are shown with dots at the appropriate ends to indicate the following of the Lenz's law.
8. The reluctance depends directly on the length of the magnetic path hence its value will also be doubled.
9. The region of influence around a stationary electric charge is termed as the electric flux whereas the region of influence around a moving electric charge (i.e. current) is called magnetic field.
10. The ease with which an electric field is able to penetrate a medium is called the permittivity of the medium. The ease with which a magnetic field is able to penetrate a medium is called the permeability of the medium.

Problems

1. **(a)** 0.965 Tesla; **(b)** 180°
2. **(a)** 5 Ω, 6.5 Ω and 8.2 Ω; (b) 1125 W
3. η = 95.03%, load = 89.44%
4. Iron loss (which remains same at half and full load = 1075.17 W, Cu loss (at full load) = 2660.8 W, Cu loss (at half full load) 664.2 W
5. **(a)** 70.7% full load, **(b)** 97.24%
6. 200 AT
7. 0.2 Wb/m^2
8. 1.086 A
9. 0.1989 × 106 AT/Wb, 0.1574 A

5 Electromechanical Energy Conversion

LEARNING OBJECTIVES

After a careful study of the chapter, you would be able to understand:

- General construction and principle of operation of rotating machines.
- Derivation of emf and torque equations for DC machines.
- Operating principle and performance of three-phase induction motors.
- Various types of single-phase induction motors.
- Basics of synchronous machines: alternators and synchronous motors.
- Characteristics and applications of various rotating machines.

5.1 Principles of Electromechanical Energy Conversion

The basic phenomena occurring in the operation of electric motors and generators are the electromechanical energy conversion, which involves the transfer and conversion of electrical energy into mechanical energy and vice versa, in the presence of a coupling magnetic field. This interchange of the two energies is not 100% efficient, as some of the energy gets lost in the core and windings of the machines as well as in the friction in rotating parts. Depending upon whether the electrical system involved is AC or DC, the electromechanical conversion device is called *AC motor/generator* or *DC motor/generator,* respectively.

The fundamental principles of torque and emf production remain absolutely the same in both AC and DC forms of machines. The difference in the expressions of torque and emf between AC and DC machines exists due to the differences in the mechanical constructions of the two machine types.

Since the electromechanical energy conversion pertains to the electrical and mechanical energies, the quantities involved are torque and speed (for mechanical system) and voltage and current (for electrical system). Figure 1 shows the relationship between the two.

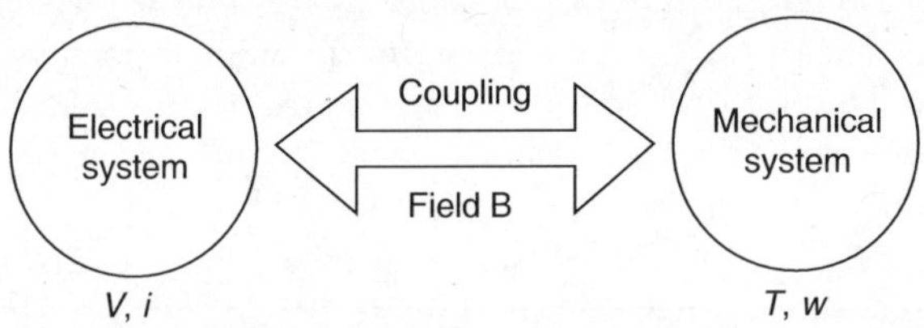

Figure 1 Electromechanical energy conversion sequence.

In order to understand the principles of electromechanical energy conversion, the knowledge of basic construction of rotating machines is required. Basically, all rotating machines comprise a stator (the static part) and a rotor (the rotating part). The rotor is a cylindrical body, carrying winding on its outer periphery.

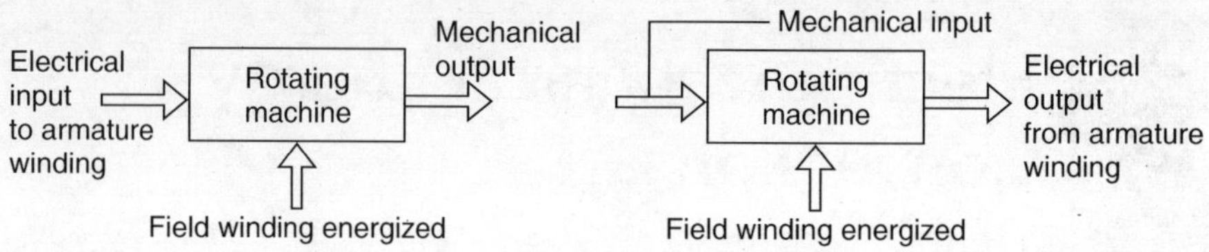

Figure 2 General operating patterns of a rotating machine. (a) A motor and (b) a generator.

The rotor is mounted on a shaft supported by bearings. It is a free to rotate inside a hollow cylindrical stator. The stator also carries a winding on its inner periphery. The basic role of rotating machine is to work either as a generator or as a motor. Figure 2 explains the difference between the two. Figure 2 shows that

1. There are two major windings in the rotating machines: *armature winding* and *field winding*.
2. Both in the case of generator as well as motor, the electrical energy is supplied to the field winding. In the case of generators, the mechanical energy is given as input (through a prime mover, such as steam turbine, hydroelectric turbine, diesel engine, gas turbine, etc., which is mechanically coupled to the shaft of the rotor). The electrical output is collected through the armature winding, to which any electrical load (resistance, inductance, etc.) can be connected.
3. In the case of motors, the electrical energy is given as input to the armature winding and the mechanical output is obtained from the rotor shaft, to which any rotating mechanical load (pump, fan, etc.) can be connected.
4. Thus, the function of a field winding is always to develop a magnetic field, both in generators as well as in motors. The role of armature winding is to produce electrical output (in the case of generators) and to accept electrical input (in the case of motors).
5. The placement of the armature winding and the field winding on stator or rotor varies from machine to machine.

To summarize, major *magnetic parts* in a rotating machine are: (a) stator (a fixed structure) and (b) rotor (a rotating body, to absorb or produce mechanical energy). Major *electrical windings* are: (a) armature winding (to absorb or produce electrical energy) and (b) field winding (to develop coupling magnetic field).

The following sequence of events occurs for successful motoring action to take place:

1. The electrical energy is the input to the rotating machine, which when applied causes the flow of current through the conductors that are placed in an already existing magnetic field.
2. The combined presence of magnetic field and electric current leads to the development of a force on each conductor, and if the mechanical construction is conducive to movement, the conductor rotates with a particular angular velocity.
3. On the application of a voltage, a current flows, and along with the magnetic field the current causes the production of a torque (force in rotational motion) leading ultimately to the conductor rotation at an angular velocity. The electrical power $v \times i$ gets converted to mechanical power $T \times \omega$.
4. Along with the motoring action, described above, the revolving conductors also cut through the magnetic field, thus causing an emf to get induced in them. This is a sort of "retaliatory" or "reaction" emf, similar to the E_1 developed in the transformer primary, as discussed before. This reaction emf, called *back-emf*, is responsible for maintaining a stable value of torque in the motor.

In the generating action, the reverse processes take place:

1. A prime mover (water or steam turbine) moves the rotor in the presence of a magnetic field, thus causing an emf to get induced across the armature winding terminals in accordance with the Faraday's law of electromagnetic induction.

2. When an electrical load (i.e. any combination of resistance, inductance or capacitance) is applied, the current starts flowing, and so long as the mechanical energy is continued to be supplied, the electrical output is maintained.

Fleming's right-hand and Fleming's left-hand rules explain the concept of mechanical to electrical energy conversion in the case of generators, and electrical to mechanical energy conversion in the case of motors, respectively. In both the rules, the main finger (thumb), forefinger and the central (or middle) finger are used. In both the rules:

1. The **F**orefinger indicates the **F**ield.
2. The **C**entral finger indicates the **C**urrent.
3. The **M**ain finger indicates the **M**otion.

In Fleming's right-hand rule, the directions of motion (indicated by the thumb) and field are known, and thus the middle finger indicates the direction of the (induced) current, which is orthogonal (i.e. perpendicular) both to the field and the motion. Thus, this rule is applicable for the generators.

In Flemings left-hand rule, the directions of current (indicated by the central finger) and field are known, and thus the main finger (thumb) indicates the direction of the motion, which is orthogonal (i.e. perpendicular) both to the field and the current. Thus, this rule is applicable for the motors.

To summarize, Figure 3 indicates the two rules. If field is F, motion is M and current is C, we have

1. Fleming's left-hand rule: For motor action F, C are known, M is found.
2. Fleming's right-hand rule: For generator action F, M are known, C is found.

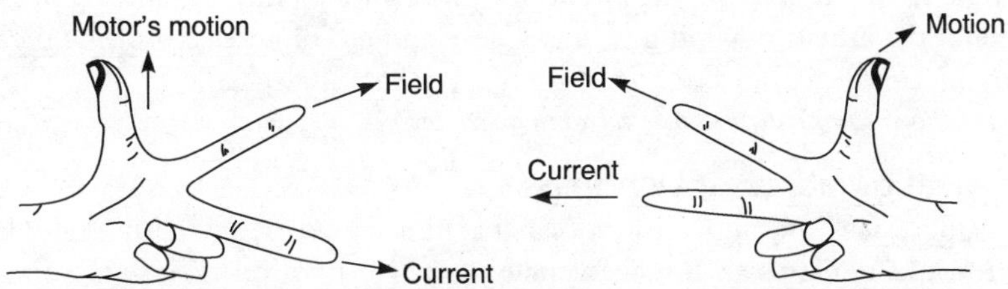

Figure 3 Fleming's left-hand and Fleming's right-hand rules.

The above concept can be further elaborated by the following explanation: As shown in Figure 4, if a conductor is moved upwards in the presence of a field, as shown, it will have an emf induced in such a direction that the current flows into the paper (which is indicated by a cross). According to the Faraday's law of electromagnetic induction and Lenz's law, the direction of the induced current (in the conductor) is such that it leads to the production of a force in a direction so as to oppose the movement of the conductor. This can be verified by applying Fleming's right-hand rule.

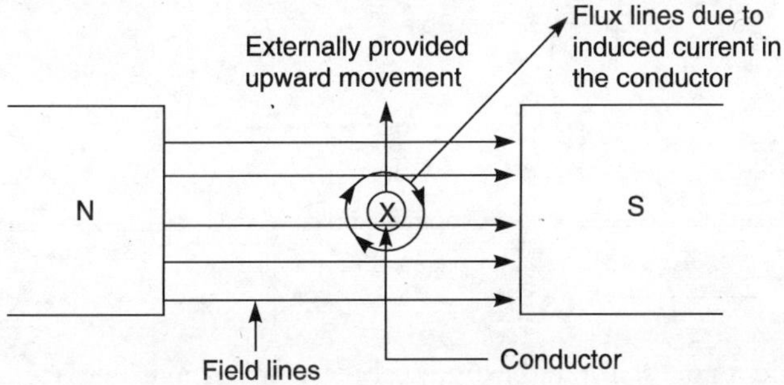

Figure 4 Illustration of the principle of generator.

The right-hand Ampere's rule states that if the thumb of the right hand is put in the direction of the current, then the remaining four fingers, if closed (to form a thumbs-up sign), indicate the direction in which the flux due to that current will be produced. The opposite of this is also true. That is, if the closed fingers are indicating the direction of current (or currents), then the thumb of right-hand indicates the direction of the flux.

As can be seen, the flux (for inward current) will be clockwise encircling the conductor. This flux will strengthen the already existing field flux (due to the magnet N–S) at the top of the conductor and oppose the field at the bottom of the conductor. This will result in the accumulation of extra flux lines at the top and dearth of flux lines at the bottom of the conductor. This results in a catapult-like action on the conductor which thus experiences a force in the downward direction. Now since the conductor is being forced to move in the upward direction (by external means), we can say that the Lenz's law is satisfied. Hence our assumed current direction (inward) is correct. Now in Figure 4, if in the presence of the magnetic field (due to the magnet N–S), we provide the current (say, into the plane of paper), we will see that the conductor experiences force in the downward direction and will move accordingly. This is the principle of a motor. Fleming's left-hand rule will verify the downward motion of the conductor. Here, the electromagnetic torque T_e and the actual mechanical torque T_m are in the same direction. Thus, we conclude the following.

1. In a generator, the electromagnetic torque (T_e) is produced due to interaction of the induced current in the armature winding and the field. Due to Lenz's law, it opposes the direction of motion of the rotor (which is moved by a prime mover). The prime mover torque, that is the mechanical torque (T_m), thus keeps on opposing T_e. Thus *the direction of rotation and the direction of electromagnetic torque are opposite in a generator.*
2. In a motor, the supply is given to the armature winding too. Thus, the interaction of the fluxes due to the field and the armature windings causes development of an electromagnetic torque (T_e). The current-carrying conductors placed on the rotor experience a mechanical torque (T_m). In motors, *the direction of rotation is same as that of electromagnetic torque.*

5.1.1 Various Arrangements of Windings

The working principle of a rotating machine can be understood better with slight knowledge of its construction. The magnetic parts can be salient-pole type or cylindrical type. Accordingly, the electrical windings are concentric DC winding or distributed AC winding, respectively. (Some of the typical arrangements are explained through Figure 5 and Figure 6.) Table 1 shows the winding arrangement in a DC machine, induction machine and synchronous machine.

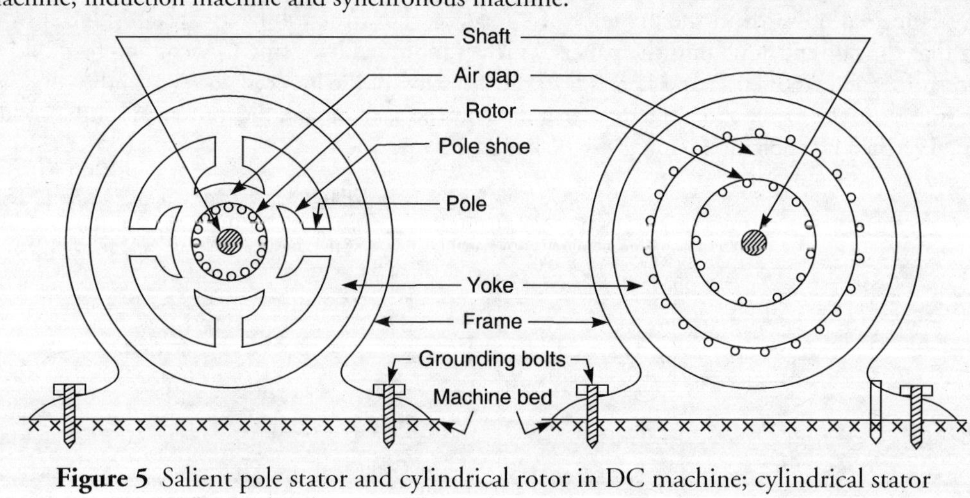

Figure 5 Salient pole stator and cylindrical rotor in DC machine; cylindrical stator and cylindrical rotor in three-phase induction machines.

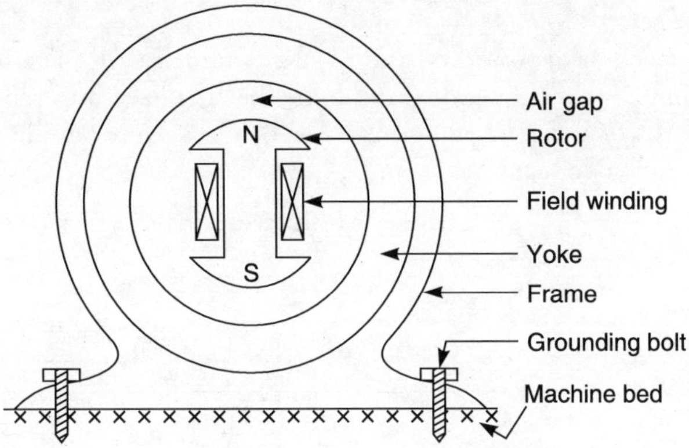

Figure 6 Cylindrical stator and salient pole rotor in three-phase synchronous machines.

Table 1 Winding arrangement

DC machine	*Induction machine*	*Synchronous machine*
Salient pole stator (houses DC concentric field winding)	*Cylindrical stator* (houses AC distributed field winding)	*Cylindrical stator* (houses AC distributed armature winding)
Cylindrical rotor (houses AC distributed armature winding)	*Cylindrical rotor* (houses AC distributed armature winding)	*Salient pole/cylindrical rotor* (houses DC concentric field winding)

5.2 DC Machines

In DC machines, the stator contains salient (horse shoe magnet type) poles to carry the field winding and to produce the magnetic field. The armature winding of DC machine is placed on the rotor, which cuts the magnetic field produced by the field winding placed on the stator. The current in armature winding is AC, which is rectified into DC by means of the commutator-brush arrangement situated over the rotor shaft. The windings of DC machine are represented by the symbols shown in Figure 7.

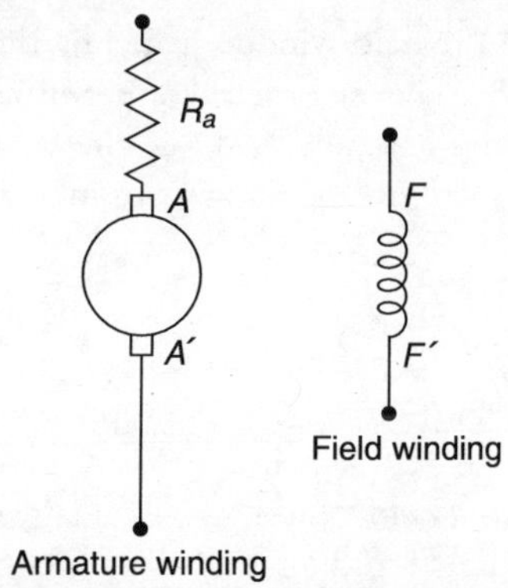

Figure 7 Circuit representation of the armature and field windings of a DC machine. Here R_a is the armature winding resistance.

5.2.1 Types of DC Machines

The classification of DC machines is based on the *type of excitation* used. The term *excitation* means the energization of field winding. The field winding can either be energized by a separate supply arrangement (called *separately-excited field*) or it can be energized by its own armature winding (called *self-excited field*). Accordingly, the classification is done, as shown in Figure 8.

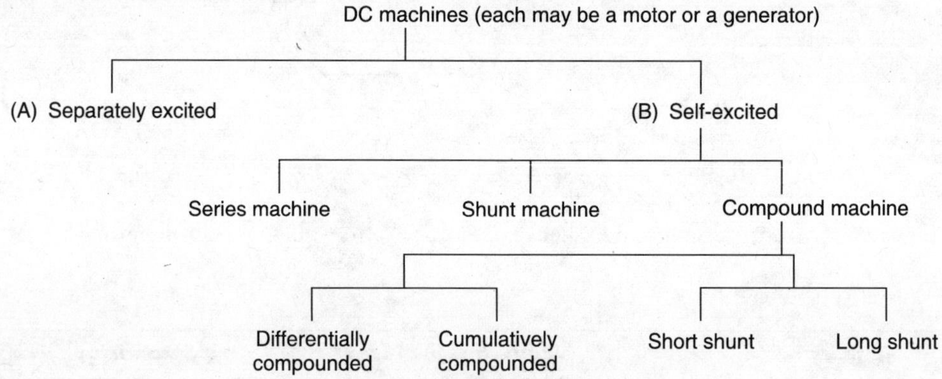

Figure 8 Classification of DC machines.

Thus, there are two major types of DC machines.

1. **Separately excited DC machine:** A variable DC supply arrangement for field winding is made separately, as shown in Figure 9.

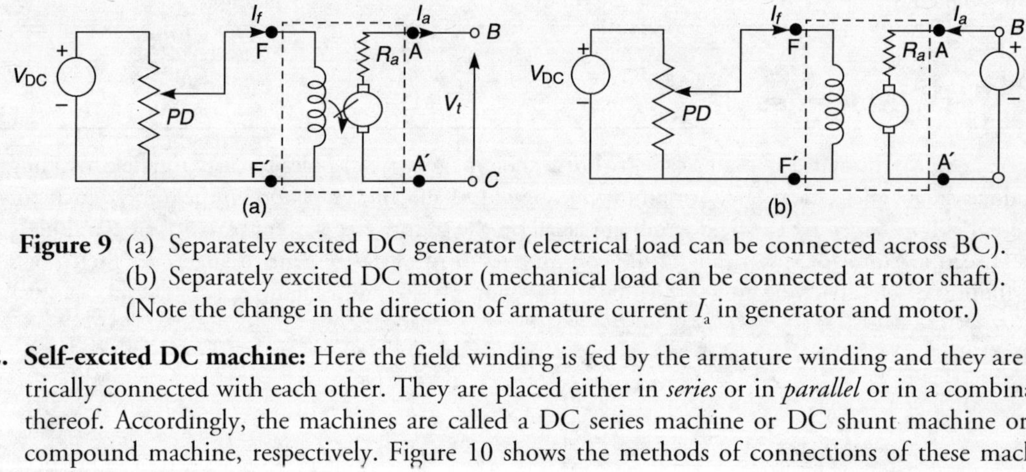

Figure 9 (a) Separately excited DC generator (electrical load can be connected across BC).
(b) Separately excited DC motor (mechanical load can be connected at rotor shaft).
(Note the change in the direction of armature current I_a in generator and motor.)

2. **Self-excited DC machine:** Here the field winding is fed by the armature winding and they are electrically connected with each other. They are placed either in *series* or in *parallel* or in a combination thereof. Accordingly, the machines are called a DC series machine or DC shunt machine or DC compound machine, respectively. Figure 10 shows the methods of connections of these machines working as motors.

The connections of these machines working as generators can be drawn by replacing the DC supply by electrical load (i.e. load resistance) and reversing the direction of load current.

The DC shunt motor takes the line current I_L from the supply, out of which the current I_a goes to the armature circuit and the current I_f goes to the field circuit. The value of the armature resistance R_a is usually less than 1 Ω, whereas the resistance of the field winding R_f is usually greater than 100 Ω for the shunt field. In the series field, the resistance of the field winding is comparable to that of the armature winding, that is it is around 1 Ω or less. In series motors, all the three currents, I_L, I_a and I_f, are same as can be seen from Figure 10.

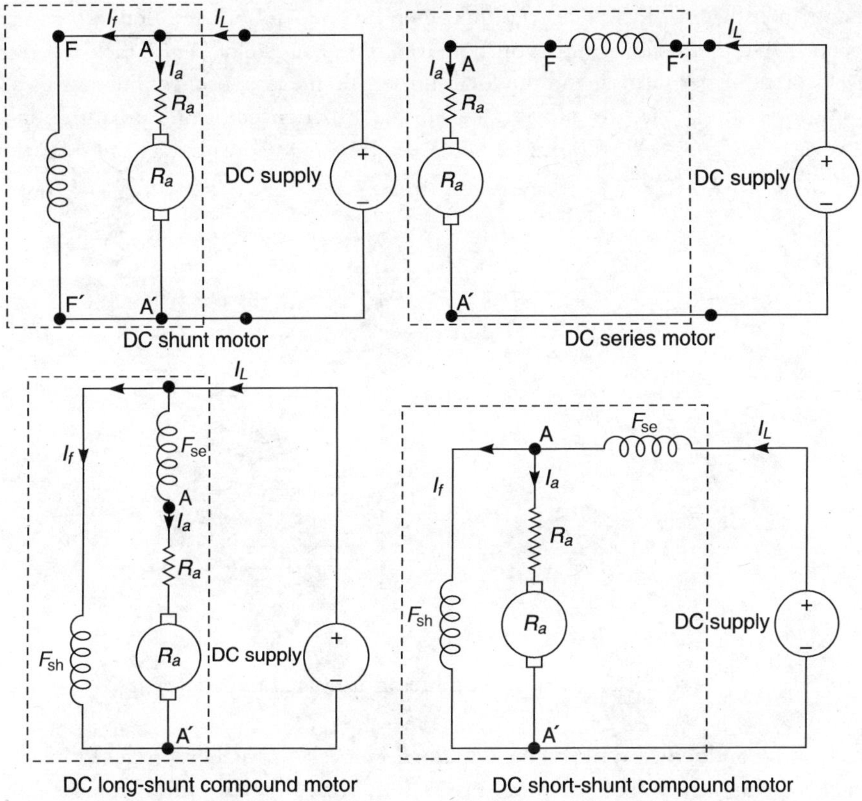

Figure 10 Various types of self-excited DC motor configurations.

In compound motors, two configurations are possible. If the shunt field winding is long enough to encompass both the armature and the series field winding, it is called the long-shunt DC compound motor. If the shunt field winding is only covering the armature across it, the configuration is called the short-shunt DC compound motor.

If the flux due to the series field winding is assisting the one due to the shunt field winding, it is called a *cumulatively-compounded DC motor*, whereas if the series field winding is wound in such a direction that its flux is opposing the one due to shunt field winding, it is called the *differentially-compounded DC motor*.

Since the armature also carries winding, an emf (e) is induced in it on application of an input voltage (in any of these motor configurations). This emf opposes the applied voltage. Hence it is also called the *back-emf* e_b. The value of e_b is equal to the applied voltage (V_t) minus the voltage drop that takes place in the armature winding due to the flow of the armature current, I_a. Mathematically,

$$e_b = V_t - I_a R_a \tag{5.1}$$

5.2.2 EMF Equation of DC Generators

In the case of generators, the volt–ampere relationship would be

$$e = V_t + I_a R_a$$

Let there be P number of poles on stator, each developing flux equal to ϕ webers. Let Z number of armature conductors be placed in 'a' number of parallel paths on the rotor (i.e. armature winding carries A parallel paths), and the rotor rotates at N rpm. Each armature conductor (when makes one rotation), sweeps past all the poles and cuts their fluxes. Thus, it cuts $P\phi$ flux when it completes one rotation. Now the armature (or the rotor) is making N rotations in a minute or $N/60$ rotation in a second. Hence each armature conductor also makes $N/60$ rotations per second, and the total flux cut by the armature conductor in a second is thus $P\phi N/60$ Wb. Thus, we can say that the rate of change of flux experienced by one armature conductor would be $P\phi N/60$ Wb/sec. Hence, $e = P\phi N/60$ can be said to be the *induced emf* due to one armature conductor (see Figure 11).

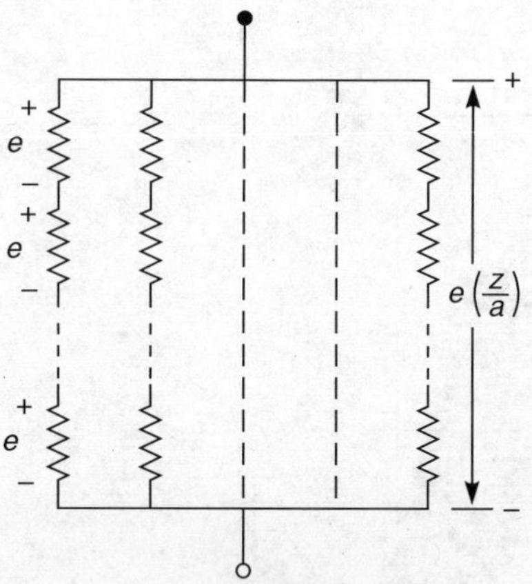

Figure 11 Armature conductors arranged in parallel.

Now there are Z such conductors in the whole armature. Hence, the total emf induced in all the armature conductors will be $Z \times e$ volts. However, as can be seen from Figure 11, the output voltage is available at the ends only at the brushes. Hence the factor $Z \times e$ needs to be divided by the number of parallel paths a, to get the voltage induced at the ends of the armature, that is at the brushes. Thus, the expression of induced emf can be written as

$$e_b = e\frac{Z}{a}\ \text{V} \quad \text{or} \quad e_b = \frac{P\phi NZ}{60a}\ \text{V} \tag{5.2}$$

There are two types of armature winding patterns: (i) lap winding and (ii) wave winding. The number of parallel paths $a = 2$ for wave winding and $a = P$ for lap winding of the armature.

5.2.3 Torque Equation of DC Motors

In general, the voltage–current relationship of a DC motor can be expressed as

$$V_t = e_b + I_a R_a$$

where e_b = induced back emf in armature winding, I_a = armature current and R_a = resistance of armature winding. Multiplying both sides by I_a, we get the power balance equation

$$V_t I_a = e_b I_a + I_a^2 R_a$$

that is

Power input = Power output + Losses in armature winding

The power output is the mechanical power available at the rotor shaft (neglecting the rotational and other losses). Now,

$$\text{Power} = \text{Torque (in N-m)} \times \text{Speed (in mechanical radians/sec)}$$

Here the motor speed being N rpm means $2\pi N/60$ rad/sec. Hence,

$$\text{Output power, } e_b I_a = \frac{2\pi N T_e}{60} \text{ W}$$

where T_e is the torque developed by the rotor (in Nm). Therefore,

$$T_e = \frac{60 \times e_b I_a}{2\pi N} \text{ Nm}$$

Substituting for e_b, we get

$$T_e = \frac{60}{2\pi N} \times \frac{P\phi NZ}{60\ a} I_a = \frac{P\phi Z I_a}{2\pi a} \text{Nm}$$

or

$$T_e = k\phi I_a \text{ Nm} \tag{5.3}$$

where $k = PZ/2\pi a$ is constant.

5.2.4 Characteristics of DC Generators

There are mainly three characteristics of DC generators:

1. Magnetizing characteristic
2. External characteristic
3. Internal characteristic

Magnetizing characteristic (no-load or open-circuit characteristic)

This characteristic is the plot of induced emf (E) against field current (I_f). This is also known as *Open Circuit Characteristic* (OCC). The induced emf, $E \propto N\phi$. For a given speed, the variation of E can be plotted against ϕ (where ϕ is proportional to I_f).

In a separately excited DC generator as the one shown in Figure 12(a), the field is present due to the voltage source separately connected across the field winding. The curve, in Figure 12(b), shows that if the rotor is rotated at the same speed, and even if there is no current in the field winding, a small emf equal to OA gets developed in the generator. This is because of the residual magnetism present in the stator poles.

With the increase in I_f from zero to higher values, the induced emf rises nearly linearly, depicted by the region AB. Point B is called the *knee* of the curve. An increase in I_f beyond this causes magnetic saturation of the stator poles, and hence ϕ tends to become constant. This saturation is seen in the region BC of the curve, where an increase in I_f does not lead to corresponding increase in induced emf E.

In a self-excited DC generator, as the one shown in Figure 13(b), the external supply to the field is absent. Hence the rotation of the rotor, without any magnetic field to cut, is a fruitless exercise, as Faraday's law cannot get satisfied. In fact, the emf that gets induced due to the rotation of the rotor in the field is the only factor responsible to excite the field winding to produce magnetic field. Thus, in self-excited DC generators, the field current and the induced emf are interdependent.

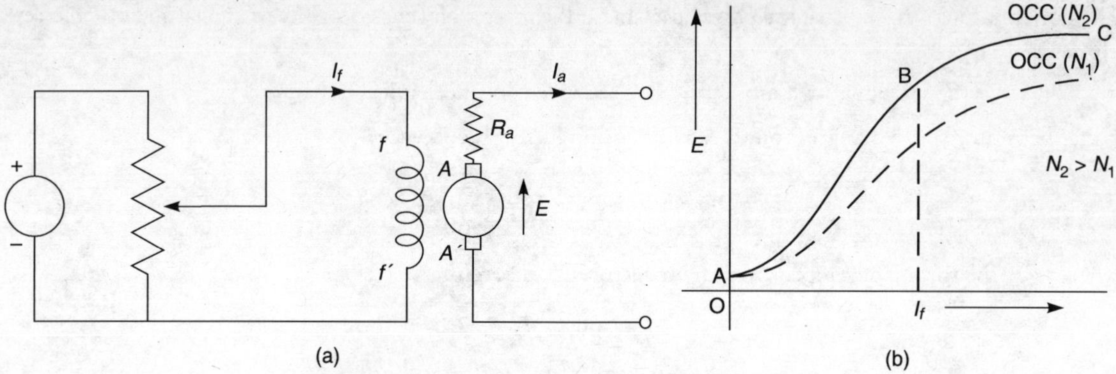

Figure 12 (a) Circuit arrangement; (b) OCC.

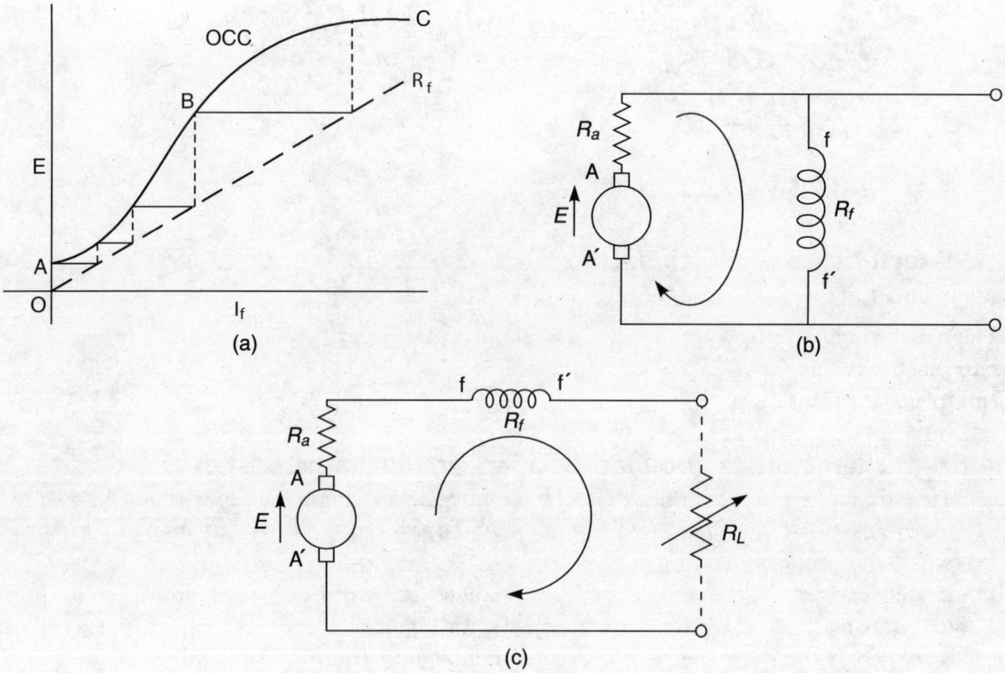

Figure 13 Building up of emf in self-excited (a) DC generators; (b) DC shunt generator; (c) DC series generator.

Thus, the voltage build up at the generator output terminals is a complex phenomenon in self-excited DC generators. Building up of emf requires certain prerequisite conditions to be fulfilled. The following four preconditions need to be attained in order to ensure a smooth voltage build up in self-excited DC generators:

1. **There should be certain "residual magnetism" present in the field poles:** This residual magnetism is the same that was discussed in the **B–H** curve in earlier chapter. Thus, if a small amount of magnetism is present, that is a small number of field lines are present, the rotor when rotated, cuts these lines, causing the induction of a small emf. This small emf circulates small current in the field winding. The magnetic field thus gets strengthened. The rotor continues to move in this increased

field and thus a higher emf is induced, causing a higher current which is (as before) again fed to the field winding, thus further strengthening it. Thus, the voltage gradually builds up. This increase in the voltage continues till the field winding reaches the saturation stage, that is when further increase in the field current does not increase the magnetic field (or in other words, the flux does not increase on increasing the field current). Figure 13(a) shows the voltage build up till saturation.

2. **The direction of the field winding should be proper:** This means that the manner in which the field winding has been wound should be such that when the current due to residual magnetism flows through it, the direction of the flux produced should be in such a way as to enhance the magnetic field due to residual magnetism in order to be able to further strengthen the field flux. It must not oppose the same
3. **The voltage build up takes place till the OCC intersects with the given field resistance line:** The slope of the OCC is the critical (maximum possible) value of field resistance. If the value of the resistance of field winding is greater than this critical value of the field resistance, the generator would not generate any emf.
4. **Further, for a given value of field resistance to which the OCC (for a particular minimum speed) coincides or intersects, building up of emf is possible:** Since $E \propto N\phi$, for same value of I_f (i.e. ϕ), the induced emf is lower for lower speed. For lower speeds, the OCC may not intersect with the field resistance, thereby prohibiting the building up of emf in the generator.

The armature winding, the field winding and the load are all series-connected in a DC series generator. Hence, for the development of emf in DC series generator, that is for a current to flow through the field winding, it is necessary that an electrical load be connected across the generator, failing which the field current will not circulate in the field winding. The field winding of DC series generator is required to carry the armature current; hence, less number of turns of thicker wires is used to make the field winding. This results in a low-field winding resistance which leads to smaller voltage drop across the field winding. However, at the same time as the armature current is large, the magnetic field due to field winding gets saturated fast. The use of DC series generators is less.

External (or load) and internal characteristics of DC generators

When a DC generator is loaded, its terminal voltage drops down. Variation of the generator voltage with the change in load current is called the *external characteristic of DC generators.*

Separately-excited DC generator

As can be seen in Figure 14(a), the KVL around the loop containing the armature and the load gives,

$$E = V_t + I_a r_a$$

Thus with the addition of extra load (i.e. more number of parallel resistances like R_L), the current I_L (and hence I_a) increases. This results in an increased value of $I_a R_a$ drop.

According to KVL, the voltage equation for armature circuit is

$$V_t = E - I_a r_a$$

But due to the phenomenon of armature reaction (i.e. reduction in stator main field due to demagnetizing effect of current in armature winding), the terminal voltage actually obtained is

$$V_t = E - I_a r_a - \text{Drop due to armature reaction}$$

The external characteristic is plotted by gradually increasing the load current I_L, and obtaining the corresponding terminal voltage V_t, as shown by the chain line in the characteristics plot. The armature resistance drop $I_a r_a$ is also plotted in the same graph by using the same scales for current and voltage. This

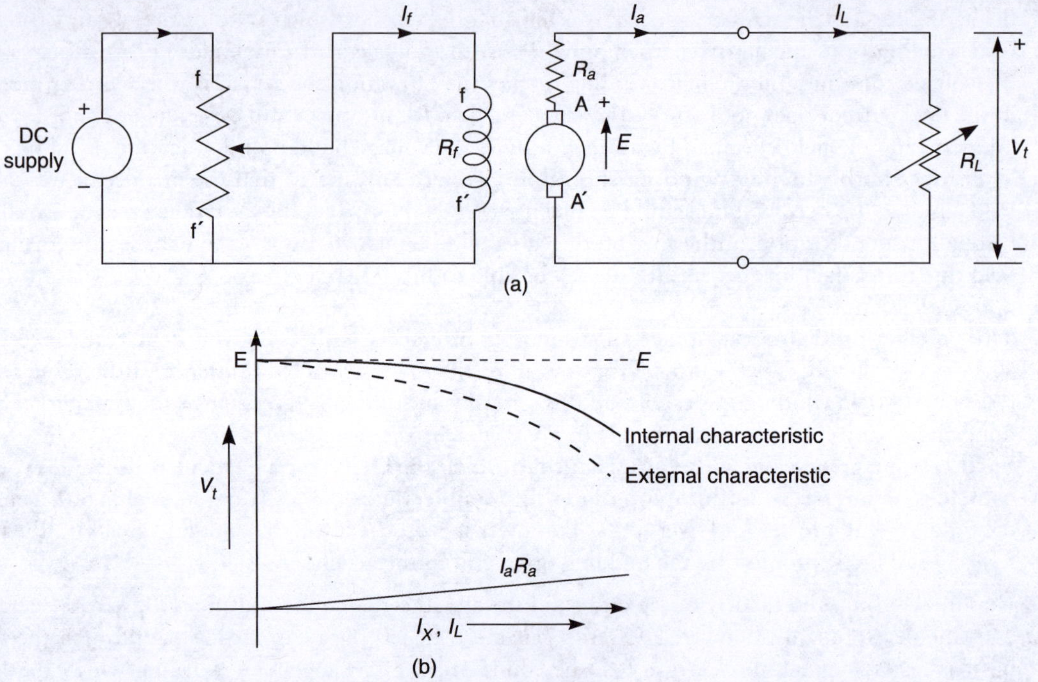

Figure 14 Separately-excited DC generators: (a) external (or load characteristic); (b) internal characteristics.

drop is added to the external characteristic to obtain the "internal characteristic" shown by the solid line in the characteristic plot. In fact, the gap between the induced EMF E (shown by dotted line) and the internal characteristic is the voltage drop due to armature reaction, which cannot be measured directly. Hence, the internal characteristic can only be obtained by using the external characteristic, as discussed.

Self-excited DC generator

In the case of self-excited DC shunt generators, as shown in Figure 15, the terminal voltage drops down rapidly with the increase in the load current I_L. This is shown by chain line in the characteristic plot.

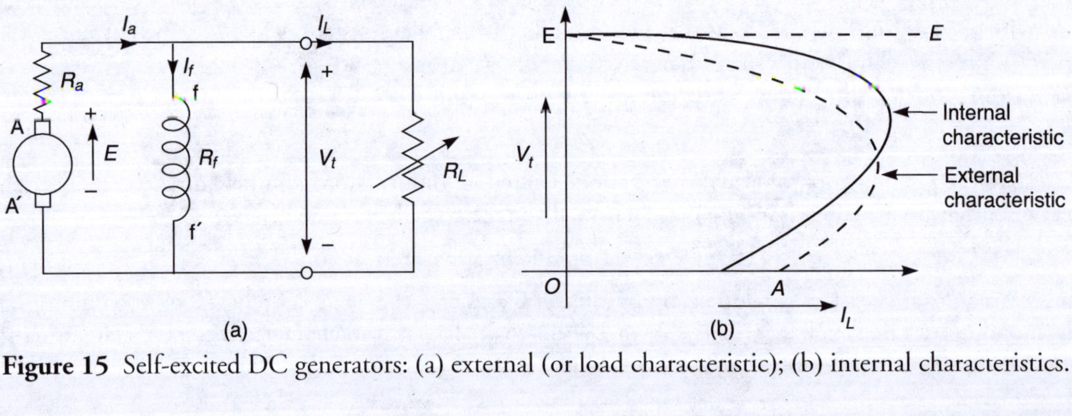

Figure 15 Self-excited DC generators: (a) external (or load characteristic); (b) internal characteristics.

Beyond certain value of I_L, if the load resistance is further reduced, it causes a drop in V_t, which results in reduction in I_f (field weakening). This leads to reduction in E, which further reduces V_t. The process goes on, and the operation becomes unstable till the point of operation reaches A, where the induced emf (E) drops down to zero. The generator stops working.

The internal and external characteristics of a DC series generator are plotted, as shown in Figure 16. Figure 16(c) shows the external characteristics of compound generators. Depending upon the extent and the nature of assistance (i.e. enhancement of flux) provided by the series field winding to the shunt field winding, three types of cumulatively compounded generators are designed: over-compounded, flat (or level) compounded and under-compounded generators. Also, the characteristic for the differentially-compounded generator is shown in which the magnetic fields due to series and shunt field windings oppose each other. In differentially-compounded generators, if the load current becomes large, the series field gets saturated, and the counter shunt field reduces the flux, thereby reducing the induced emf by the generator. This results in decrease in load current, which further leads to weakening of the field. The process continues to give a reverse slope characteristic for the differentially-compounded generator as shown in Figure 16.

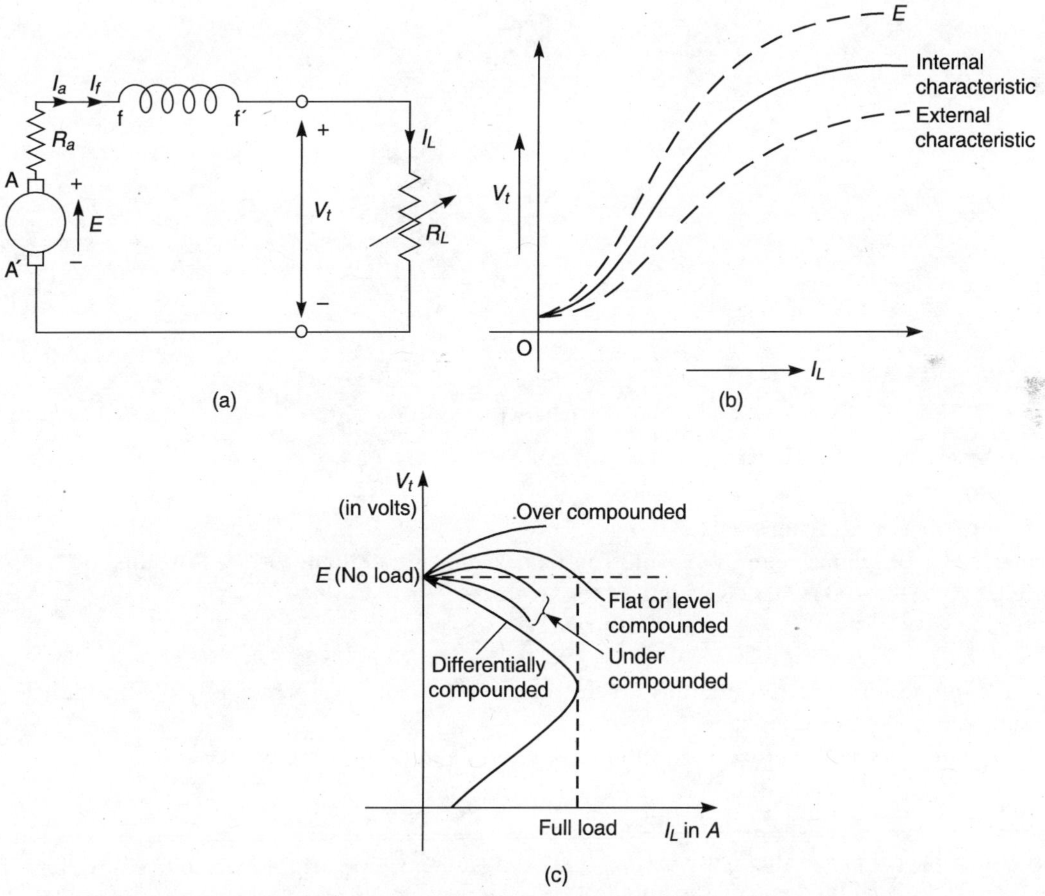

Figure 16 (a) DC series generators; (b) external and internal characteristics; (c) external characteristics of compound generators.

5.2.5 Characteristics of DC Motors

The expression for torque of a DC motor is given by

$$T = \frac{P\phi Z I_a}{2\pi a} \tag{5.4}$$

or

$$T = k_t \phi I_a, \quad \text{where the constant } k_t = \frac{PZ}{2\pi a} \tag{5.5}$$

The voltage–current relationship of a DC motor is given by

$$e_b = V_t - I_a R$$

where

$$R = \begin{cases} R_a & \text{for DC shunt motor} \\ R_a + R_f & \text{for DC series motor} \end{cases}$$

But,

$$E = \frac{P\phi NZ}{60a}$$

or

$$\frac{P\phi NZ}{60\,a} = V_t - I_a R$$

or

$$N = \frac{60\,a}{PZ}\,\frac{V_t - I_a R}{\phi}$$

or

$$N = K_n \frac{V_t - I_a R}{\phi}, \quad \text{where the constant } K_n = \frac{60\,a}{PZ} \tag{5.6}$$

DC shunt motor characteristics

In the case of a DC shunt motor, ϕ remains constant as long as the supply voltage is maintained constant. Hence Eq. (5.5) for torque of a DC shunt motor becomes

$$T \propto I_a$$

Here, the torque varies nearly linearly with the armature current. It drops slightly due to the effect of armature reaction as shown in Figure 17(a).

For the same reason of ϕ being constant in a DC shunt motor, Eq. (5.6) for speed becomes

$$N \propto V_t - I_a R_a$$

This characteristic is plotted in Figure 17(b). Here also the ideal characteristic, shown by dotted line, drops down (shown by solid line) due to armature reaction.

Superimposing the characteristics shown in Figures 17(a) and (b), the speed–torque characteristic of DC shunt motor is plotted as shown in Figure 17(c).

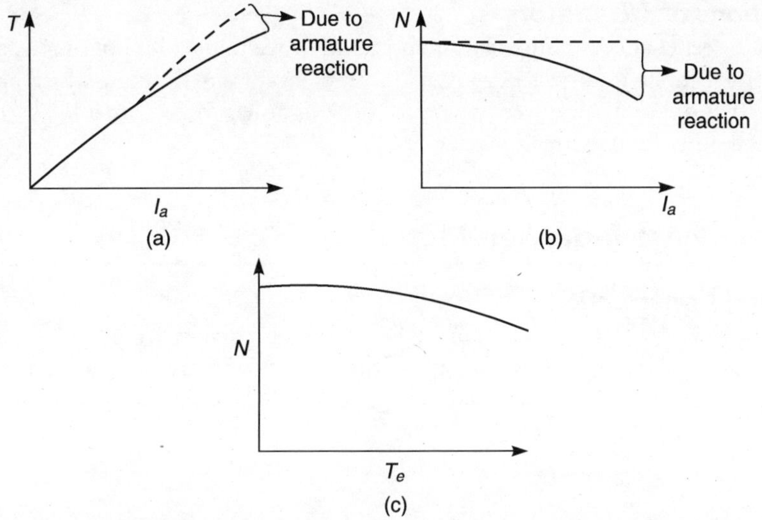

Figure 17 DC shunt motor: (a) torque–current characteristic; (b) speed–current characteristic; (c) speed–torque characteristic.

DC series motor characteristics

For DC series motors, Eqs. (5.5) and (5.6) can be rewritten with the condition that in this motor the field winding is kept in series with the armature winding. Hence $I_f = I_a$. Also, we know that $I_f \propto \phi$. Therefore, in DC series motors, $\phi \propto I_a$. Hence Eqs. (5.5) and (5.6) can be written as

$$T \propto I_a^2 \quad \text{and} \quad N \propto \frac{V_t - I_a(R_a + R_f)}{I_a}$$

It is clear from the equation for torque that the T–I_a characteristic is a parabola, as shown in Figure 18(a). In DC motors, the value of field winding resistance is kept much higher than that of the armature winding. The reason for this is that the current flowing through the field winding should be small so that the core does not get saturated. Now, since the current drawn by the armature winding is large (which has to be the same as the field winding current in DC series motors), the magnetic core gets saturated. The effect of magnetic saturation changes this characteristic from parabola to nearly a straight line, especially towards higher values of the armature current. The N–I_a characteristic is plotted in Figure 18(b). The value of R_f is sufficiently large making the voltage drop across ($R_a + R_f$) large enough to oppose the terminal voltage. Thus, the shape of the characteristic is mainly governed by the denominator in the expression of speed N. Hence the N–I_a characteristic is more or less like a rectangular hyperbola. Superimposing Figures 18(a) and (b), we get the speed–torque characteristic for DC series motor as given in Figure 18(c).

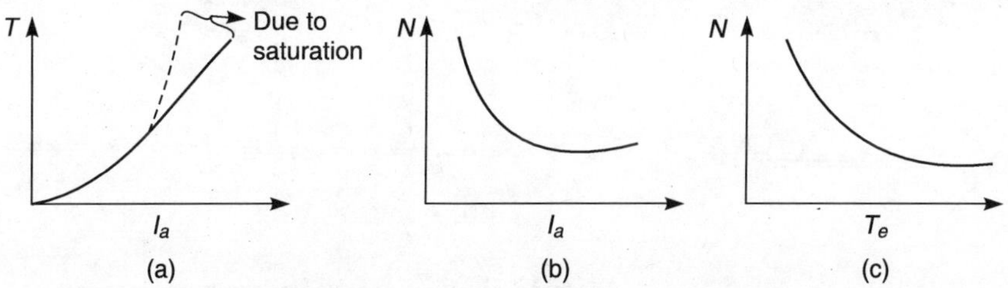

Figure 18 Characteristics of the DC series motor.

5.2.6 Applications of DC Motors

DC shunt motors are generally constant-speed motors, and hence find wide applications in the steel rolling mills and certain other similar industrial purposes. DC series motors are the ones, which posses high starting torque, and hence they are useful in places where a dead weight has to be started like electric locomotives, cranes, elevators and other (mainly traction) devices.

5.3 Three-Phase Induction Motor

5.3.1 Types and Construction

A three-phase induction motor (3-ϕ IM) consists of a stator (containing the field winding) and a rotor (containing the armature winding). The stator, unlike that of a DC motor, is a continuous winding structure (i.e. no salient poles are here). Rotor winding in this type of motor is always a closed circuit. Rotors are of two types:

1. Squirrel cage rotor
2. Wound rotor (or slip ring rotor)

The squirrel cage rotor is a solid cylindrical structure with grooves made at regular intervals throughout the length of the cylinder. In these grooves, molten conducting material is poured (which gets hardened) which takes the shape of strips that are actually called the winding of the squirrel cage rotor. Since the basic material of the rotor is not electrically conducting, these conducting strips are connected electrically by two conducing rings, which are put at the two ends of the rotor.

Thus, every conducting strip gets electrically connected to the neighboring and every other strip and we can say that "A squirrel cage rotor winding is electrically a closed circuit." The second type of rotor is the wound rotor, which has a construction similar to that of the stator (i.e. continuous windings situated in the slots).

Thus like the stator, here also three open ends emerge out, which are connected to three resistances via three slip rings, as shown in Figure 19. Hence this type of motor is also called *slip ring induction motor*. The three nonslip ring ends of the resistances are connected together, thus again ensuring that the rotor here is an electrically closed circuit too.

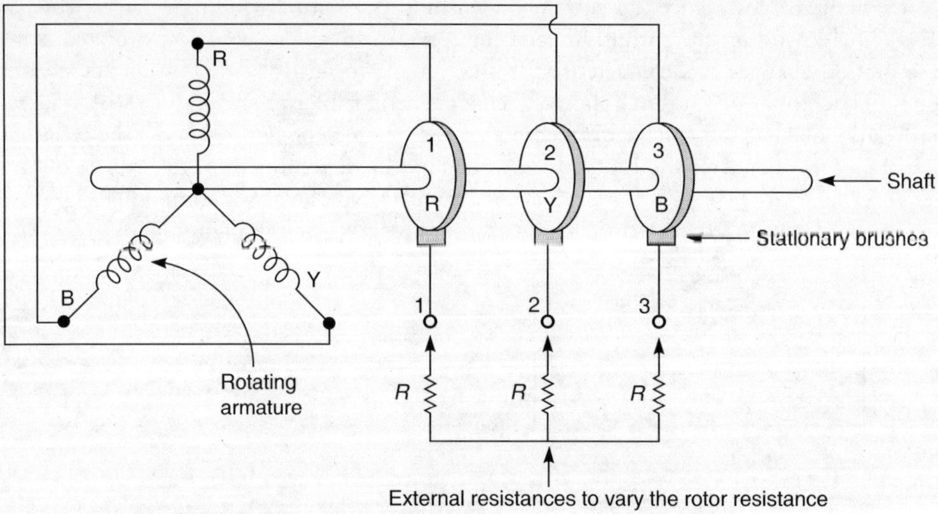

Figure 19 Layout of a wound rotor three-phase induction motor.

5.3.2 Principle of Operation

The principle of operation of a 3-ϕ IM can be understood as follows:

1. When a balanced three-phase supply is given to the stator of a 3-ϕ IM, a rotating magnetic field (RMF) gets produced in the air gap between the stator and the rotor. The speed (in rpm) of the RMF is named the *synchronous speed* and is denoted by the symbol N_s where $N_s = 120 \times f/p$ (rpm) with f the frequency of the supply voltage and P the number of poles for which the stator (and compulsorily the rotor) has been wound.
2. This RMF cuts the stationary rotor conductors. The accompanying rate of change of flux causes emf to be induced in the rotor conductors (according to Faraday's law of electromagnetic induction).
3. Since the rotors of both types of 3-ϕ IM rotors are electrically closed circuits (as mentioned earlier), the induced emf facilitates the flow of current in the rotor windings.
4. Now all the four necessary ingredients for the production of electromagnetic torque, that is number of turns (N), area (A), magnetic field (B) and current (I), are available in the rotor. Hence the torque $T_e = NABI$ is produced and the rotor starts moving.
5. Since the torque has been produced entirely by the induced voltage and current, according to Lenz's law, it opposes the cause that produces it.
6. The rotor speed (in rpm) is N_r (say). The difference between the RMF speed (N_s) and rotor (i.e. conductor) speed (N_r) is precisely the reason behind the induction of emf in the rotor conductors. This difference ($N_s - N_r$) is also called *slip speed.*
7. According to Lenz's law, the torque is produced in such a direction so as to oppose (and hence reduce) the value of ($N_s - N_r$). This results in the rotor moving in the direction of the RMF.
8. Although the rotor speed (N_r) can never ever reach N_s (due to losses), even if it hypothetically does so, the value of ($N_s - N_r$) will vanish and the induced emf will become zero. The rotor (which was getting its entire energy through induction) will stop getting energy to sustain its motion, and hence will slow down. Since all this happens almost instantaneously, we can say that a 3-ϕ IM rotor can never ever run at a synchronous speed.

Slip

We define the term *slip* as

$$s = \frac{N_s - N_r}{N_s} \text{ and Percentage slip} = 100 \times s\% \tag{5.7}$$

The slip at the time of starting (i.e. at $N_r = 0$) is one, that is 100%. This means that the emf is trying to move the rotor along with itself, but completely fails to do so (i.e. getting a hundred percentage slippage). Similarly, if (hypothetically speaking) $N_r = N_s$, slip = 0. This means that here both N_s and N_r are moving together (with no slippage in between them). Normally, 3-ϕ IM runs between 2% and 10% slip values. Thus, a four-pole, 50 Hz, 3-ϕ IM, running at 4% slip, actually runs at

$$N_r = (1-s)N_s = (1-0.04)\times\frac{120\times 50}{4} = 1440 \text{ rpm}$$

Rotor frequency

We have

$$N_s = \frac{120f}{P}$$

So,

$$f = \frac{PN_s}{120} \tag{5.8}$$

Now the frequency of the rotor-induced emf and current is

$$f_r = \frac{P(N_s - N_r)}{120}$$

The term $N_s - N_r$ has been used because emf (and hence current) gets induced in the rotor due to this factor (also called *slip speed*) only. By multiplying the numerator and denominator by N_s in above expression we get

$$f_r = \frac{P(N_s - N_r)}{N_s}\,\frac{N_s}{120} = sf \tag{5.9}$$

Hence, the rotor frequency is slip (absolute value) times the supply frequency. Thus, for a motor running at 4% slip with 50 Hz supply frequency, $f_r = 0.04 \times 50 = 2$ Hz.

5.3.3 Slip–Torque Characteristics

To derive and plot the slip–torque characteristics of a 3-ϕ IM, we need to draw the electrical equivalent circuit (on a per-phase basis) similar to that of a single-phase transformer. The primary will be stator and secondary the rotor side. The circuit is shown in Figure 20.

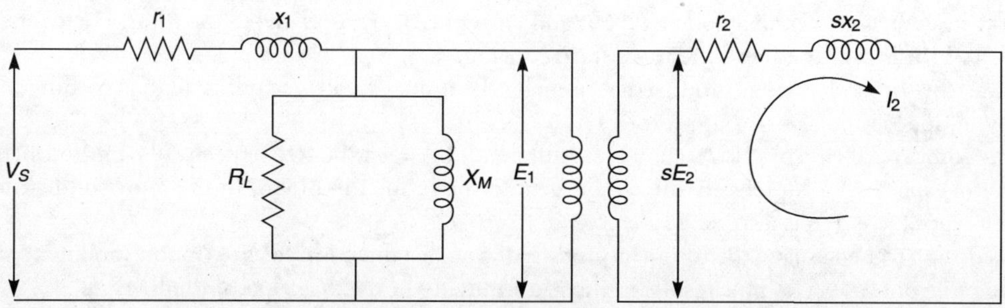

Figure 20 Equivalent circuit of a 3-ϕ IM on a per-phase basis.

As discussed before, since the rotor frequency is sf, every rotor quantity (dependent on frequency) will have the term s multiplied to it. sE_2 and sx_2 are thus the secondary-induced emf and rotor leakage reactance, respectively. r_2 remains unaffected as it does not depend on frequency. Now,

$$\text{Rotor current, } I_2 = \frac{sE_2}{\sqrt{r_2^2 + (sx_2)^2}}$$

This can be written as

$$I_2 = \frac{E_2}{\sqrt{(r_2/s)^2 + x_2^2}}$$

Now since r_2/s can be written as

$$\frac{r_2}{s} + r_2 - r_2 = r_2 + r_2\left(\frac{1}{s} - 1\right)$$

the equivalent circuit can be modified to, as shown in Figure 21.

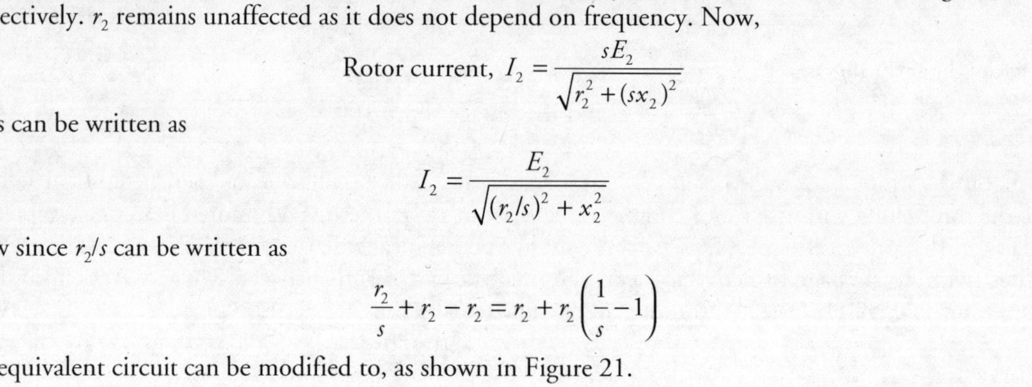

Figure 21 Equivalent circuit of a 3-ϕ IM with electrical equivalent of the mechanical load.

The circuit now looks a lot more closer to that of the transformer with the "electrical equivalent of the mechanical load of a 3-ϕ IM on a per-phase basis" being

$$r_2\left(\frac{1}{s}-1\right)$$

Initially, when the slip s is 1, the load is zero, that is a short-circuit resulting in a high starting current value. If s becomes zero (i.e. $N_r = N_s$), the load is infinity, that is $I_2 = 0$, indicating the stoppage of the energy transfer to rotor.

Power flow

Figure 22(a) shows the power flow diagram. The electrical power fed to a stator flows to the rotor through air gap. The copper loss in the stator winding and the iron loss (at supply frequency) in the stator core take place. The rest of the power is transferred through air gap to the rotor. The copper loss in the rotor winding takes place. The frequency of rotor current is very small (as $f_r = sf$; where the range of s varies from 0.01 to 0.05). Hence the iron loss in the rotor core is neglected. Cu loss in rotor = $I_2^2 r_2$.

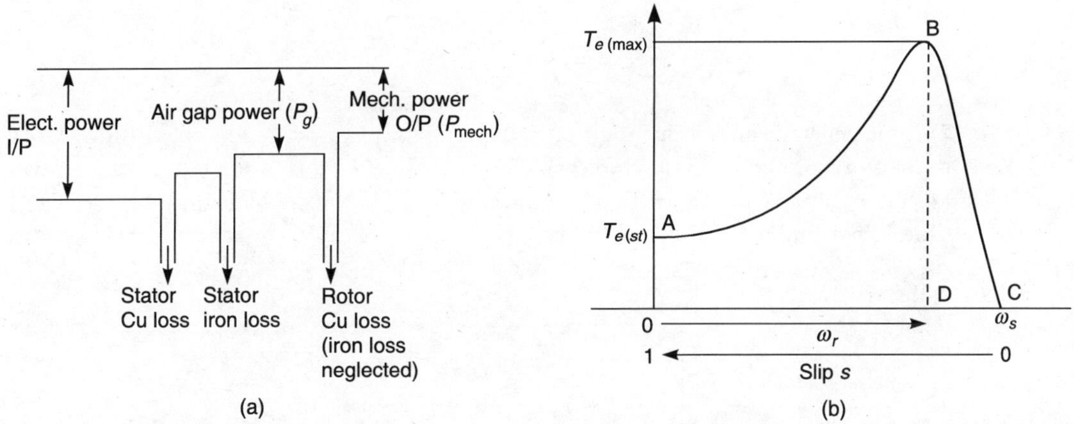

Figure 22 (a) Power flow diagram of 3-ϕ IM; (b) torque-slip (and hence torque-speed) characteristic of a 3-ϕ IM.

The power relationship in Figure 22(a) is as follows:

$$P_g : P_{mech} : \text{Rotor Cu loss} :: 1 : (1-s) : s \tag{5.10}$$

where P_g = per-phase air gap power, P_{mech} = mechanical power output = $I_2^2 r_2[(1/s) - 1]$. For the three-phase induction motor, we have

$$P_{mech} = T_e \omega_r$$

where T_e = electromagnetic torque produced and ω_r = rotor speed in rad/sec. Then,

$$T_e = \frac{P_{mech}}{\omega_r} \tag{5.11}$$

From Eq. (5.10),

$$P_{mech} = (1-s)P_g$$

Also, we know that

$$\omega_r = (1-s)\omega_s \text{ and gap power } P_g = \frac{I_2^2 r_2}{s}$$

where ω_s = synchronous speed of the RMF in red/sec. On substituting for P_{mech} and T_e in Eq. (5.11), we get

$$T_e = \frac{(1-s)P_g}{(1-s)\omega_s} = \frac{P_g}{\omega_s}$$

Now from the power relationship of Eq. (5.10), we have

$$P_g = \frac{3\phi \text{ rotor Cu loss}}{s}$$

Using the single-phase equivalent circuit of the three-phase induction motor, we get

$$P_g = \frac{3\text{-}\phi \text{ rotor Cu loss}}{s} = 3I_2^2\frac{r_2}{s} = \frac{3s^2E_2^2}{r_2^2+s^2x_2^2}\frac{r_2}{s} = \frac{3sE_2r_2}{r_2^2+s^2x_2^2}$$

So

$$T_e = \frac{3}{\omega_s}\frac{sE_2^2r^2}{r_2^2+s^2x_2^2} \tag{5.12}$$

If in Eq. (5.12) we put $s = 1$, we get the starting torque:

$$T_{(est)} = \frac{3}{\omega_s}\frac{E_2^2r_2}{r_2^2+x_2^2}$$

Equation (5.12) is the general equation for the electromagnetic torque produced by the 3-ϕ IM. Usually, $x_2 >> r_2$ (in both transformers and induction motors). Now, initially the value of slip is one. It gets down to 0.99, 0.98 and so on, at initial (starting) period. Thus, $(sx_2)^2 >> r_2^2$ in the denominator of Eq. (5.21) and hence r_2^2 can be neglected, and with r_2, E_2 and W_S being constants, we can say that at initial values of slip, the torque

$$T_e = \frac{3}{\omega_s}\frac{E_2r_2}{s^2x_2^2}$$

or

$$T_e \propto \frac{1}{s} \quad \text{(Region AB in Figure 22)}$$

This is a rectangular hyperbola ($xy = c$).

Later when the motor runs at its highest speed (with values of slip close to zero), the term $(sx_2)^2$ in the denominator becomes very small (say for 4% slip, it becomes $0.0016\,x_2^2$), and hence it can be neglected in front of r_2^2. This results in the value of torque $T_e \propto s$ [Region BC in Figure 22(b)]. Thus ABC is the torque-slip (or torque-speed) characteristic of a 3-ϕ IM. (We know that the resistance of a squirrel cage rotor cannot be varied. The resistance of a wound rotor can indeed be changed as per Figure 18.)

Now, if Eq. (5.12) is differentiated with respect to slip s and the answer equated to zero, we get the "condition of the occurrence of maximum torque":

$$\frac{dT_e}{ds} = \frac{3}{w_s}E_2^2r_2\left[\frac{(r_2^2 + s^2x_2^2) - s(2sx_2^2)}{(r_2^2+s^2x_2^2)^2}\right] = 0$$

On solving, we get $r_2 = sx_2$. That is, when the rotor resistance in ohms is equal to the rotor leakage reactance (in ohms), the maximum value of torque occurs. (Of course, the stator impedance has been neglected in this analysis.) The slip at this point is called the "slip at maximum torque (s_{mt})." Mathematically, it is given by

$$s_{mt} = \frac{r_2}{x_2}$$

Putting this value in the torque expression, Eq. (5.12), we get

$$T_{e(\max)} = \frac{3}{\omega_S}\frac{s_{mt}E_2^2 r_2}{r_2^2 + s_{mt}^2 x_2^2} = \frac{3}{\omega_S}\frac{E_2^2}{2x_2}$$

Thus, we conclude that "the maximum torque of a 3-ϕ IM is independent of the rotor resistance. But, the slip, at which it occurs, does depend on the rotor resistance. In Figure 23, the region to the left of the torque axis will have slip value greater than one. Now, slip = $(N_s - N_r)/N_s$ can be greater than one only when N_r is having an opposite polarity (or direction of rotation of N_r is against that of N_s).

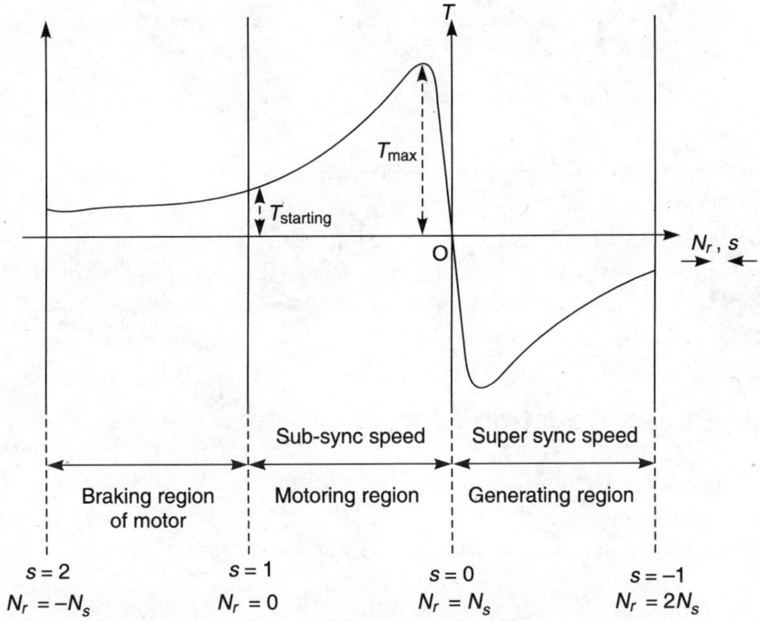

Figure 23 Showing the three regions of operation of a 3-ϕ IM.

A normal 3-ϕ IM operation does not allow this. If the motor is installed in a vehicle, which is going down a slope, it will try to rotate downward (due to gravity), and if the direction of rotation of rotating magnetic field (RMF) is against it, the rotor will disobey the principle of operation (of a 3-ϕ IM) and continue moving downward (due to gravity of the whole system). But in trying to move the rotor in its own direction (i.e. upward, in this case), the RMF will apply a sort of "braking" torque to the downward moving rotor. Thus, the region to the left of the torque axis, where the value of slip ranges from 1 to 2, is called the *braking region.*

The region to the right of the point N_s (i.e. point C) in Figure 22 is of the negative slip. The negative value of the slip means $N_s < N_r$. This can only be possible if the rotor is powered externally. This thus means *generator operation*, and the negative slip region is also called the *generating region.* In Figure 22, the actual region of operation of a 3-ϕ IM is the linear one, that is BC. Thus the operating slip is between s_{mt} and $s = 0$.

If the rotor resistance (obviously of the wound rotor motor) is now increased, the following happens:

1. The value of $s_{mt} = r_2/x_2$ increases, causing the position of occurrence of the maximum torque to shift towards left (at a higher slip, as shown in Figure 24).
2. The value of the starting torque increases.

Thus with $r_{2D} > r_{2C} > r_{2B} > r_{2A}$, we have the following T_e-slip characteristic: When the rotor resistance becomes equal to x_2, the $s_{mt} = r_{2D}/x_2 = 1$, and thus the maximum torque occurs at the starting itself.

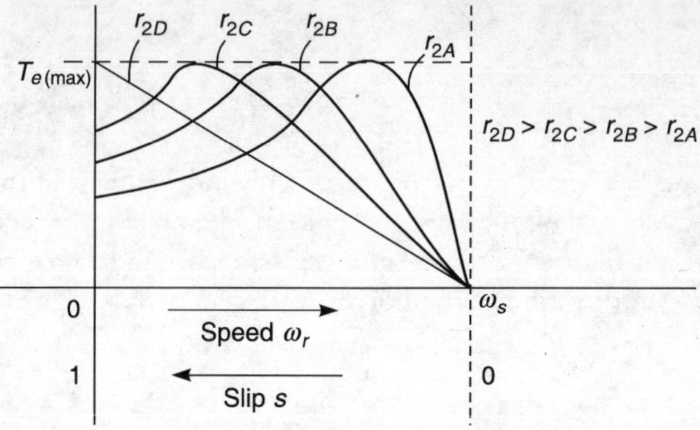

Figure 24 Variation of the torque-speed characteristic with the rotor resistance.

5.3.4 Applications of Three-Phase Induction Motor

A squirrel cage-type 3-ϕ IM is extremely rugged, and hence usually finds applications in a number of industrial applications like water pumps, mills and manufacturing. Also, a significant application of the 3-ϕ IM is in the electric trains (or locomotives), from where it is fast replacing the conventionally-used DC series motor. This is happening because of the continuous increase in the efficiency of the 3-ϕ IMs with better controlling methods coming up.

5.4 Single-Phase Induction Motors

The construction of the stator of a single-phase induction motor (1-ϕ IM) is similar to that of a 3-ϕ IM with continuous winding throughout the circular periphery (for a particular number of poles). Two terminals, one for phase and the other for neutral, are there in the stator winding. The rotors may be squirrel-cage type or single-phase wound-rotor type.

5.4.1 Principle of Operation

Single-phase supply is given to the stator winding, as shown in Figure 25. As a result of giving a single-phase alternating supply, a pulsating magnetic field gets produced in the air gap between the stator and the rotor.

1. Since the average value of a pulsating (or alternating) quantity is zero over a cycle, the average torque produced has a zero value, and hence a 1-ϕ IM is not a self-starting motor.
2. But a peculiar phenomena is observed, whenever a push is given (externally) to the rotor of 1-ϕ IM, in either direction—it builds up its speed in that direction similar to a 3-ϕ IM.

These two features of the 1-ϕ IM are well explained by the *double revolving field theory*. The theory states that when a single-phase supply is given to the stator of a 1-ϕ IM, two RMFs similar to that in the 3-ϕ IM get produced in the air gap. These two RMFs move in opposite (forward and backward) directions as shown in Figure 27. The phasor value of each RMF is equal to half of the actual flux ϕ_m in the air gap.

The five different locations of the RMF phasors are shown in Figure 26. It can be seen that the resultants follow a definite sinusoidal pattern (ϕ_m, $\phi_m/2$, 0, $-\phi_m/2$, $-\phi_m$, and so on), while the individual RMF phasors rotate in opposite directions.

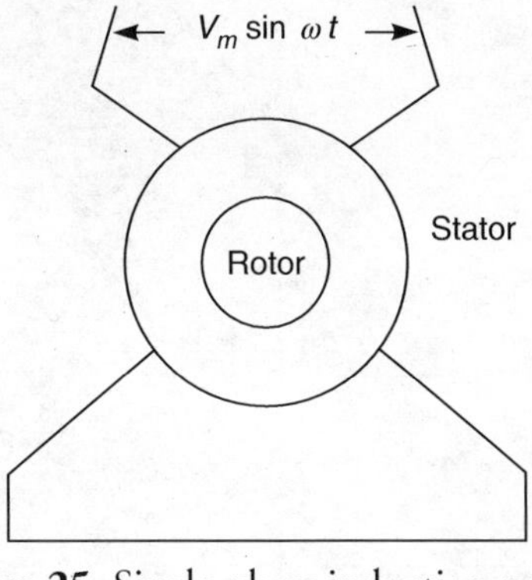

Figure 25 Single-phase induction motor.

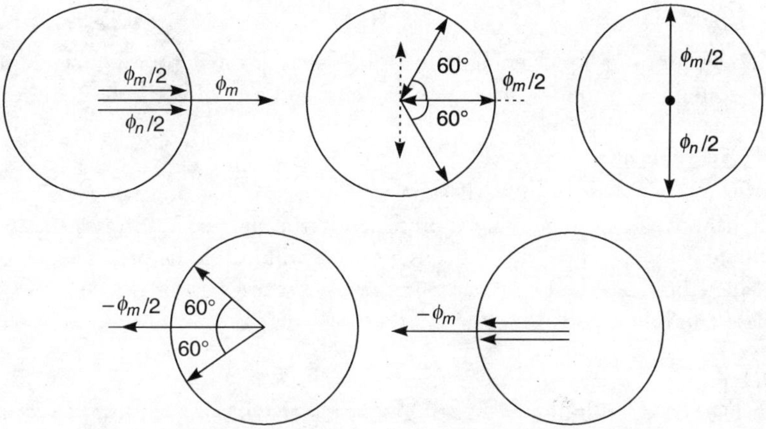

Figure 26 The phasor resultants of the two rotating magnetic fields moving in opposite directions.

Now since the actual field in the air gap (due to single-phase AC supply) is also sinusoidal, it proves that the double revolving field theory hypothesis is correct. Now, once it has been proved to be true, the hypothesis helps in explaining the two features of a 1-ϕ IM mentioned before.

Figure 27 shows the torque-slip (and hence torque-speed) characteristics of the two RMFs. One of the RMF is said to be rotating in forward direction (with torque T_{fwd}) and the other in the backward direction (with torque T_{rev}).

As can be seen at the time of starting, when the slip s is 1 (i.e. $s_F = s_B = 1$), the forward torque component (T_{fwd}) and the reverse torque component (T_{rev}) are equal in magnitude and opposite in direction. Hence the net torque at this point will be zero. This explains the nonself-starting behavior of a 1-ϕ IM. Next, if a push is given to the rotor in either direction (say, forward direction), the torque in that direction will become greater than that for the backward direction, and hence the net torque will be positive (AB > BC), as shown in Figure 27 (in that direction), and the rotor speed builds up in forward direction. The same explanation holds good for the reverse direction.

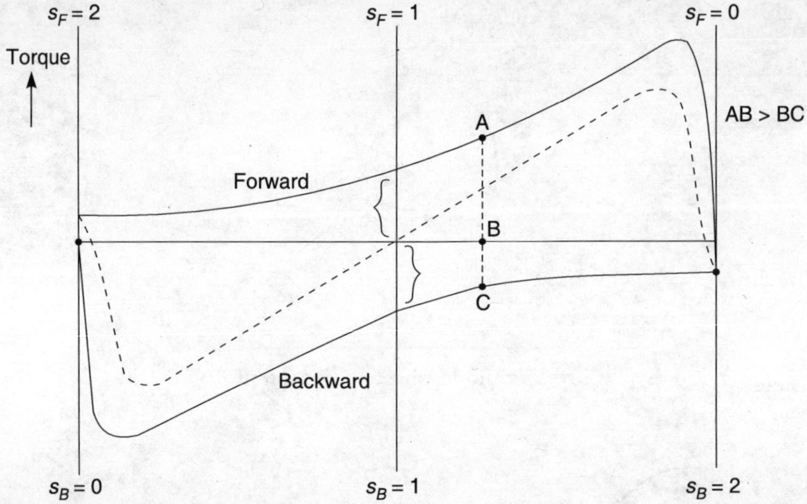

Figure 27 The torque-slip characteristic of the two oppositely rotating magnetic fields, along with the torque-slip characteristic of a 1-ϕ IM (in the chain line).

5.4.2 Methods of Starting

The main reason for a 1-ϕ IM being nonself starting is the fact that on giving a single-phase supply, a pulsating magnetic field of zero average value is produced in the air gap. If, instead, the single-phase supply is split into two different phases somehow, a starting torque will indeed be produced whose value depends on the sine of the angle between the two phase currents. There are various methods of splitting the single phase into two phases, and these constitute the methods of starting of 1-ϕ IM.

Reactance start

Here, as shown in Figure 28, an inductive starting winding (having high resistance R_{st} and reactance X_{st}) is connected in parallel to the stator's main winding (having low resistance R_m and reactance X_m). Thus the angle $\tan^{-1}(X_{st}/R_{st})$ of the starting winding is deliberately kept lower than that of the main winding, as shown in the phasor diagram of Figure 28. In this figure, R_{st} and X_{st} are the starting winding resistance and reactance; R_m and X_m are the main (stator) winding resistance and reactance; and CS is the centrifugal switch.

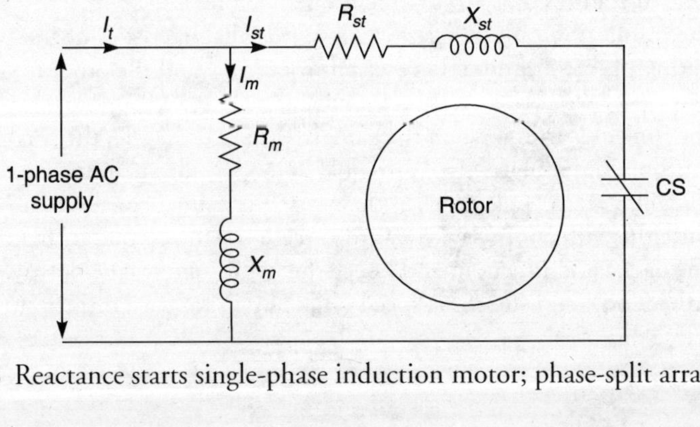

Figure 28 Reactance starts single-phase induction motor; phase-split arrangement.

We also have

$$\phi_{st}\left(=\tan^{-1}\frac{X_{st}}{R_{st}}\right)<\phi_m\left(=\tan^{-1}\frac{X_m}{R_m}\right)$$

The angle ϕ between I_{st} and I_m phasors equals $\phi_m - \phi_{st}$. The starting torque $T_{est} \propto \sin\phi$ gets created because now a RMF is produced (based on the fact that an RMF gets produced for two or more phase supply), and thus the principle of operation of a 3-ϕ IM gets applied for the forward and backward fields.

As the motor attains around 70% of its rated speed, the starting winding is no more required in the circuit. If it is allowed to remain in the circuit, it accounts for the real power loss, thus reducing the motor efficiency. Hence, the centrifugal switch is set to open to disconnect the starting windings. Figure 29 shows the phasor relationship between the applied voltage phasor, starting winding current phasor and the main stator winding current phasor.

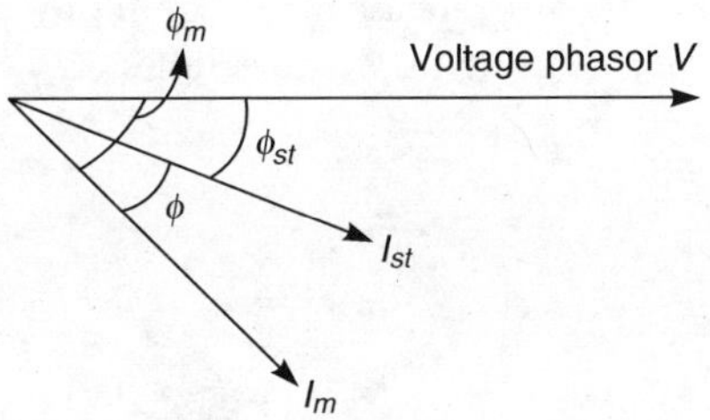

Figure 29 Phasor diagram showing the phase splitting in the reactance 1-ϕ IM.

Figure 30 shows the percentage full-load torque versus speed characteristic for the reactance start motor.

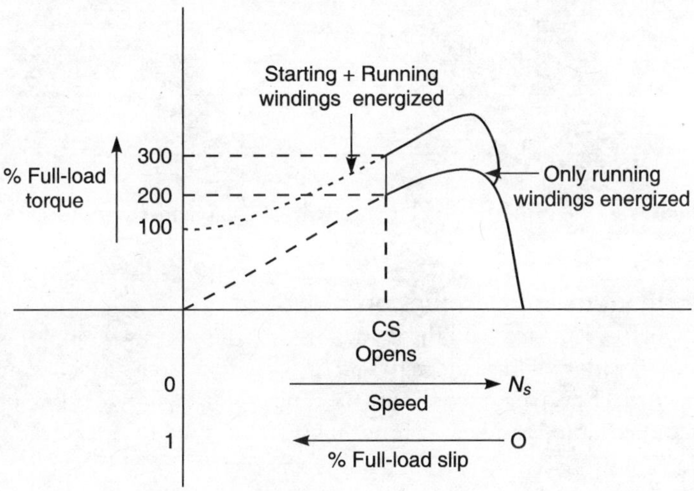

Figure 30 Reactance start 1-ϕ IM; % full-load torque versus speed characteristic.

Split-phase, capacitor start motor

Since the starting torque depends upon the sine of the angle ϕ between the two winding currents (I_{st} and I_m), its maximum value can be achieved if the angle ϕ is 90°. To make $\phi = 90°$, a capacitor is connected in the starting winding, as shown in Figure 31(a). The corresponding phasor diagram is shown in Figure 31(b).

In the reactance start motor, the angle between I_m and I_{st} is approximately 25–30°, whereas in the capacitor start motor, the angle is close to 90°. Hence the starting torque here is around sin 90°/sin 25° ~ 2.4 times more. The torque-slip characteristic shown in Figure 32 reveals the same.

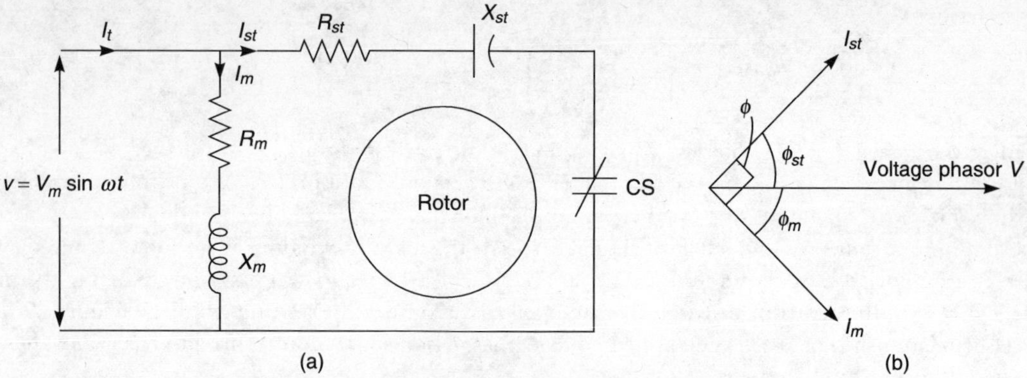

Figure 31 (a) Capacitance start single-phase induction motor; (b) phasor diagram showing the phase splitting in the capacitor-start 1-ϕ IM.

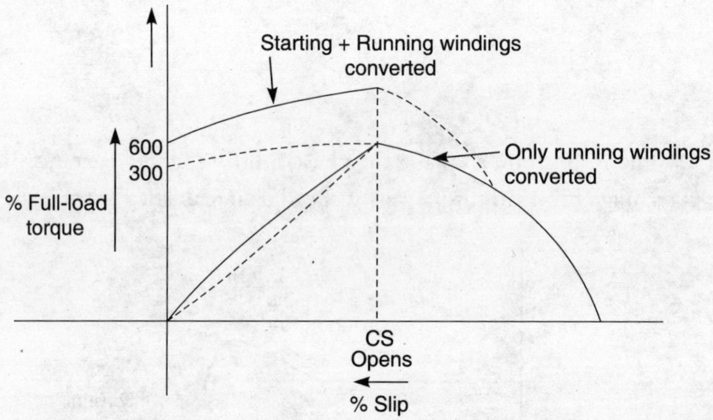

Figure 32 Torque-speed characteristic of a capacitor start 1-ϕ IM.

Capacitor run (permanent split) induction motor

Here the centrifugal switch is removed, and the capacitor (and the starting winding) remains in the circuit throughout the motor operation. Thus, the capacitor used here is designed for continuous duty and is preferably an oil-filled type. The capacitor current at the time of starting is rather low and the starting torque is also lower than the earlier motors studied. A reversing switch is used to reverse the direction of rotation, if required (see Figure 33).

Applications of the single-phase induction motor

As can be thought of, 1-ϕ IMs find applications in a lot of places which primarily require low powers. Ceiling and table fans, small (fractional horse power) water pumps, refrigerators, blowers, etc. are some low-power applications which make use of the 1-ϕ IM. The ceiling fans usually make use of the capacitor-start and run-type of motor configuration. This is the reason why most of the times the failure of a ceiling fan is related with the condenser getting damaged and needs to be changed.

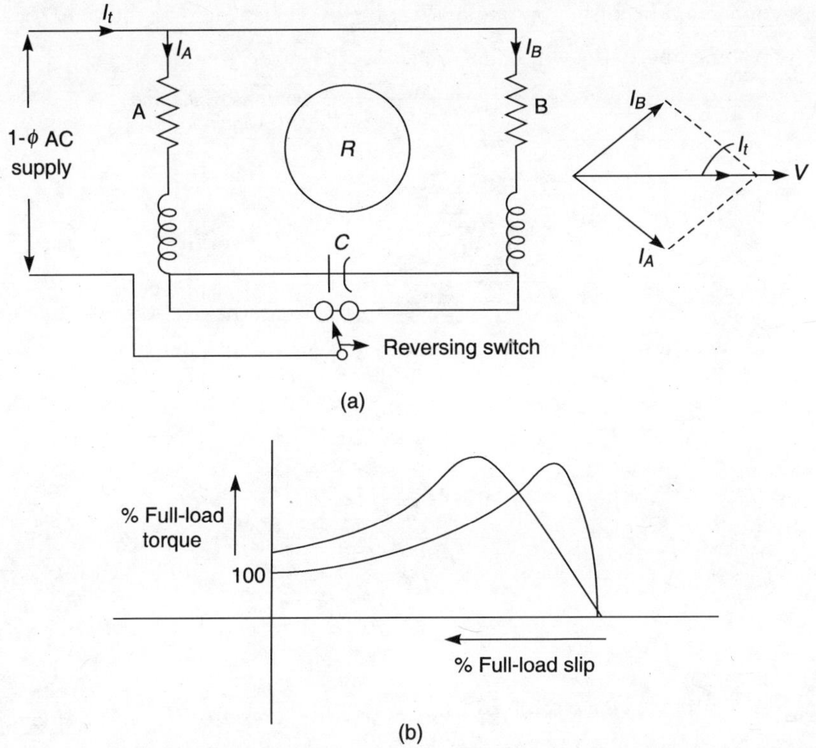

Figure 33 1-ϕ IM: (a) Capacitor-run type; (b) torque-slip characteristic of capacitor-run type.

5.5 Three-Phase Synchronous Machines

The name of *synchronous machine* is based on the fact that the rotor of this machine rotates at a synchronous speed (N_s rpm), where

$$N_s = \frac{120 f_s}{P}$$

Here f_s is frequency of armature current in Hz and P is the number of stator poles. As mentioned before, synchronous machines are of two types: *cylindrical rotor* and *salient pole rotor*. If the prime mover is a high-speed turbine (such as steam turbine used in thermal and nuclear power plants), cylindrical rotor with two poles is used. However, if the prime mover is a low-speed turbine (such as water turbine used in hydroelectric plants), salient pole rotor with larger number of poles is used.

The stator houses a three-phase AC armature winding, while a DC field winding is placed on the rotor. This arrangement of having the winding is reverse to that in the DC machines. The armature winding is placed on the stator because it operates at high voltages (as high as 11 kV). It would be difficult to place the armature winding on the rotor, because

1. High voltage requires larger insulation, which would make the rotor bulky, causing a reduction in the torque/weight ratio.

2. Taking out connection from the rotating body at high voltage would be difficult.

The synchronous machine houses the windings, as shown in Figures 34 and 35.

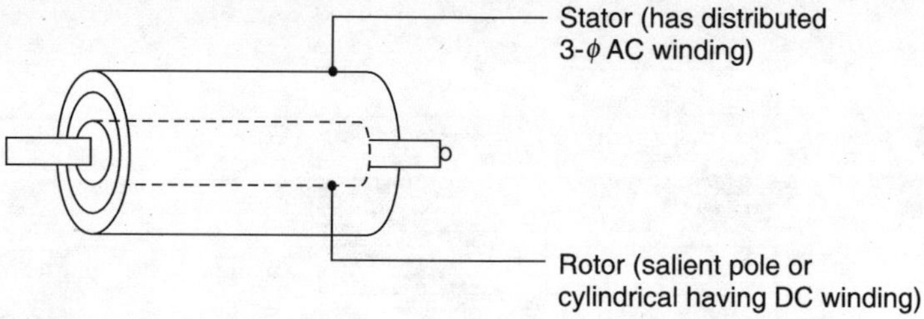

Figure 34 Placement of windings in synchronous machines.

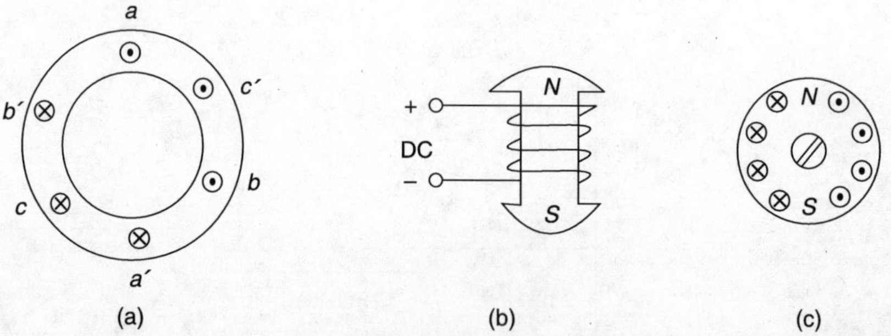

Figure 35 (a) Three coils of the three-phase winding on stator; (b) salient pole rotor (used for low-speed machines); (c) cylindrical rotor (used for higher-speed machines).

The AC armature winding is placed on the inner periphery of the stator as shown in Figure 35(a). The rotor carries DC field winding. The rotor design depends on speed on which the machine is required to run. The salient pole rotor is shown in Figure 35(b) in which concentric winding makes electromagnetic poles. For higher speed machines, the rotor is made cylindrical as shown in Figure 35(c). The distributed field winding is placed over the outer periphery of the rotor. The sense of the winding is such that it produces electromagnetic poles on the rotor.

5.5.1 *Principle of Operation of Alternator*

The role of alternator, that is the synchronous generator, is to convert mechanical input to three-phase electrical output. The DC field winding, placed on the rotor, develops fixed electromagnetic poles in it. The construction of both machine types is such that when DC is given to the rotor, a sinusoidal field is created in the air gap between the stator and the rotor. This is achieved by the typical shape of the pole shoes as shown in Figure 36(a). The rotor is rotated at synchronous speed by a mechanically coupled prime mover (turbine/engine), and when the rotor cuts the sinusoidal field, an emf is induced in the 3-ϕ armature windings. The distribution of 3-ϕ AC winding on the stator and the shapes of the rotor poles are such that the emf induced in each armature coil is sinusoidal. Figure 36(b) shows developed diagram of the machine. The flux distribution in the air gap due to the rotor electromagnets (shown by "N" and "S" poles) is nearly sinusoidal as shown in Figure 36(b). Maximum flux is cut by the armature conductor facing the "N" or the "S" poles, thereby generating maximum currents (in opposite directions). The adjoining armature conductors cut less flux (due to sinusoidal distribution of flux) and generate less currents. Figure 36(a) shows single coils in phase the three phases of armature winding placed 120° apart from each other. They are a-a′, b-b′ and c-c′.

Figure 36(b) shows the three situations when the rotor poles ("N" and "S") shift in space and occupy next consecutive position. This is how with rotation of rotor, the sinusoidal currents get induced in the three phases of the armature windings. A closer look at any one of the armature conductors (say, "a") in the three figures of Figure 36 clearly illustrates this.

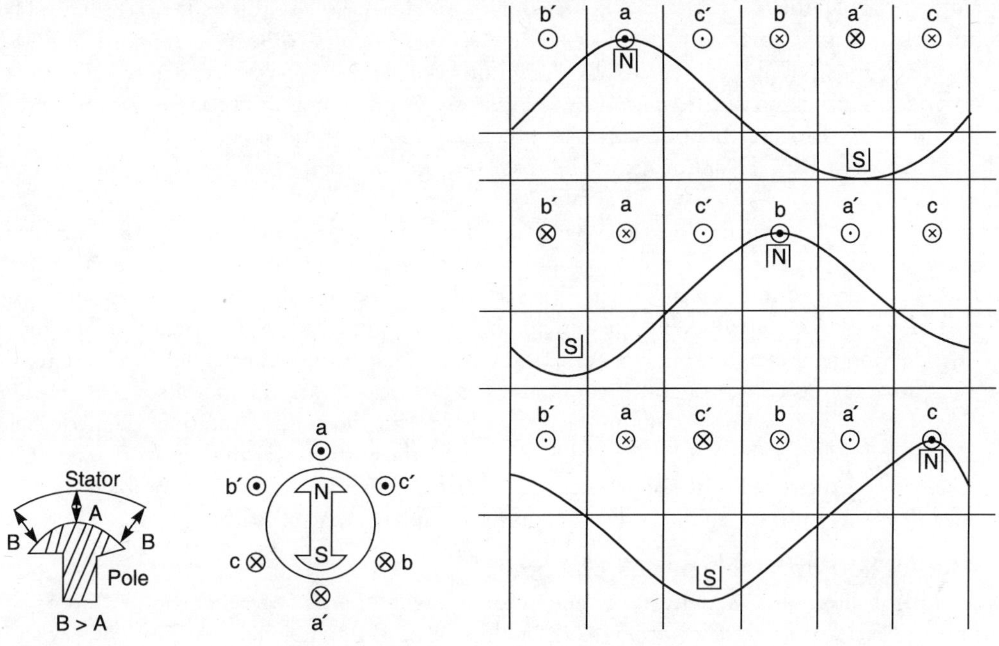

Figure 36 Generation of AC current in armature winding due to rotation of rotor.

5.5.2 Principle of Operation of Synchronous Motor

The 3-ϕ AC armature winding, placed at stator, when energized by a 3-ϕ AC supply, develops a rotating magnetic field in the air gap. This rotating magnetic field is such as if P number of fictitious poles rotate in the air gap at synchronous speed (see Figure 37). The rotor contains DC field winding, which creates electromagnetic poles on the rotor. The rotor poles, when aligned and locked with the stator fictitious poles, make the rotor rotate in the direction of the stator's rotating magnetic field.

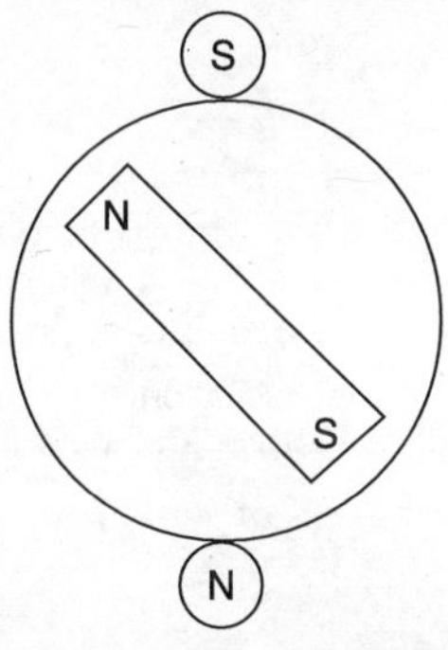

Figure 37 The interaction of the fictitious stator poles with the fixed polarity rotor poles in a synchronous motor.

The synchronous motors are inherently *nonself-starting* motors. At the time of starting, when the rotor is stationary, the fictitious poles due to stator field change their polarity f times per second (where f is the frequency of the input AC supply to the armature). Thus, with rapidly changing polarities (100 times for a 50 Hz supply), the stator poles seem to be sweeping over the rotor poles. When a stator's S pole passes over the rotor's N pole, the rotor tries to align itself with it. The stationary rotor's inertia is high. Hence it takes some time to begin movement. By this time, the stator's N poles come over the same N pole of the rotor. Now this rotor's N pole gets repelled by the stator's N pole. In a very short span of time, the rotor's N pole experiences attraction and repulsion by the stator's S and N poles, respectively. Thus, alternately the N and S poles of the stator field quickly pass over the stationary rotor poles, and the rotor fails to get aligned and locks itself with the stator field. The rotor thus fails to rotate, making the synchronous motor nonself-starting.

In order to make it self-starting, a special winding called the *damper winding* is put in the rotor (along with the already existing DC winding). This special winding is made up of copper bars that are short circuited at both ends (and thus are closed circuit electrically). Hence it helps start the motor like a 3-ϕ IM. The rotor field winding is initially kept open circuited. The supply is fed to the armature winding, which results in the development of a rotating magnetic field in the air gap. On the principle of induction motor, a current starts flowing in the short-circuited copper bars at the rotor. Thus, the motor starts as an induction motor and the rotor poles soon get rotating at a speed close to the stator's rotating magnetic field (i.e. the synchronous speed). At this point, if the DC supply to the rotor field winding is switched ON, the rotor quickly catches the speed of the stator field and gets locked with it. Once the rotor starts rotating at synchronous speed, it continues to do so throughout for complete range of loads.

5.5.3 Application of Synchronous Machines

The operation of alternator (synchronous generator), as well as that of synchronous motor, greatly depends on the excitation (i.e. the supply to the field winding) of the machine. The power factor and the magnitude of the armature current depend on the level of excitation of the field winding. The excitation, at which the armature current is at unity power factor, is called *normal excitation.* Excitation below the level excitation is called *under excitation*, and above that is called normal excitation. The variation of armature current (i.e. the load current) with variation in excitation is represented by V curves of the machine, as is shown in Figure 38.

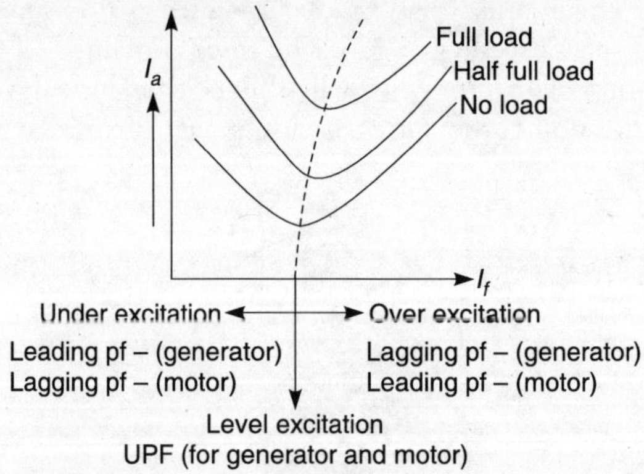

Figure 38 V curves of synchronous machine.

The V curves shown in Figure 38 give the following important information:

1. At any load (ranging from no load to full load), the armature current drawn is minimum at normal excitation.

2. The armature current increases with the increase in excitation (i.e. over excitation) or decrease in excitation (i.e. under excitation).

Application of alternators

Alternators are invariably used by all the conventional power plants for generation of electricity. The excitation (refer to V curves) of alternators is adjusted in order to meet the reactive power requirement of electrical load on them.

Application of synchronous motors

Synchronous motors are the constant-speed motors. The following are the applications of synchronous motors:

1. Motors having speeds below 500 rpm are used in electroplating, reciprocating compressors, DC generators, centrifugal pumps, rolling mills, paper mills, etc.
2. Motors having speeds above 500 rpm are used in fans, blowers, frequency changers, etc.
3. Over-excited synchronous motors, running at no load, draw leading power factor current (refer to V curves). Hence they are called *synchronous condensers*. These are used at electrical substations for improving the power factor of power system.

Solved Examples

Example 1

The armature winding of a DC separately-excited machine has resistance of 0.15 Ω, and is connected to a 230 V supply. Calculate the emf across armature terminals, if the machine is running as a: (a) generator supplying 75 A and (b) motor drawing 65 A.

Solution

The machine is separately-excited and nothing is given about its field winding. Hence it is assumed that the field current, and hence the flux, is maintained constant. Given: $R_a = 0.15\ \Omega$, $V_t = 230$ (terminal voltage). In the case of (a), by applying KVL [see Figure 39(a)], we get

$$E_g = V_t + I_a R_a = 230 + 75 \times 0.15 = 241.25\ \text{V}$$

In the case of (b), by applying KVL, we get

$$E_b\ (\text{back-emf}) = V_t - I_a R_a = 230 - 65 \times 0.15 = 220.25\ \text{V}$$

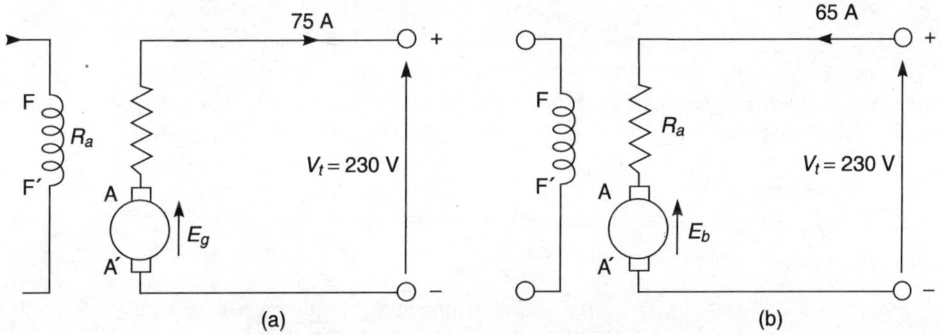

Figure 39 Figure for Example 1.

Example 2

An eight-pole lap wound DC generator has 450 armature turns. It operates at 0.02 Wb flux per pole and runs at 1000 rpm at no load. Find the emf induced by it.

Solution

Given $P=8$, $a=8$ (for lap winding, $a=P$), $Z=450\times 2$ (every turn has 2 conductors) $=900$, $\phi=0.02$ Wb/pole and $N=1000$ rpm. The emf induced is

$$E_g = \frac{P\phi NZ}{60a} = \frac{8\times 0.02\times 1000\times 900}{60\times 8} = 300\text{ V}$$

Example 3

A six-pole lap wound DC generator has 720 conductors. A flux of 80 m Wb/pole is driven at 1000 rpm. Find the generated emf.

Solution

Given number of poles, $P=6$; number of parallel paths, $a=6$ (as in lap winding, $a=p$); flux, $\phi=80\times 10^{-3}$ Wb; speed, $N=1000$ rpm; number of conductors, $Z=720$. The expression of emf is given by

$$E = \frac{P\phi NZ}{60a}$$

Substituting the values, we get

$$E = \frac{6\times 80\times 10^{-3}\times 1000\times 720}{60\times 6}$$

Therefore, $E=960$ V.

Example 4

A 6-pole wave wound DC generator has 720 conductors. A flux of 80 mWb/pole is driven at 1000 rpm. Find the generated emf.

Solution

Given number of poles, $P=6$; number of parallel paths, $a=2$ (as in wave winding, a is always 2); flux, $\phi=80\times 10^{-3}$ Wb; speed, $N=1000$ rpm; number of conductors, $Z=720$. The expression of emf is given by

$$E = \frac{P\phi NZ}{60a}$$

Substituting the values, we get

$$E = \frac{6\times 80\times 10^{-3}\times 1000\times 720}{60\times 2} = 2880\text{ V}$$

Example 5

A separately excited DC generator develops an induced emf of 220 V, while running at 900 rpm and with a flux of 0.05 Wb/pole. What would be the required speed if the emf of 200 V is to be generated with a flux of 0.045 Wb?

Solution

Given $E_{g1} = 220$ V, $N_1 = 900$ rpm, $\phi_1 = 0.05$ Wb, $E_{g2} = 200$ V, $\phi_2 = 0.045$ Wb. We have to find N_2. We know that

$$E_g = \frac{P\phi NZ}{60a}$$

Since P, Z and a are constants, we have

$$\frac{E_{g1}}{E_{g2}} = \frac{\phi_1 N_1}{\phi_2 N_2} \quad \text{or} \quad N_2 = \frac{\phi_1 N_1}{\phi_2}\frac{E_{g2}}{E_{g1}}$$

Substituting the values, we get

$$N_2 = \frac{0.05 \times 900}{0.045}\frac{200}{220} = 909\,\text{rpm}$$

Example 6

An eight-pole wave wound 250 V DC shunt generator has armature resistance of 0.2 Ω and field resistance of 300 Ω. It supplies to 5 kW heater and 3 kW lighting loads, while running at 1000 rpm. Find the flux per pole. Allow 1 V drop per brush and neglect armature reaction. The generator has 200 conductors.

Solution

Given load = 5 + 3 = 8 kW, $V = 250$ V, $P = 8$, $a = 2$ (wave winding has $a = 2$), $R_a = 0.2\ \Omega$, $R_f = 300\ \Omega$, $N = 1000$ rpm, brush drop = 2 V (1 V per brush) and $Z = 200$. From the given load and voltage, the load current can be obtained as

$$I_L = \frac{8 \times 1000}{250} = 32\,\text{A} \quad \text{and} \quad I_f = \frac{V_t}{R_f} = \frac{250}{300} = 0.83\,\text{A}$$

Hence from Figure 40, applying KCL it can be seen that

$$I_a = I_L + I_f = 32.83\ \text{A}$$

and by KVL we get

$$E_g = V_t + I_a R_a = 250 + 32.83 \times 0.2 = 256.566\ \text{V}$$

We know that

$$E_g = \frac{P\phi NZ}{60a}$$

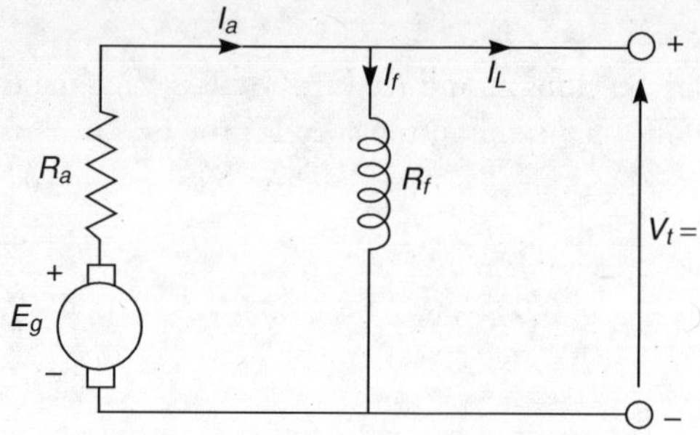

Figure 40 Figure for Example 6.

Substituting the values, we get

$$256.566 = \frac{8 \times \phi \times 1000 \times 200}{60 \times 2}$$

or

$$\phi = \frac{256.566 \times 60 \times 2}{8 \times 1000 \times 200} = 0.0192 \text{ Wb per pole}$$

Example 7

An eight-pole DC generator has 500 armature conductors and a useful flux of 0.05 Wb. What will be the emf generated if it is lap-connected and runs at 1200 rpm? What must be the speed at which it is to be driven to produce the same emf, if it is wave wound?

Solution

Given $P = 4$, $Z = 500$, $\phi = 0.05$ Wb. For lap wound armature, $a = p = 4$. For wave wound armature, $a = 2$. $N_{\text{lap}} = 1200$ rpm, $N_{\text{wave}} = ?$ Induced emf is given by

$$E_b = \frac{P\phi NZ}{60a}$$

For lap winding,

$$E_{b(\text{lap})} = \frac{4 \times 0.05 \times 1200 \times 500}{60 \times 4} = 500 \text{ V}$$

Given that

$$E_{b(\text{wave})} = E_{b(\text{lap})} = 500 \text{ V}$$

Hence

$$500 = \frac{4 \times .05 \times N_{\text{wave}} \times 500}{60 \times 2} \quad \text{or} \quad N_{\text{wave}} = \frac{500 \times 60 \times 2}{4 \times 0.05 \times 500} = 600 \text{ rpm}$$

Thus, for the same induced emf, the speed will be 600 rpm with wave winding.

Example 8

A four-pole DC shunt motor working on 220 V DC supply takes a line current of 3 A at no-load, while running at 1500 rpm (see Figure 41). Determine the speed when the motor takes a line current of 50 A. Assume armature and field resistances as 0.2 Ω and 400 Ω, respectively. **(UPTU December 2003)**

Solution

Given $V_t = 220$ V, $I_{L_1} = 3$ A, $N_1 = 1500$ rpm, $I_{L_2} = 50$ A, $R_a = 0.2\ \Omega$, $R_f = 400\ \Omega$. Then $N_2 = ?$. Here

$$I_f = I_{f_1} = I_{f_2} = \frac{220}{400} = 0.55 \text{ A}$$

$$I_{a1} = I_{L1} - I_f(\text{at no-load}) = 3 - 0.55 = 2.45 \text{ A}$$

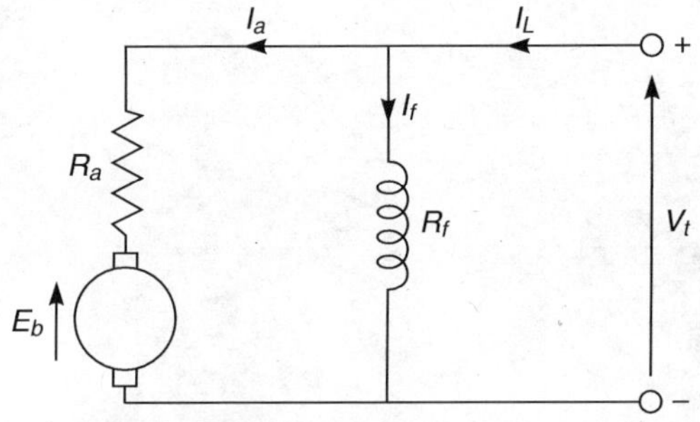

Figure 41 Figure for Example 8.

Similarly,

$$I_{a2} = I_{L2} - I_f = 50 - 0.55 = 49.45 \text{ A}$$

Also,

$$E_1 = V_t - I_{a1}R_a = 220 - (2.45 \times 0.2) = 219.51 \text{ V}$$

and

$$E_2 = V_t - I_{a2}R_a = 220 - (49.45 \times 0.2) = 210.11 \text{ V}$$

Since $E \propto N\phi$ and ϕ is constant (because I_f is constant), we have

$$\frac{E_2}{E_1} = \frac{N_2}{N_1} \quad \text{or} \quad N_2 = N_1\frac{E_2}{E_1}$$

Hence

$$N_2 = 1500 \times \frac{210.11}{219.51} = 1435.77 \text{ rpm}$$

The new speed of motor is 1435.77 rpm.

Example 9

The combined armature and field winding resistance of a DC series motor is 0.5 Ω. When supplied by a 230 V source, it draws a current of 100 A and runs at 960 rpm. Find the speed, if the motor draws a current of 150 A.

Solution

Given $R_a + R_f = R = 0.5\ \Omega$, $V_t = 230$ V, $I_{a1} = 100$ A, $I_{a2} = 150$ A, $N_1 = 960$ rpm. We have to find N_2. Refer to Figure 42. We know that, $E_g \propto N\phi$. Here, $\phi \propto I_a$. Then,

$$E_g \propto NI_a \tag{1}$$

Also,

$$E_g = V_t - I_aR \tag{2}$$

Then

$$\frac{N_1 I_{a1}}{N_2 I_{a2}} = \frac{V_t - I_{a1}R}{V_t - I_{a2}R}$$

or

$$N_2 = \frac{V_t - I_{a2}R}{V_t - I_{a1}R}\frac{I_{a1}}{I_{a2}}N_1 = \frac{230 - 150 \times 0.5}{230 - 100 \times 0.5} \times \frac{100}{150} \times 960 = 551 \text{ rpm}$$

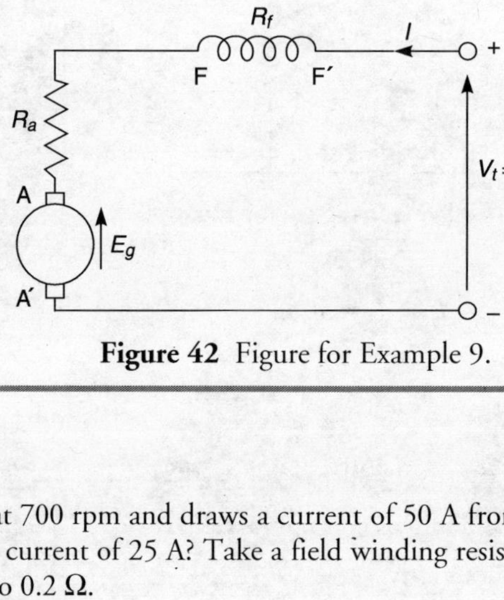

Figure 42 Figure for Example 9.

Example 10

A DC shunt motor runs at 700 rpm and draws a current of 50 A from the 230 V supply. At what speed would it run, if it draws a current of 25 A? Take a field winding resistance equal to 150 Ω and armature winding resistance equal to 0.2 Ω.

Solution

Given $R_a = 0.2\ \Omega$, $R_f = 150\ \Omega$, $I_{L1} = 50$ A, $N_1 = 700$ rpm, $I_{L2} = 25$ A. To find N_2, refer to Figure 43.

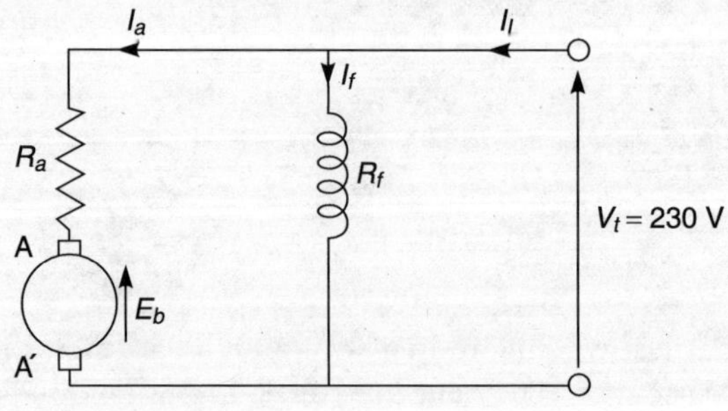

Figure 43 Figure for Example 10.

Applying KCL in the circuit, we get

$$I_L = I_a + I_f \quad (1)$$

Applying KVL in the circuit, we get

$$V_t = I_f R_f = E_b + I_a R_a \quad (2)$$

In the case of Eq. (1), $I_{L1} = 50$ A, $N_1 = 700$ rpm. Then,

$$I_{f1} = \frac{V_t}{R_f} = \frac{230}{150} = 1.53\text{ A}$$

Therefore,

$$I_{a1} = I_{L1} - I_{f1} = 50 - 1.53 = 48.467\text{ A}$$

Hence,

$$E_{b1} = V_f - I_{a1}R_a = 230 - (48.467 \times 0.2) = 220.3\text{ V}$$

In Eq. (2), $I_{L2} = 25$A, $N_2 = ?$ Hence,

$$I_{f2} = I_{f1} = 1.53$$

as there is no change in the field circuit. Then

$$I_{a2} = I_{L2} - I_{f2} = 25 - 1.53 = 23.47\text{ A}$$

Hence

$$E_{b2} = V_t - I_{a2}R_a = 230 - (23.47 \times 0.2) = 225.3\text{ V}$$

Also, we know that

$$E_b = \frac{P\phi NZ}{60a} = K\phi N \quad \left[\text{where } K = \frac{PZ}{60a}\right]$$

Therefore,

$$\frac{E_{b1}}{E_{b2}} = \frac{N_1}{N_2} \quad [\text{as } \phi\text{, i.e. } I_f\text{, is also same}]$$

Substituting the values from Eqs. (1) and (2), we get

$$N_2 = \frac{E_{b2}}{E_{b1}} N_1 = \frac{225.3}{220.3} \times 700 = 716\text{ rpm}$$

Example 11

A 230 V DC shunt motor has the armature resistance of 0.2 Ω. It draws armature current of 50 A while running at 700 rpm. Find the new speed of the motor, if it is running at a constant torque and its field is reduced by 20%.

Solution

Given $V_t = 230$ V, $R_a = 0.2\ \Omega$, $N_1 = 700$ rpm, $I_a = 50$ A. Refer to Figure 44. Since,

$$E_b = \frac{P\phi NZ}{60a}$$

Therefore,

$$E_b = K\phi N \tag{1}$$

or

$$\frac{E_{b1}}{E_{b2}} = \frac{\phi_1 N_1}{\phi_2 N_2} \quad \text{and} \quad E_b = V_t - I_a R_a \tag{2}$$

Also,

$$T = \frac{P\phi Z I_a}{2\pi a} \quad \text{or} \quad T \propto \phi I_a$$

For constant torque,

$$\phi_1 I_{a1} = \phi_2 I_{a2}$$

or

$$I_{a2} = \frac{\phi_1}{\phi_2} I_{a1} = \frac{\phi_1}{0.8\phi_1} I_{a1} = \frac{1}{0.8} I_{a1} \text{ (As the field is reduced by 20\%)}$$

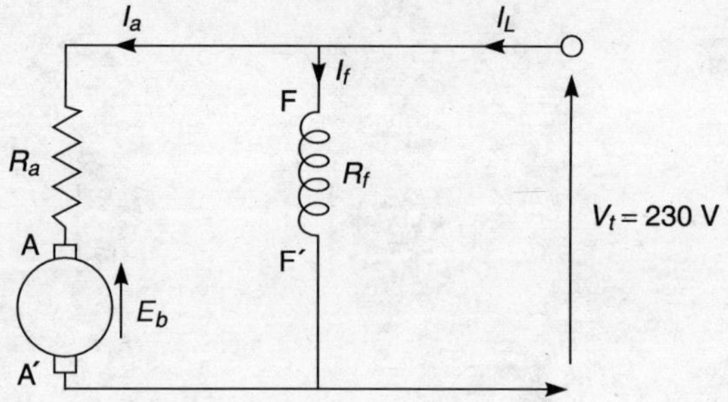

Figure 44 Figure for Example 11.

Then

$$I_{a2} = \frac{50}{0.8} = 62.5\ \text{A}$$

Using Eqs. (1) and (2), we get

$$\frac{\phi_1 N_1}{\phi_2 N_2} = \frac{V_t - I_{a1} R_a}{V_t - I_{a2} R_a}$$

Hence

$$N_2 = \frac{V_t - I_{a2} R_a}{V_t - I_{a1} R_a} \frac{\phi_1}{\phi_2} N_1 = \frac{230 - 62.5 \times 0.2}{230 - 50 \times 0.2} \times \frac{1}{0.8} \times 700 = 865 \text{ rpm}$$

Example 12

A 140 V DC shunt motor has an armature resistance of 0.2 Ω and a field resistance of 70 Ω. The full-load line current is 40 A and the full-load speed is 1800 rpm. If the brush contact drop is 3 V, find the speed of the motor at half load. **(UPTU June 2010)**

Solution

The circuit diagram for the given problem is shown in Figure 45.

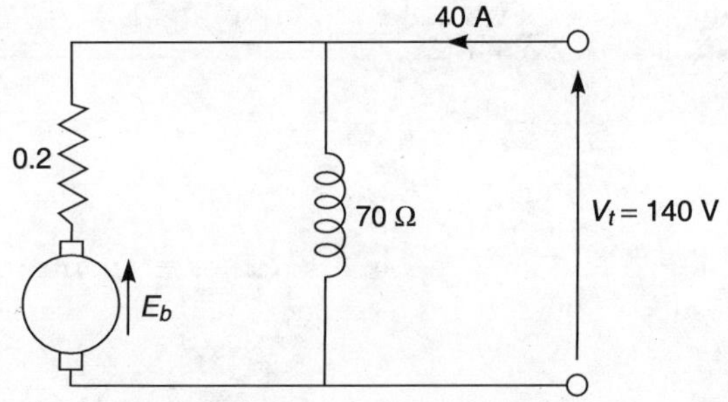

Figure 45 Figure for Example 12.

Given, $R_a = 0.2\ \Omega$, $R_f = 70\ \Omega$, $I_{L_1} = 40$ A, $N = 1800$ rpm, $V_{\text{brush}} = 3$ V. We know that there are two brushes in a DC motor. Then brush drop $= 2 \times 3 = 6$ V. Also, we know that the back-emf is given by

$$E_b = \frac{P\phi NZ}{60a} \quad \text{or} \quad E_b \propto N\phi$$

In the given problem, the voltage is not changed, hence ϕ will remain constant. Therefore, we can say that

$$\frac{E_{b_1}}{E_{b_2}} = \frac{N_1}{N_2} \tag{1}$$

The back-emf can be calculated as

$$E_{b1} = (V_t - I_{a1}R_a - V_{\text{brush}})$$

But

$$I_{a1} = I_{L1} - I_f$$

where

$$I_f = \frac{140}{70} = 2 \text{ A}$$

Then

$$I_{a1} = 40 - 2 = 38 \text{ A}$$

Hence

$$E_{b1} = (140 - 38 \times 0.2 - 6) = 126.4 \text{ V} \tag{2}$$

Next, the load current (i.e. I_L) is halved. Hence $I_{L2} = 20$ A. So,

$$I_{a2} = I_{L2} - I_f = 20 - 2 = 18 \text{ A}$$

Thus,

$$E_{b2} = (140 - 18 \times 0.2 - 6) = 130.4 \text{ V} \tag{3}$$

Substituting Eqs. (2) and (3) in Eq. (1), we get

$$N_2 = \frac{130.4}{126.4} \times 1800 = 1857 \text{ rpm}$$

Thus, the speed of the motor at half load would be 1857 rpm.

Example 13

A DC shunt machine connected to 230 V supply has resistance of armature as 0.115 Ω and of field winding as 115 Ω. Find the ratio of the speed as a generator to the speed as a motor with the line current in each case being 100 A. **(UPTU June 2003)**

Solution

Given $R_a = 0.115\ \Omega$, $R_f = 115\ \Omega$, $I_L = 100$ A, $V_t = 230$ V. The circuit diagram for the machine, working as a generator and as a motor, is given as shown in Figures 46 and 47, respectively.

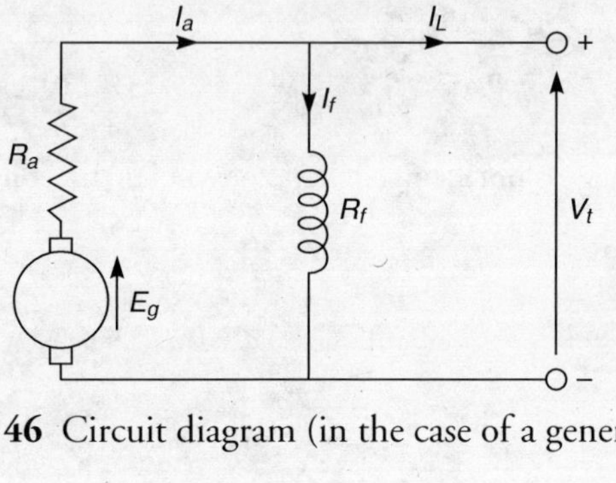

Figure 46 Circuit diagram (in the case of a generator).

(a) In the case of a generator,

$$E_g = V_t + I_a R_a, \qquad I_f = \frac{V_t}{R_f}, \qquad I_a = I_L + I_f$$

Hence

$$I_f = \frac{230}{115} = 2 \text{ A} \quad \text{and} \quad I_a = 100 + 2 = 102 \text{ A}$$

Therefore,

$$E_g = 230 + (102 \times 0.115) = 241.73 \text{ V}$$

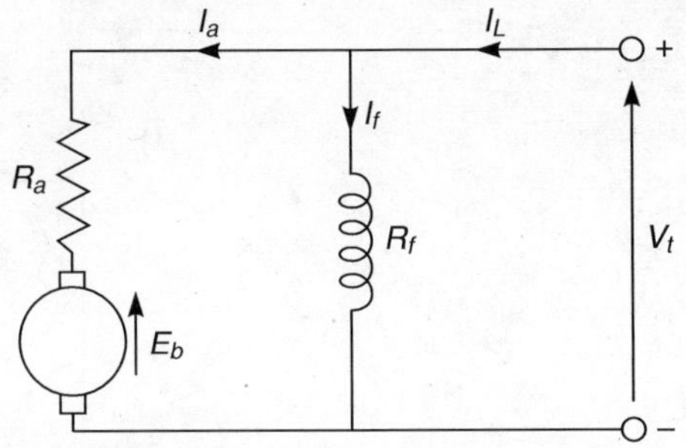

Figure 47 Circuit diagram (in the case of motor).

(b) Figure 47 shows the circuit diagram in the case of motor.
In the case of motor,

$$E_b = V_t - I_a R_a, \qquad I_f = \frac{V_t}{R_f}, \qquad I_a = I_L - I_f$$

Hence

$$I_f = \frac{230}{115} = 2 \text{ A} \quad \text{and} \quad I_a = 100 - 2 = 98 \text{ A}$$

Therefore,

$$E_b = 230 - (98 \times 0.115) = 218.73 \text{ V}$$

We know that in a shunt machine, I_f, and hence ϕ, is constant for constant terminal voltage. Therefore, $E \propto N$. Hence,

$$\frac{E_g}{E_b} = \frac{N_g}{N_m}$$

or

$$\frac{N_g}{N_m} = \frac{241.73}{218.73} = 1.105$$

The ratio of speed as generator to that of motor = 1.105.

Example 14

A 10 kW, 250 V DC shunt generator, having an armature resistance of 0.1 Ω and a field resistance of 250 Ω, delivers full load at rated voltage and 800 rpm (see Figure 48). The machine is now run as a motor while taking 10 kW at 250 V. What is the speed of the motor? Neglect the contact drop.

(UPTU December 2001)

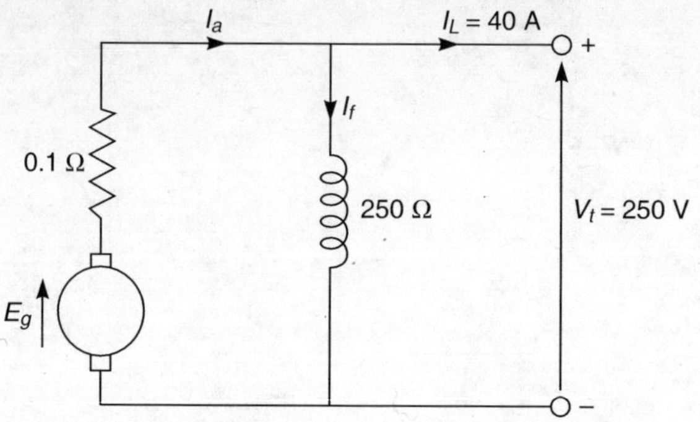

Figure 48 Figure for Example 14 (generator operation).

Solution

A 10 kW generator would draw a load current of

$$I_L = \frac{10 \times 1000}{250} = 40 \text{ A}$$

We know that

$$I_f = \frac{250}{250} = 1 \text{ A}$$

Then

$$I_a = 40 + 1 = 41 \text{ A}$$

and

$$E_g = V_t + I_a R_a = 250 + 41 \times 0.1 = 254.1 \text{ V}$$

Now, when the machine runs as a motor, the diagram is as shown in Figure 49.

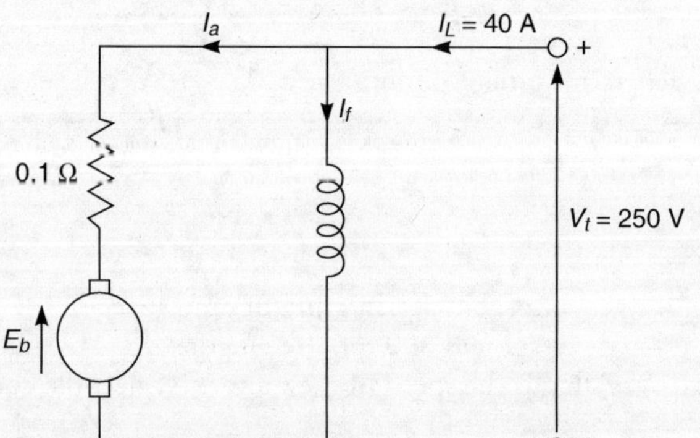

Figure 49 Circuit diagram when the machine runs as a motor.

Here $E_b = (V_t - I_a R_a)$. I_a here will be

$$I_a = I_L - I_f = 40 - 1 = 39 \text{ A}$$

and

$$E_b = (250 - 39 \times 0.1) = 246.1 \text{ V}$$

Since ϕ is constant in shunt machine,

$$\frac{E_g}{E_b} = \frac{800}{N} \quad \text{or} \quad N = \frac{800 \times E_b}{E_g} = \frac{800 \times 246.1}{254.1} = 774.8 \text{ rpm}$$

The speed of motor will be 774.8 rpm.

Induction Motor Numericals

Example 15

Find out the percentage slip of a three-phase, four-pole 50 Hz, induction motor, running at 1450 rpm.

Solution

Given $P = 4$ poles, $N_r = 1450$ rpm and $f_s = 50$ Hz. We have to find out % slip. We know that

$$N_s = \frac{120 f_s}{P} = \frac{120 \times 50}{4} = 1500 \text{ rpm}$$

Hence we get

$$\% \text{slip} = \frac{N_s - N_r}{N_s} \times 100 = \frac{1500 - 1450}{1500} \times 100 = 3.33\%$$

Example 16

A three-phase, four-pole slip ring induction motor has a rotor frequency of 2 Hz. The motor is supplied by a three-phase, 50 Hz supply. Determine the percentage slip and the rotor speed.

Solution

Given $P = 4$ poles, $f_r = 2$ Hz and $f_s = 50$ Hz. We have to find out % slip and N_r. We know that

$$f_r = s f_s$$

Hence

$$\text{Slip } s = \frac{f_r}{f_s} = \frac{2}{50} = 0.04$$

Also we know that $N_r = (1 - s)N_s$. Then,

$$N_s = \frac{120f_s}{P} = \frac{120 \times 50}{4} = 1500 \text{ rpm}$$

Therefore,

$$\text{Rotor speed, } N_r = (1 - 0.04) \times 1500 = 1440 \text{ rpm}$$

Example 17

Find out the frequency of rotor current when a three-phase, four-pole, 50 Hz induction motor is run at 1050 rpm.

Solution

Given $P = 4$ poles, $f_s = 50$ Hz and $N_r = 1050$ rpm. We have to find f_r. We know that $f_r = sf_s$. Also we know that

$$s = \frac{N_s - N_r}{N_s}$$

where

$$N_s = \frac{120f_s}{P} = \frac{120 \times 50}{4} = 1500 \text{ rpm}$$

Hence

$$s = \frac{1500 - 1050}{1500} = 0.3$$

Also

$$f_r = 0.3 \times 50 = 15 \text{ Hz}$$

Example 18

A three-phase slip ring induction motor has four poles. The motor frequency is found to be 2.5 Hz while connected to a 400 V, three phase, 100 Hz supply. Determine the speed of the stator's rotating magnetic field and the rotor speed in rpm.

Solution

Given number of poles = 4; supply frequency, $f_s = 100$ Hz; rotor frequency, $f_r = 2.5$ Hz. From the given data, the speed of the stator's rotating magnetic field is

$$N_s = \frac{120f_s}{P} = \frac{120 \times 100}{4} = 3000 \text{ rpm}$$

We know that $f_r = 2.5$. Also $f_r = sf_s$. Therefore,

$$s = \frac{2.5}{100} = 0.025$$

The rotor speed is

$$N_r = (1 - s)\,N_s = (1 - 0.025)\,3000 = 2925 \text{ rpm}$$

Example 19

A three-phase, four-pole induction motor runs with 4% slip. The supply frequency is 50 Hz. Find out the speed of induction motor. **(UPTU December 2010)**

Solution

Given slip, $s = 4\% = 0.04$; number of poles, $P = 4$; supply frequency, $f_s = 50$ Hz. We know that the synchronous speed N_s is given by

$$N_s = \frac{120 f_s}{P}$$

Therefore,

$$N_s = \frac{120 \times 50}{4} = 1500 \text{ rpm}$$

Also, we know that the slip is given by

$$s = \frac{N_s - N_r}{N_s} \quad \text{or} \quad N_r = (1 - s)\,N_s$$

Substituting the values, we get the speed of the motor as

$$N_r = (1 - 0.04) \times 1500 = 1440 \text{ rpm}$$

Example 20

At a certain load, a three-phase, 50 Hz, eight-pole induction motor runs with a slip of 2.67%. Find the following:

(a) The speed of rotor with respect to stator
(b) The speed of rotor magnetic field with respect to stator
(c) The speed of rotor magnetic field with respect to stator magnetic field
(d) The speed of rotor magnetic field with respect to rotor
(e) The speed of stator magnetic field with respect to rotor

Solution

Given $f_s = 50$ Hz, slip $s = 2.67\%$, $P = 8$. Then

$$N_s = \frac{120 f_s}{P} = \frac{120 \times 50}{8} = 750 \text{ rpm}$$

Also,

$$N_r = (1 - s)\, N_s = (1 - 0.0267)\, N_s = 730 \text{ rpm}$$

Figure 50 explains the relative speeds with regard to the stationary frame of reference. The rotor rotates at N_r rpm. The magnetic field due to rotor, ϕ_r, rotates at synchronous speed N_s rpm. The magnetic field due to stator ϕ_s rotates at synchronous speed N_s rpm. The stator is stationary. Keeping all these in mind:

(a) The speed of rotor with respect to stator = N_r rpm = 730 rpm
(b) The speed of rotor magnetic field with respect to stator = N_r rpm = 750 rpm
(c) The speed of rotor magnetic field with respect to stator magnetic field = $(N_s - N_s)$ rpm = 0 rpm
(d) The speed of rotor magnetic field with respect to rotor = $(N_s - N_r) = sN_s$ (called *slip speed*) = (750 − 730) = 20 rpm
(e) The speed of stator magnetic field with respect to rotor = $(N_s - N_r)$ = 20 rpm

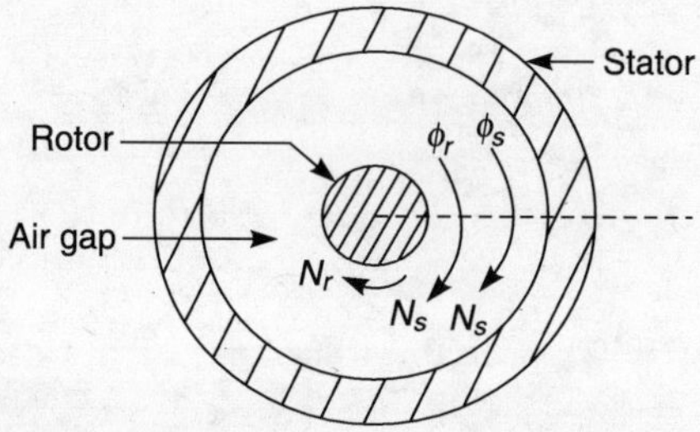

Figure 50 Figure for Example 20.

Example 21

A three-phase, 50 Hz, 440 V induction motor runs at 1470 rpm. Find

(a) Numberius of poles
(b) Speed of rotating magnetic field
(c) Frequency of the rotor current

Solution

Given $f_s = 50$ Hz and $N_r = 1470$ rpm. We have to find out P, N_s and f_r.

(a) We know that

$$N_s = \frac{120 f_s}{P}$$

Also, we know that the rotor slips behind the synchronous speed N_s by a very small amount. For instance, let us assume, $N_r = N_s$. Hence,

$$1470 = \frac{120 \times 50}{P} \quad \text{or} \quad P = \frac{120 \times 50}{1470} = 4.0816$$

Since the number of poles must be an integer, we have $P = 4$.

(b) The speed of rotor magnetic field

$$N_s = \frac{120 f_s}{P} = \frac{120 \times 50}{4} = 1500 \text{ rpm}$$

(c) The frequency of rotor current $f_r = sf_s$. Here,

$$\text{Slip } s = \frac{N_s - N_r}{N_s} = \frac{1500 - 1470}{1500} = 0.02$$

Hence, $f_r = 0.02 \times 50 = 1$ Hz.

Example 22

A three-phase, 50 Hz induction motor has a full-load speed of 960 rpm. Calculate

(a) Slip
(b) Frequency of rotor induced emf
(c) Number of poles
(d) Speed of rotor field with respect to rotor structure
(e) Speed of rotor field with respect to stator field

(UPTU June 2010)

Solution

Given $N_r = 960$ rpm, $f_s = 50$ Hz.

(a) In order to find the slip, it is required to know the synchronous speed of the field (i.e. N_s). We know that

$$N_s = \frac{120 f_s}{P}$$

From this,

$$P = \frac{120 f_s}{N_s}$$

Say $N_s \approx N_r$. Therefore,

$$P \approx \frac{120 \times 50}{960} = 6.25$$

Since the number of poles cannot be fractional, $P = 6$ [answer to part (c)].

Now with $P = 6$, the value of N_s can be obtained as

$$N_s = \frac{120 f_s}{P} = \frac{120 \times 50}{6} = 1000\,\text{rpm}$$

Hence, as per the definition of slip

$$s = \frac{N_s - N_r}{N_s} = \frac{1000 - 960}{1000} = 0.04$$

Thus, slip = 0.04.

(b) The frequency of rotor-induced emf is given by the expression, $f_r = sf_s$. Therefore, the frequency of rotor-induced emf for the given rotor speed would be

$$f_r = (0.04)(50) = 5\text{ Hz}, \quad f_s = 2\text{ Hz}$$

(c) Number of poles has been obtained as a part of answer of (a), as $P = 6$.

(d) The rotor field rotates at synchronous speed (i.e. at 1000 rpm in this core). In the problem, the speed of rotor is given as 960 rpm. Hence, the slip speed, that is the speed of rotor field with respect to rotor structure, is $sN_s = (N_s - N_r) = (1000 - 960) = 40$ rpm.

(e) There is no relative speed between the fields due to stator and rotor. Hence, the speed of rotor field with respect to stator field (i.e. $N_s - N_s$) is zero.

Example 23

A motor generator set used for producing variable frequency AC supply consists of a three-phase, 10-pole synchronous motor and a 24-pole three-phase synchronous generator. The motor generator set is fed from a 25 Hz, three-phase AC supply. A 6-pole, three-phase induction motor is electrically connected to the terminals of the synchronous generates and runs at a slip of 5%. Determine:

(a) Frequency of the generated voltage of the synchronous generator.

(b) Speed at which the induction motor is running.

Solution

The given formation is shown in Figure 51.

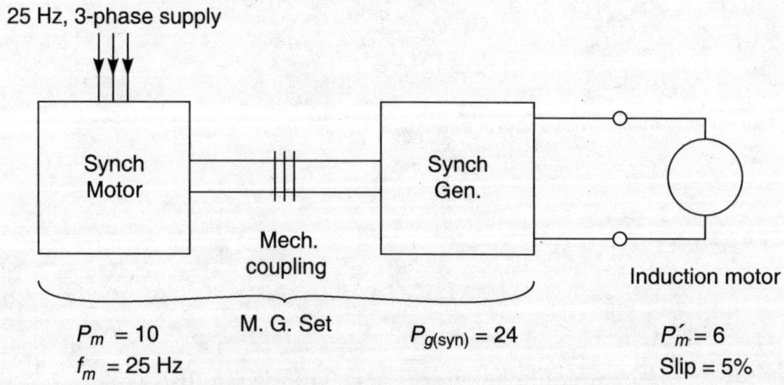

Figure 51 Figure for Example 23.

(a) The synchronous speed for synchronous motor is

$$N_m = \frac{120 f_m}{P_m} = \frac{120 \times 25}{10} = 300 \text{ rpm}$$

Since the synchronous motor is mechanically coupled with the synchronous generator, the shaft of the generator also rotates at 300 rpm. Hence, $N_g = 300$ rpm. We know that the rotor of synchronous machine rotates at synchronous speed. Therefore, for synch generator, $N_{S_g} = 300$ rpm. Hence the frequency of the current generated by the synch generator is given by

$$f_g = N_g \times \frac{P_g}{120} = 300 \times \frac{24}{120} = 60 \text{ Hz}$$

Hence, the frequency of the generated voltage of synch generator is 60 Hz.

(b) Next, for induction motor we know that

$$N_{s(\text{Induction Motor})} = 120 \times \frac{f_g}{P_M'} = \frac{120 \times 60}{6} = 1200 \text{ rpm}$$

Also,

$$N_r = (1 - s)\, N_s = (1 - 0.05)\, 1200 = 1140 \text{ rpm}$$

Thus, the speed of the induction motor is 1140 rpm.

Summary

The chapter focuses mainly on the energy conversion principles and their applications in various types of electrical machines, like the DC machine, single- and three-phase induction machines and three-phase synchronous machine. The constructions of various machine types have been elaborated. The various machine parts, that is stator and rotor, consist of windings in which currents flow. Field winding current is provided from the outside as an input in both machine types: motor and generator. Armature current is either the input current (for a motor) or induced current (for a generator) which is produced as a result of the movement of the rotor according to the Faraday's law of electromagnetic induction (for a generator). It is seen that a DC motor and a three-phase induction motor are self-starting motors. The three-phase synchronous motor and the single-phase induction motor are nonself-starting motors and certain arrangements (like damper windings and split phase windings, respectively) need to be done in order to make them start at the switch getting on.

Short Answer Questions

1. Which of the DC motors is used in purposes requiring high starting torque?
2. Which of the two winding types in DC motors is suited for handling higher armature voltage?
3. Among a motor and a transformer, whose efficiency is higher and why?
4. What is the frequency of the rotor-induced emf in a three-phase induction motor at the time of starting?

5. In a three-phase induction motor, the frequency of the supply is 50 Hz, number of poles is 4 and the motor is squirrel age and running at 1440 rpm. What is the slip?
6. What are the ranges of slip in motoring, braking and generating regions?
7. Under full load running conditions, the core losses occurring in a three-phase induction motor rotor are negligible. Explain.
8. At which slip does a three-phase synchronous motor always run?
9. Which is the best method of starting a single-phase induction motor?
10. Which are the doubly excited motors and which are the singly excited motors?

Review Questions

1. Explain the basic principles of electromechanical energy conversion. How does a motor and a generator work?
2. What are the names and functions of the various parts of the electric machine? Draw a neat sketch of the DC motor explaining its major parts.
3. Describe the working principle of DC series motor, and draw its various characteristics. **(UPTU June 2011)**
4. Discuss the working principle of a three-phase synchronous motor. Also, differentiate synchronous motor from induction motor. **(UPTU June 2011)**
5. Explain the squirrel-cage rotor and phase-wound rotor in induction motor. **(UPTU June 2011)**
6. Discuss the principle of operation of single-phase induction motor. Also write its applications. **(UPTU June 2011)**
7. Explain the torque-slip characteristic of a three-phase induction motor. **(UPTU June 2011)**
8. Derive the expression of torque for a DC motor. Also, discuss the characteristics of a DC shunt motor. **(UPTU June 2011)**
9. Explain the starting methods of single-phase induction motors. **(UPTU December 2010)**
10. Sketch and explain the speed current characteristics of the following DC motors:
    (a) Series motor
    (b) Shunt motor
    (c) Cumulatively compounded motor
    (d) Differentially compounded motor
11. Discuss the working principle of a three-phase induction motor. **(UPTU December 2002)**
12. Derive the expression of emf of a DC generator. What are the requisite conditions for building up of voltage in a self-excited DC generator?

Problems

DC Machines

1. A six-pole DC generator with wave connected armature has 41 slots, with 12 conductors/slots. Armature resistance is $R_a = 0.5\ \Omega$. Shunt resistance is $R_{sh} = 200\ \Omega$. Flux per pole is 165 mWb. Speed in rpm = 1250. Find the terminal voltage across a load resistance of 12.5 Ω connected across the armature terminals.
2. A wave connected 10-pole DC generator has 600 armature conductors and a useful flux of 80 mWb. Find the induced emf when it runs at 1250 rpm. At what speed should it be run to produce the same emf if the generator is lap wound?

3. The induced emf of a four-pole, lap-connected, 1000 rpm DC generator is 1800 V with flux per pole being 90 mWb. Find the number of slots if there are 12 conductors per slot.
4. A 200 V DC shunt motor takes 22 A at a speed of 1000 rpm. Given $R_a = 0.5\ \Omega$, $R_f = 100\ \Omega$. Find the speed if the current taken is 40 A.
5. A four-pole DC shunt motor with rated input voltage of 200 V takes a line current of 4 A at no-load, while running at 1250 rpm. Find the speed at full-load. The input power (electrical) at full-load is 10 kW. Given $R_a = 0.5\ \Omega$, $R_f = 200\ \Omega$.

Three-phase Induction Motor

6. A four-pole, 50 Hz, three-phase induction motor runs at 1440 rpm. Find: (a) the speed of the rotating magnetic field, (b) percentage slip, (c) rotor frequency and (d) the motor speed, if the slip is doubled.
7. A six-pole, 50 Hz, three-phase induction motor has a full-load speed of 940 rpm. Find: (a) the slip speed, (b) rotor induced emf frequency and (c) the speed at and beyond which the rotor induced emf becomes zero.
8. The speed of a 3-ϕ IM with supply frequency of 50 Hz is 1470 rpm. Find the suitable number of poles in the stator of the motor.
9. A three-phase, 50 Hz induction motor has its rotor emf making 180 oscillations per minute at a certain load. If the rotor moves at 940 rpm for that load, find: (a) the slip, (b) the number of stator poles, (c) speed of the stator's rotating magnetic field with respect to the rotor.
10. In Problem 9, find the speed of rotor's rotating magnetic field with respect to: (a) stator and rotor and (b) stator's rotating magnetic field.

Answers

Short Answer Questions

1. The DC series motor is used for high starting torque purposes as in it the torque varies as the square of the armature current. So whenever the mechanical load increases, the armature current increases and the torque increases sharply.
2. The number of parallel paths in the wave-type armature winding is restricted to two, hence it is able to carry less current. So it is designed to handle higher voltage as against the lap-type winding.
3. The efficiency of a transformer is higher than that of a motor because the rotational losses occurring in motors are not present in the transformer.
4. The frequency of the rotor-induced emf at the time of starting is same as that of the supply because the relative difference between the rotating magnetic field and the rotor at starting is zero.
5. The synchronous speed of the rotating magnetic field is $120 \times 50/4 = 1500$ rpm and the slip will be $(1500 - 1440)/1500 = 0.04 = 4\%$.
6. Motoring region: 0 to 1; Braking region: 1 to 2; Generating region: 0 to −1.
7. The usual values of slip at which the motor runs is around 5%, i.e. $s = 0.05$. The frequency of the rotor-induced emf is slip times the supply frequency. Hence the rotor frequency is very low at full load. Thus the hysteresis and eddy current losses which depend upon 1.6 and 2 powers of frequency, respectively, are very less and are hence neglected.
8. The three-phase synchronous motor always runs at synchronous speed and hence at zero slip.
9. The capacitor start and run method is the best as not only it provides a high starting torque but it improves the overall running power factor too.
10. Three-phase synchronous motor and DC motor are doubly excited motors (as supplies are given to both stator and rotor circuits). Moreover, three-phase induction motor and single-phase induction motor are singly excited motors as supply is given only to the stator here.

Problems

1. 4867 V
2. 5000 V; 6250 rpm
3. 100 slots
4. 952.63 rpm
5. 1105.2 rpm
6. **(a)** 1500 rpm, **(b)** 4%, **(c)** 2 Hz, **(d)** 1380 rpm
7. **(a)** 60 rpm, **(b)** 3 Hz, **(c)** 1000 rpm
8. 4
9. **(a)** 6%, **(b)** 6, **(c)** 60 rpm
10. **(a)** 1000 rpm, 60 rpm; **(b)** zero rpm

6 Miscellaneous Topics in Electrical Engineering

LEARNING OBJECTIVES

After a careful study of the chapter, you would be able to understand:

- The basics of Ohm's law.
- Dependent sources and their applications in DC circuits.
- Reciprocity and Millman's theorems for DC circuits.
- Superposition theorem, Thevenin's theorem, Maximum power transfer theorem and Norton's theorem for AC circuits.
- The concept of time constant for inductive and capacitive circuits.
- The aspects of energy storage in inductors and capacitors.
- The analysis of unbalanced three-phase systems.
- Cathode Ray Oscilloscope (CRO) and its applications.
- The measurement of insulation resistance.
- The laws of electromagnetism with the concept of energy storage and the analysis of linear and non-linear magnetic circuits.
- Basics of the Per Unit system and its calculations.
- Phasor diagram of a single-phase transformer.
- Open-circuit and short-circuit tests of a single-phase transformer.
- Voltage regulation of a single-phase transformer.
- Losses and efficiency calculations of DC motors.
- Starting and speed control of DC motors.
- Starting of three-phase induction motors.
- Stepper motor and Universal motors.

6.1 Analysis of DC Circuits

Starting from Ohm's law, the concept of DC circuits is developed, including the dependent sources. Energy storage in inductors and capacitors is also discussed in the text to follow. Application of reciprocity and Millman's theorems is also described.

6.1.1 Ohm's Law

Ohm's law states that current I flowing through a conductor results in a proportionate voltage drop V across it, provided temperature and other physical conditions of the conductor remain the same. In other words, $V \propto I$. The constant of proportionality is represented by the resistance R and its unit is Ohm (Ω). Hence, for the circuit shown in Figure 1.

$$V = IR$$

$$\Rightarrow R\,(\text{ohm}) = V\,(\text{volt})/I\,(\text{ampere})$$

Resistance is given by

$$R = \frac{\rho l}{A}$$

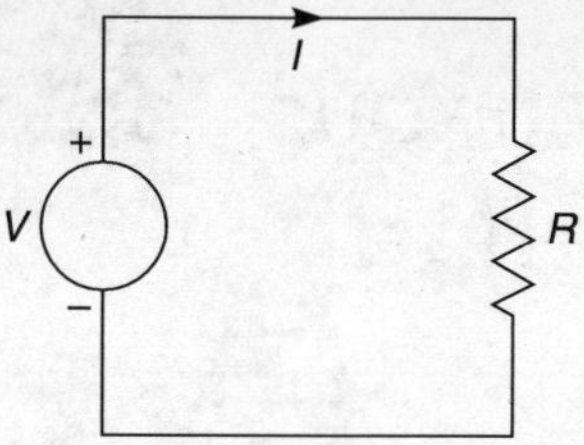

Figure 1 Circuit explaining Ohm's law.

where ρ (in Ω m) is the resistivity of the conductor material (which varies from material to material), l is the length (in m) of the conductor and A (in m^2) is the cross-sectional area of the conductor. A resistance dissipates energy in it. This energy dissipation is given by I^2R (Watts).

6.1.2 Dependent Sources and Their Application in DC Circuits

The voltage and current sources, used in electrical circuits, can be of two types – independent and dependent sources. When the value of voltage or current generated by a source does not depend on any other factor in the circuit, it is called an **independent source** – for example, a battery, an alternator, etc., which deliver power independent of any circuit conditions. However, if the value of voltage or current generated by a source depends on the value of current or voltage at any other part in the circuit, the source is known as a **dependent source**.

Dependent sources are represented by diamond-shaped symbols. The four types of dependent sources are shown in Figures 2(a)–(d).

1. **Voltage-dependent voltage source:** The circuit diagram shown in Figure 2(a) shows that the voltage source $V_2 = av_0$ is dependent on the voltage drop v_0 elsewhere in the circuit. Here "a" is a dimensionless constant.
2. **Current-dependent voltage source:** The circuit diagram of Figure 2(b) shows a current-dependent voltage source. The voltage source develops a voltage which is equal to rI_0, where I_0 is the flowing through R_4 in the circuit. Here r is an ohmic constant.
3. **Voltage-dependent current source:** The circuit diagram of Figure 2(c) shows that the current source gv_0 generates a current which is proportional to voltage drop v_0 across R_4. Here g is a constant expressed in mho.
4. **Current-dependent current source:** The circuit diagram shown in Figure 2(d) shows that the voltage source $I = bI_0$ is a current (I_0)-dependent current source. Here b is a dimensionless constant.

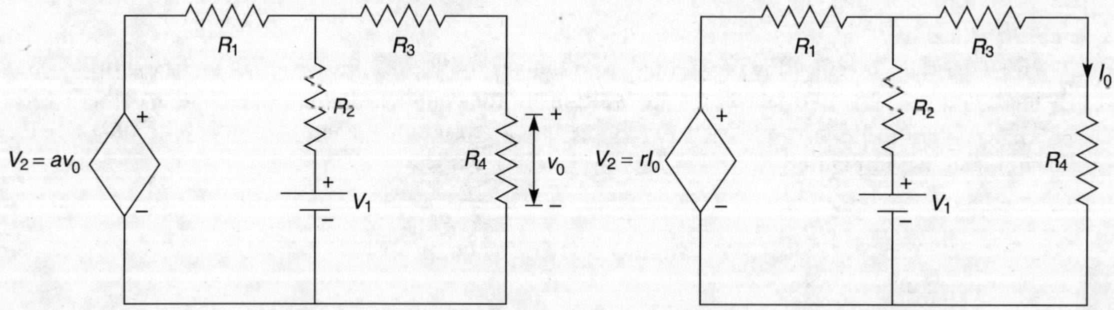

(a) (b)

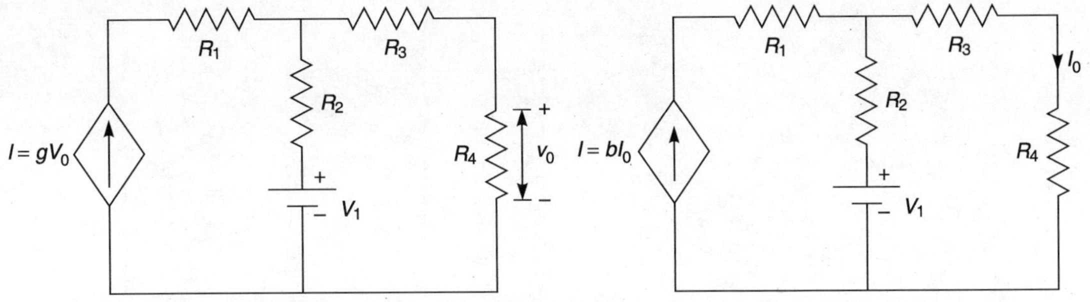

(c) (d)

Figure 2 (a) A voltage-dependent voltage source V_2 (*a* is a dimensionless constant). (b) A current-dependent voltage source V_2 (*r* is a constant in ohms). (c) Voltage-dependent current source *I* (*g* is a constant in mho). (d) Current-dependent current source *I* (*b* is a dimensionless constant).

6.1.3 Reciprocity Theorem

The reciprocity theorem is applicable to any linear and bilateral circuit. This theorem states that if a current *I* flows in a branch (say c-d) in a circuit due to a voltage source (placed say in branch a-b), then if the voltage source is shifted to branch c-d, then the current *I* will flow through branch a-b, where the voltage source was placed initially. In the circuit shown in Figure 3(a), the current in branch c-d, due to the voltage source connected between terminals a and b, is *I* amp. In Figure 3(b), the source is shifted to the branch c-d where the current *I* was flowing before. Let the current flowing through the branch a-b in Figure 3(b) be named *I*. The calculations reveal that this current is equal to the current *I* as shown flowing through branch cd in Figure 3(a).

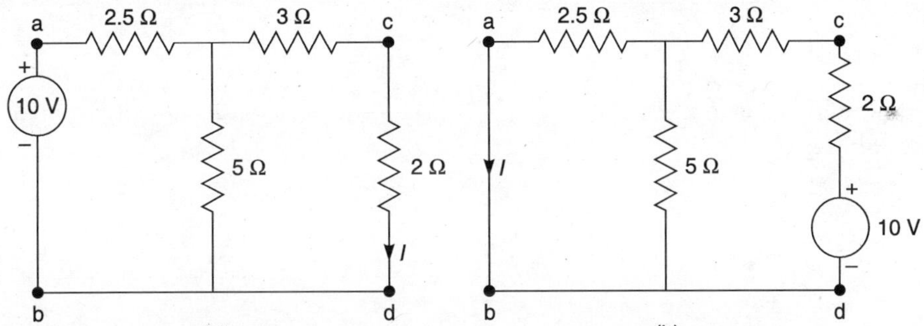

Figure 3 Reciprocity of 10 V source and the branch current *I*.

In Figure 3(a), the current supplied by the source is $10/[2.5 + (5||5)] = 2$ A. The current through branch c-d (by current divider theorem) is 1 A.

Now when the source is shifted from branch a-b to c-d, as shown in Figure 3(b), the current supplied by the source will be $10/[2 + 3 + (5||2.5)] = 1.5$ A. Thus, by current divider theorem, the current through the branch a-b in Figure 3(b) will be 1A.

6.1.4 Millman's Theorem

This theorem states that if there are a number of voltage sources (V_1, V_2, V_3, ...) having corresponding internal resistances (R_1, R_2, R_3, ...), connected in parallel in a circuit, then they all can be replaced by a single voltage source (V_{eq}) in series with an equivalent internal resistance (R_{eq}).

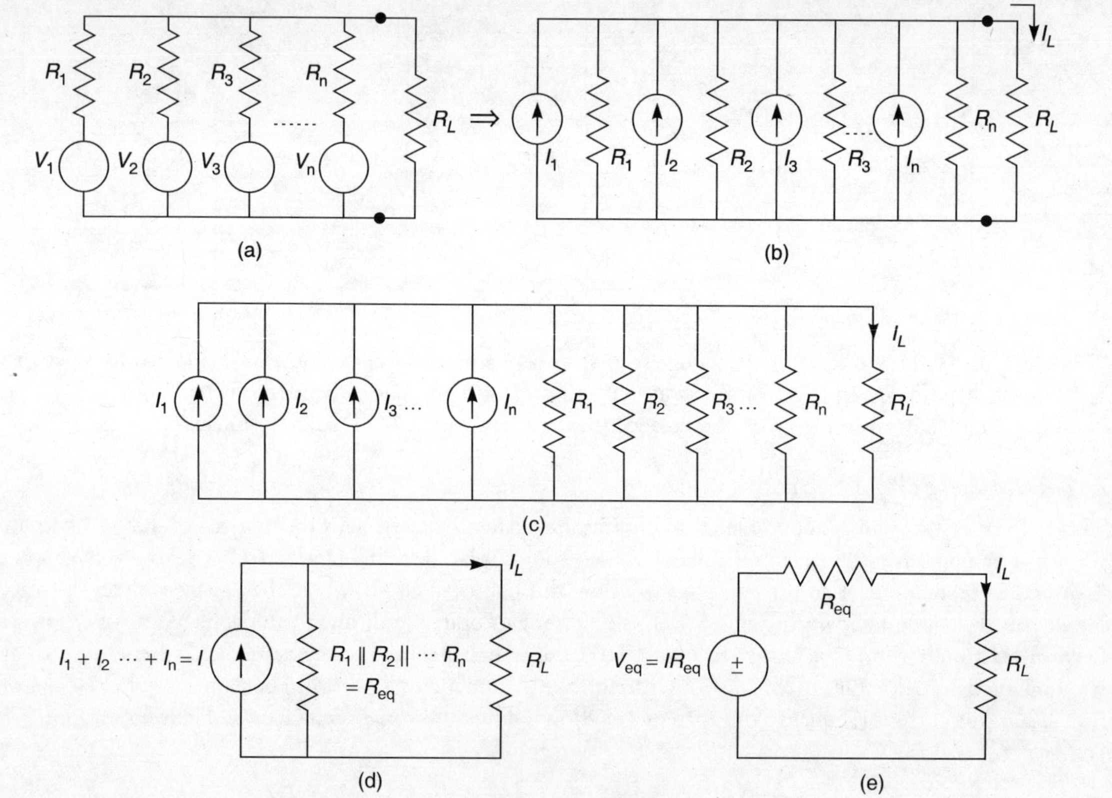

Figure 4 Circuits explaining Millman's theorem.

The circuit shown in Figure 4(a) can be represented as the circuit shown in Figure 4(b) by applying source transformation. Here, $I_L = I_1 + I_2 + I_3 + \cdots + I_n$.

$$R_{eq} = \left[\frac{1}{R_1} + \frac{1}{R_2} + \frac{1}{R_3} + \cdots + \frac{1}{R_n}\right]^{-1}$$

$$I = I_1 + I_2 + I_3 + \cdots + I_n$$

Thus, with the help of source transformation, the circuit of Figure 4(a) having multiple voltage sources can be represented by the circuit shown in Figure 4(e) having single voltage source.

6.2 Analysis of AC Circuits

The performance of the circuits is very different when they are excited by sinusoidal (AC) supply. Some related issues are discussed here.

6.2.1 Concept of Time Constant

This concept holds for first-order electrical circuits (i.e., *R-L* or *R-C* types). If a voltage *V* is applied across a series *R-L* circuit as shown in Figure 5, the differential equation for evaluating the current can be written as

$$V = L\frac{di}{dt} + Ri$$

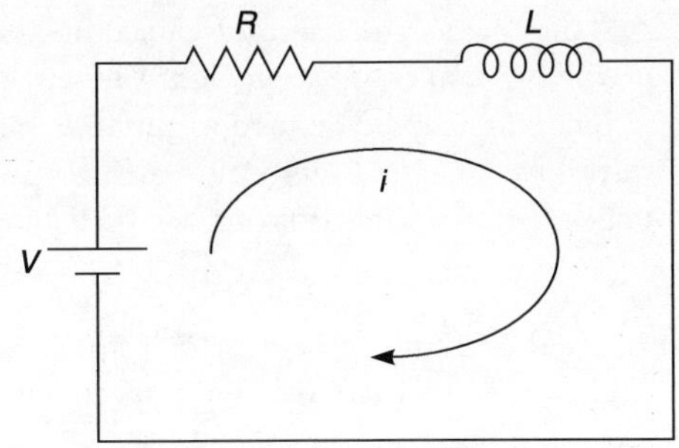

Figure 5 Illustrating the time constant for an *R-L* circuit.

or

$$\frac{di}{dt}+\frac{R}{L}i=\frac{V}{L}$$

Its solution (assuming the current i to be zero at $t = 0$) is

$$i=\frac{V}{R}[1-e^{-t/\tau}]$$

where $\tau = L/R$ is called the "time constant" of the circuit. Now if $t = \tau$, the current is given by

$$i=\frac{V}{R}[1-e^{-1}]$$

Thus $i = 0.632V/R$ or $i = 63.2\%$ of the value which it would attain at $t = \infty$ (at $t = \infty$, $i = V/R$). Similarly, the time constant can be obtained for the *R-C* series circuit shown in Figure 6.

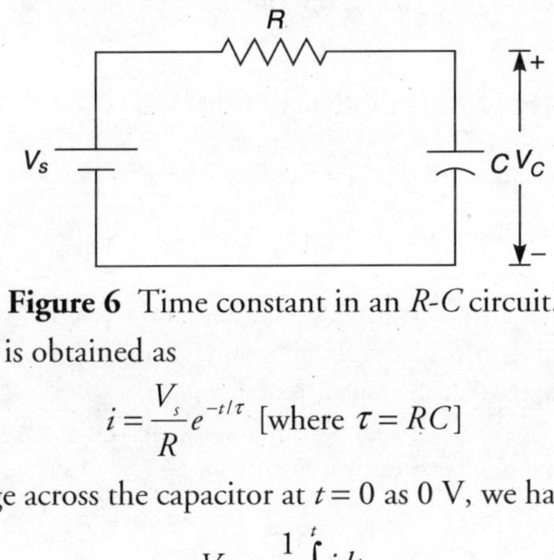

Figure 6 Time constant in an *R-C* circuit.

Here, the expression of current is obtained as

$$i=\frac{V_s}{R}e^{-t/\tau}\quad [\text{where } \tau = RC]$$

Now, taking the value of voltage across the capacitor at $t = 0$ as 0 V, we have

$$V_C=\frac{1}{C}\int_0^t i\,dt$$

On substituting the value of i into the above equation, we get

$$\begin{aligned}V_C &= \frac{1}{C}\int_0^t \frac{V_s}{R}e^{-t/RC}dt\\ &= \frac{V_s}{R}\times[-RC]\times e^{-t/RC}\Big|_0^t\\ \Rightarrow V_C &= V_s[1-e^{-t/RC}]=V_s[1-e^{-t/\tau}]\end{aligned}$$

Here also at $t = \tau$, $V_C = 0.623\ V_s$. This means the value of capacitor voltage is equal to 63.2% of the maximum possible voltage which it will attain at $t = \infty$ and whose value is V_s volts.

Thus the time constant for an *R-L* circuit may be defined as the time taken by the current through the inductor to reach to 63.2% of its prospective (at $t = \infty$) value. Similarly for an *R-C* circuit, the time constant is defined as the time taken by the voltage across the capacitor to reach 63.2% of its prospective (at $t = \infty$) value.

6.2.2 Energy Storage Aspect

A pure inductor and a pure capacitor are the circuit elements which store energy during one-half and release it in the other half of the cycle of AC current and AC voltage, respectively. They do not dissipate energy. The energy stored in an inductor L is given by

$$E_L = (1/2)LI_L^2 \text{ Watt-sec}$$

where L is the inductance in Henry and I_L is the maximum (or peak) value of current flowing through it in amperes.

When a current I_L flows through an inductor of inductance L, causing a voltage drop V_L, the integration of power gives the expression of energy stored (E_L) in the inductor. This can be understood from the following equation of energy:

$$E_L = \int P_L dt = \int V_L I_L dt = \int (LdI_L/dt)I_L dt = \int LI_L dI_L = (1/2)LI_L^2$$

Similarly, the energy stored in a capacitor C is given by

$$E_C = (1/2)\ CV_C^2 \text{ Watt-sec}$$

where C is the capacitance in Farad and V_C is the voltage across it in volts.

When a voltage V_C appears across a capacitor C_F, causing the flow of a current I_C through it, the integration of power gives the expression of energy stored in the capacitor. This can be understood from the following equation of energy:

$$E_C = \int P_C dt = \int V_C I_C dt = \int V_C (CdV_C/dt)dt = \int CV_C dV_C = (1/2)CV_C^2$$

6.2.3 Superposition Theorem

The superposition theorem for AC circuits is exactly the same as it is for DC circuits which we have discussed in Chapter 1.

If there are two or more (voltage or current) AC sources in a circuit, the response (output voltage or current) at any part of the circuit is found by algebraically summing up the contribution of each source while replacing every other source by its internal impedance.

In AC circuits, the RMS value of the source voltage (or the current, as the case may be) is used in calculations. For example, consider the circuit shown in Figure 7.

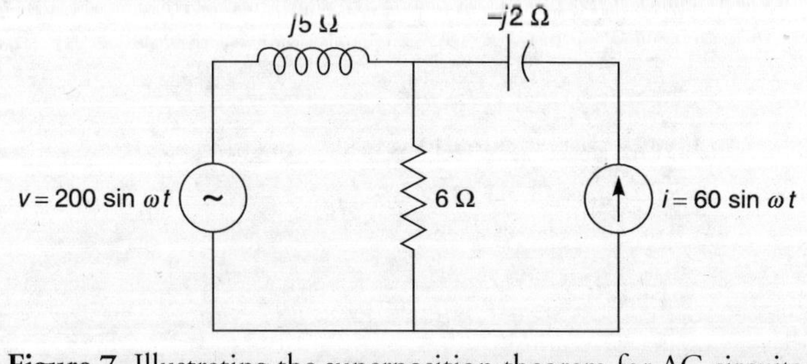

Figure 7 Illustrating the superposition theorem for AC circuits.

Here, let us say, we have to find the current through the 6 Ω resistance using the superposition theorem. As already explained in Chapter 1, we begin with the replacement of the current source by its internal impedance (i.e., infinite ohms) and find the current due to the voltage source only. We get the circuit shown in Figure 8.

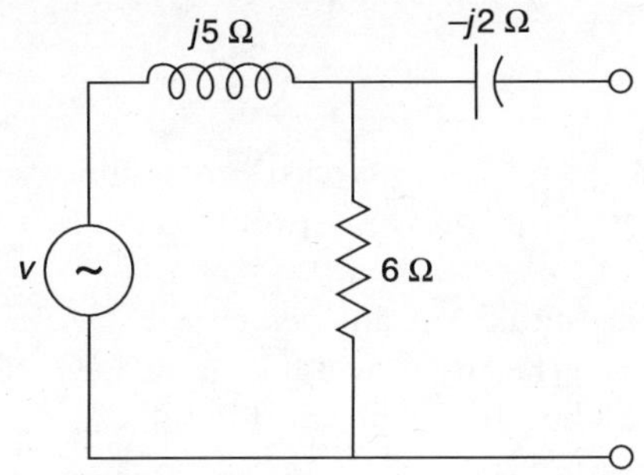

Figure 8 Circuit having only a voltage source.

The RMS value of the current through the 6 Ω resistance would be given by

$$I' = \frac{(V_m/\sqrt{2})\angle 0^\circ}{6+j5} = \frac{(200/\sqrt{2})\angle 0^\circ}{7.81\angle 39.8^\circ}$$

$$I' = 18.1\angle -39.8^\circ = 13.91 - j11.59$$

Now we find the current in the 6 Ω resistance due to the current source alone. The voltage source is replaced by its internal impedance (i.e., 0 Ω). We get the circuit shown in Figure 9.

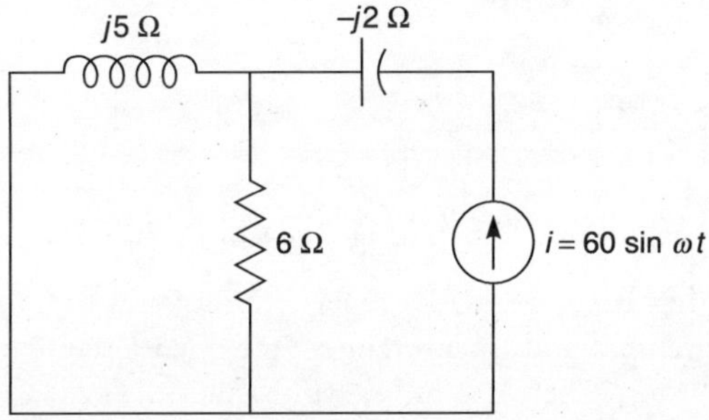

Figure 9 Circuit having only a current source.

The current i will flow in the $-j2\ \Omega$ capacitive reactance. Further it will get divided between $j5\Omega$ and 6Ω.

$$I'' = \frac{(60/\sqrt{2})\angle 0^\circ \times 5\angle 90^\circ}{6+j5}$$

$$= 27.16\angle 50.19^\circ = (17.39 + j20.8)$$

Thus, the total current is given by

$$I' + I'' = (13.91 - j11.59) + (17.39 + j20.8)$$

$$= 32.64\angle 16.51^\circ \text{A}$$

The complete expression of the current through the 6 Ω resistance will be

$$i_{(6\Omega)} = \sqrt{2} \times 32.64 \sin(\omega t + 16.51)$$
$$= 46.17 \sin(\omega t + 16.51)$$

6.2.4 Thevenin's Theorem

The principles of Thevenin's theorem for AC circuits are similar to the ones discussed for the DC circuits in Chapter 1. Let us solve the previous problem (shown in Figure 7) using Thevenin's theorem.

First, we remove the 6 Ω resistance and create an open circuit (say a-b) across its ends. The circuit looks as shown in Figure 10. The current in the circuit will flow only due to the current source i (as no current can flow due to the voltage source). Thus, starting from "b" and reaching "a" we get

$$V_{ab} = +v + j5 \times i = \frac{200}{\sqrt{2}}\angle 0° + \frac{60}{\sqrt{2}} \times 5\angle 90° = 141.42 + j212.1$$
$$V_{ab} = V_{th} = 254.95\angle 56.31°$$

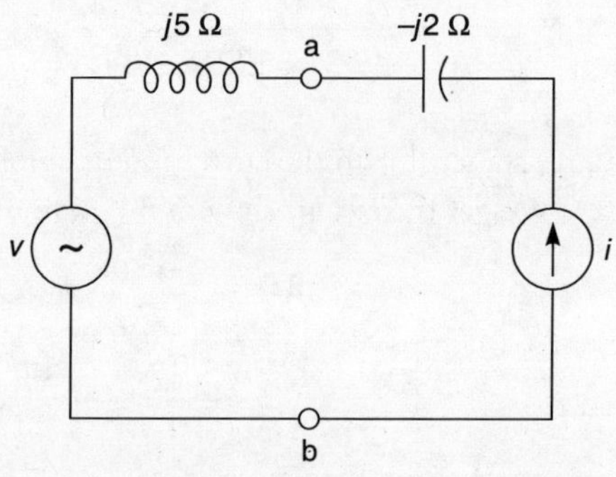

Figure 10 Illustrating Thevenin's theorem.

For finding Z_{th} (similar to R_{th} in DC circuits), we replace the sources by their internal impedances. Thus if the voltage source is short-circuited and the current source open-circuited, there remains only one path between a and b, and that is through the $j5$ Ω inductive reactance. Thus $Z_{th} = +j5\,\Omega$. Finally, Thevenin's equivalent circuit can be represented as shown in Figure 11.

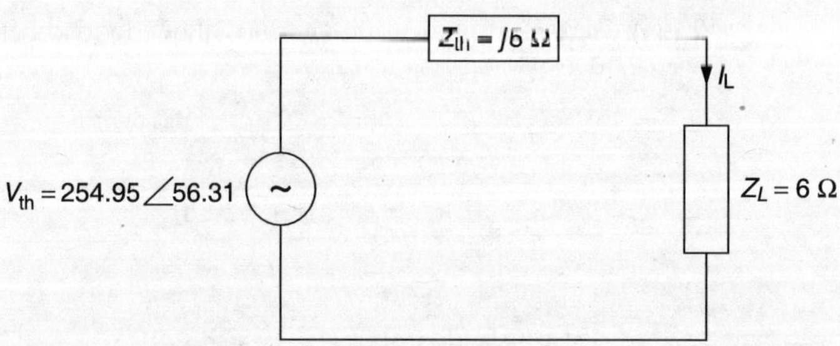

Figure 11 Thevenin's equivalent for the circuit shown in Figure 7.

The current through Z_L is

$$Z_L = \frac{254.95\angle 56.31^\circ}{6+j5}$$

$$I_L = \frac{254.95\angle 56.31^\circ}{7.81\angle 39.81^\circ} = 32.64\angle 16.5^\circ$$

Thus the complete expression for I_L will be

$$I_L = \sqrt{2}\times 32.64\sin(\omega t + 16.51)$$

$$\Rightarrow I_{(6\Omega)} = 46.17\sin(\omega t + 16.51)$$

6.2.5 Norton's Theorem

There is no difference in AC and DC analyses for Norton's theorem also. We shall solve the same problem shown in Figure 7 using Norton's theorem now.

As already stated, Z_{th} is exactly the same as Z_N. So, $Z_N = j\,5\,\Omega$. Next, we replace the 6 Ω load by a short circuit and find the current through it. The circuit looks as shown in Figure 12.

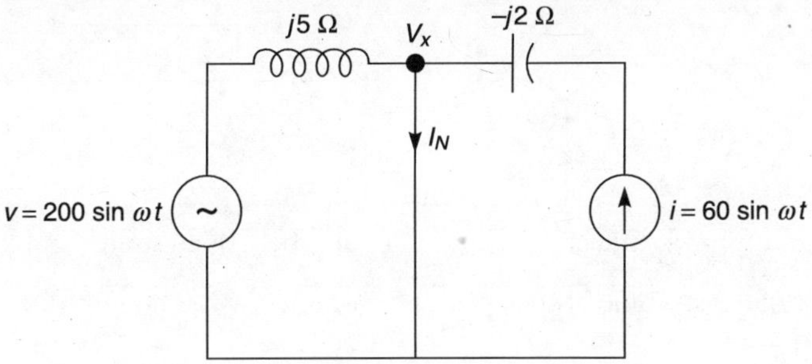

Figure 12 Finding I_N.

Applying KCL at the node having voltage V_x, we get

$$\frac{V_x - (200/\sqrt{2})\angle 0^\circ}{5\angle 90^\circ} + I_N - \frac{60}{\sqrt{2}}\angle 0^\circ = 0$$

Now $V_x = 0$ (as a result of the short circuit). So

$$I_N = \frac{40}{\sqrt{2}}\angle -90^\circ + \frac{60}{\sqrt{2}}\angle 0^\circ = 42.426 - j28.284$$

$$\Rightarrow I_N = 50.99\angle -33.69^\circ$$

Finally, Norton's equivalent circuit can be drawn as shown in Figure 13. Thus

$$I_L = \frac{I_N \times Z_N}{Z_N + Z_L} = \frac{50.99\angle -33.69^\circ \times 5\angle 90^\circ}{6+5j}$$

$$\Rightarrow I_L = \frac{50.99\angle -33.69^\circ \times 5\angle 90^\circ}{7.81\angle 39.81^\circ} = 32.64\angle 16.5^\circ$$

which also means $I_{6\,\Omega} = 46.17\sin(\omega t + 16.5)$.

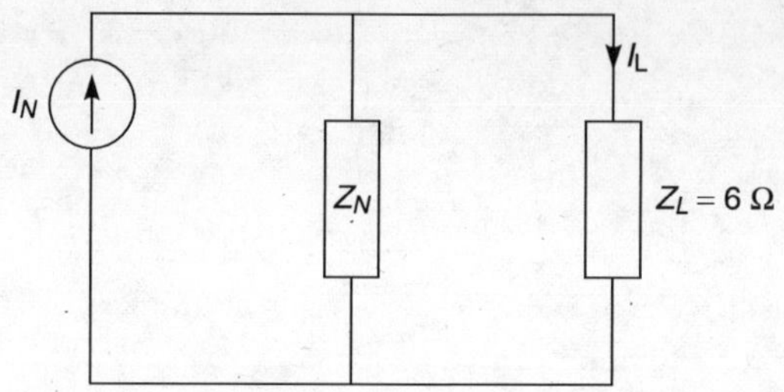

Figure 13 Norton's equivalent for the circuit shown in Figure 7.

6.2.6 Maximum Power Transfer Theorem

The maximum power transfer theorem is applicable for AC circuits with complex source impedances and complex load impedances, such as the circuit shown in Figure 14.

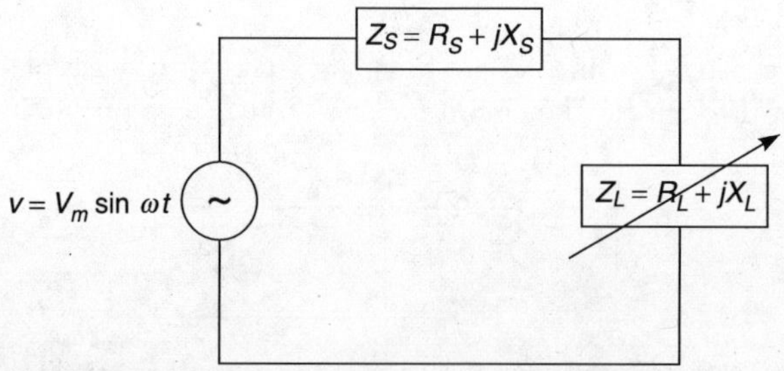

Figure 14 Maximum power transfer theorem for complex-valued source and load impedances.

As shown in Figure 14, the magnitude of the current flowing in the load is given by

$$I_L = \frac{V_m / \sqrt{2}}{\sqrt{(R_S + R_L)^2 + (X_S + X_L)^2}}$$

and the real power transferred to the load will be

$$P_L = I_L^2 R_L = \frac{V_m^2 / 2}{[(R_S + R_L)^2 + (X_S + X_L)^2]} \times R_L$$

Now if both R_L and X_L of the load Z_L are variable, we differentiate (and equate to zero) the expression of P_L with respect to X_L first and then we differentiate the answer with respect to R_L. Differentiating with respect to X_L and equating the answer to zero and solving, we get $X_L = -X_S$. Putting this value of X_L in the expression for P_L and then differentiating it with respect to R_L we get $R_L = R_S$. So, we have $Z_L = R_S - jX_S$.

Thus we conclude that if both load and source impedances are complex, the load impedance needs to be the complex conjugate of the source impedance for the maximum power to be delivered to it.

6.3 Analysis of Three-Phase Unbalanced Circuits

We have learnt about balanced three-phase circuits in Chapter 3. A three-phase system can remain balanced only if its all the three components (the generation, the transmission, and the loads) continue to remain balanced. Out of these three, it is most difficult to keep the loads balanced in a three-phase system. Hence it is important to learn about the analysis of unbalanced systems. A system becomes unbalanced when either the three phasors differ in magnitude and/or their phase differences become unequal. Usual mesh and nodal analyses are applied to solve the unbalanced three-phase systems. The analysis of a three-phase unbalanced circuit is best understood by taking some examples.

6.3.1 Problems on Analysis of Three-Phase Unbalanced Circuits

Example 1

Calculate the phase and line currents in the following three-phase delta connected load, which is supplied by a 400 V, three-phase, 50 Hz balanced supply.

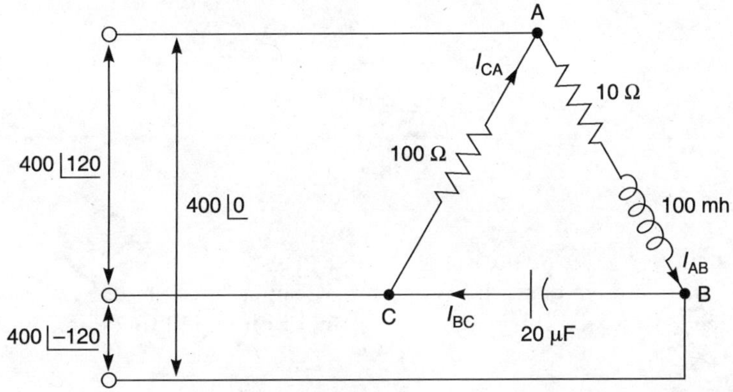

Figure 15 Unbalanced three-phase system for Example 1.

Solution

The load in this example is delta connected. In a delta connection, the line voltages are same as phase voltages. The phase impedance will be calculated with the given load's details. The phase currents will then be calculated using phase voltages and the corresponding phase impedances. Once the phase currents are obtained, KCL will be applied at nodes A, B, and C to calculate the line currents.

Calculation of phase impedances:

$$Z_{AB} = 10 + j(2\pi \times 50 \times 100 \times 10^{-3}) = 10 + j31.4 = 32.95\angle 72.33^\circ$$

$$Z_{BC} = -j\left(\frac{1}{2\pi \times 50 \times 20 \times 10^{-6}}\right) = 159.15\angle -90^\circ$$

$$Z_{CA} = 100\angle 0^\circ$$

Calculation of phase currents:

$$I_{AB} = \frac{V_{AB}}{Z_{AB}} = \frac{400\angle 0^\circ}{32.95\angle 72.33^\circ} = 12.14\angle -72.33^\circ = 3.68 - j11.56$$

$$I_{BC} = \frac{V_{BC}}{Z_{BC}} = \frac{400\angle -120^\circ}{159.15\angle -90^\circ} = 2.513\angle -30^\circ = 2.17 - j1.26$$

$$I_{CA} = \frac{V_{CA}}{Z_{CA}} = \frac{400\angle 120^\circ}{100\angle 0^\circ} = 4\angle 120^\circ = -2 + j3.46$$

Calculation of line currents:
Applying KCL at nodes A, B and C we get

$$I_A = I_{AB} - I_{CA} = (3.68 - j11.56) - (-2 + j3.46)$$
$$= 5.68 - j15.02$$
$$\Rightarrow I_A = 16.05\angle -69.28°$$
$$I_B = I_{BC} - I_{AB} = (2.17 - j1.26) - (3.68 - j11.56)$$
$$= -1.51 + j10.3$$
$$\Rightarrow I_B = 10.4\angle 98.3°$$
$$I_C = I_{CA} - I_{BC} = (-2 + j3.46) - (2.17 - j1.26)$$
$$= -4.17 + j4.72$$
$$\Rightarrow I_C = 6.29\angle 131.46°$$

Example 2

A three-phase star connected load consists of the impedances as follows: $Z_A = 50\ \Omega$, $Z_B = j20\ \Omega$, and $Z_C = -j20\ \Omega$. A balanced three-phase 400 V, 50 Hz AC supply is connected to it. Calculate the voltage drop across each branch. Also find the potential of neutral with respect to ground.

Solution

The nature of impedances given shows that the load is as shown in Figure 16.

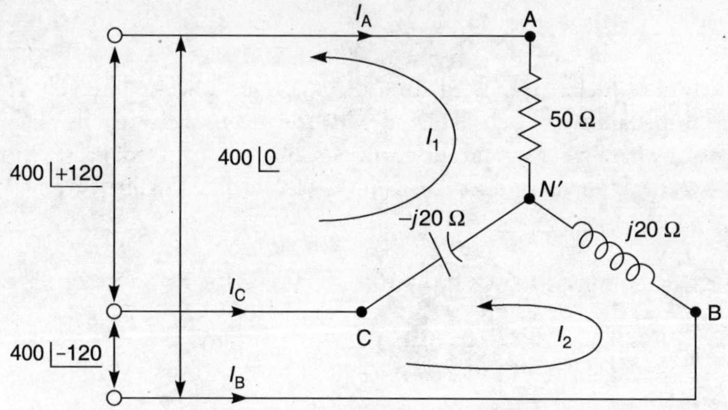

Figure 16 Unbalanced three phase system for Example 2.

Calculation of currents:

The load is star connected; hence the line currents will be the same as the phase currents. The phase currents cannot be found here directly. The balanced line voltages will not impress the balanced phase currents through the load, as the load is unbalanced. Hence KVL will be applied to find the currents.

Let us consider the two loop currents I_1 and I_2 as shown. The KVL can be written as follows.

In loop 1,

$$400\angle 120° = (50 - j20)I_1 - (-j20)I_2$$

or
$$400\angle 120° = (50 - j20)I_1 + j20I_2 \tag{6.1}$$

In loop 2,

$$400\angle -120° = -(-j20)I_1 + (-j20 + j20)I_2$$

or

$$400\angle -120° = j20I_1 = 20I_1\angle 90° \tag{6.2}$$

This implies

$$I_1 = \frac{400\angle -120°}{20\angle 90°} = 20\angle -210°$$

Substituting Eq. (6.2) in Eq. (6.1), we get

$$400\angle 120° = (50 - j20)20\angle -120° + j20I_2$$
$$400\angle 120° = (53.85\angle -21.8°)(20\angle -210°) + j20I_2\angle 90°$$
$$400\angle 120° = 1077\angle -231.8° + j20I_2$$

On solving, we get

$$I_2 = 34.17\angle -137.01°$$

Calculation of voltage of neutral (or ground):

The balanced voltage supply can be represented by the phasor diagram shown in Figure 17. It shows that potential of phase A with respect to neutral (i.e., ground) is $V_{AN}\angle -30°$. In the given problem, therefore,

$$V_{AN} = (400/\sqrt{3})\angle -30° = 230.94\angle -30° \tag{6.3}$$

For the unbalanced load given, the potential of phase A with respect to neutral N' would be 50 I_1. Therefore

$$V_{AN'} = 50(20\angle -210°) = 100\angle -210° \tag{6.4}$$

Hence the potential of neutral N' with respect to ground N will be

$$V_{N'N} = V_{AN} - V_{AN'} = 230.94\ \angle -30° - 100\angle -210° = \angle 330.94\angle -30°$$

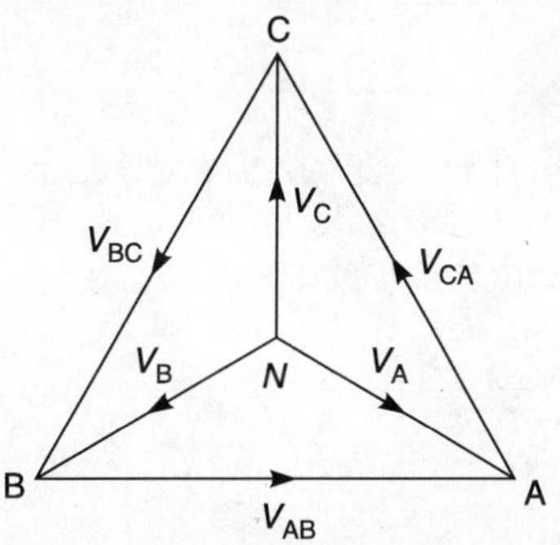

Figure 17 Phasor diagram showing neutral voltage under balanced condition.

6.4 Cathode Ray Oscilloscope and Its Applications

A Cathode Ray Oscilloscope (CRO) is a device which is used to display a voltage or current waveform on a graphical screen. It is mainly used to see the waveform, measure the voltages and the currents (both DC and AC), measure the frequency of a waveform, its phase shift, etc. It is also used for testing of various components.

A CRO basically consists of a Cathode Ray Tube (CRT) panel. The CRT is a cylindrical device which has a screen on its front. It basically consists of an electron gun to produce a beam of electrons. There is an emitting cathode, a modulator, and a focused anode in this electron gun. A beam of electrons is produced by this gun which passes through two pairs of electrodes. These sets of electrodes are placed at a certain distance and are perpendicular to each other. These are called the sets of *X* (horizontal) and *Y* (vertical) plates. The *X* plates are placed vertically so that they may deflect the beam in the horizontal direction. The basic components of a CRT are shown in Figure 18. The electron beam, starting from the electron gun and passing through the *X* and *Y* plates, strikes the screen placed at the other end of CRT. The screen is coated with a fluorescent material which illuminates the points where the electrons hit it. Thus a waveform is seen on the screen of the CRT. There is a grid made on the screen so that the measurement can be made on it.

A saw-tooth waveform (called the time base) is given to the *X* plates. This waveform is responsible for the regular horizontal deflection of the beam. The *Y* plates are placed horizontally. They are fed with the signal to be measured. They cause the vertical deflection of the beam. Thus, the deflecting arrangement (consisting of *X* and *Y* plates) facilitates the real-time movement of the input signal inside the CRT. The frequency of time base, the vertical and horizontal shifts, and the magnitude of waveform can be adjusted to get a steady-state waveform on the screen for measurement.

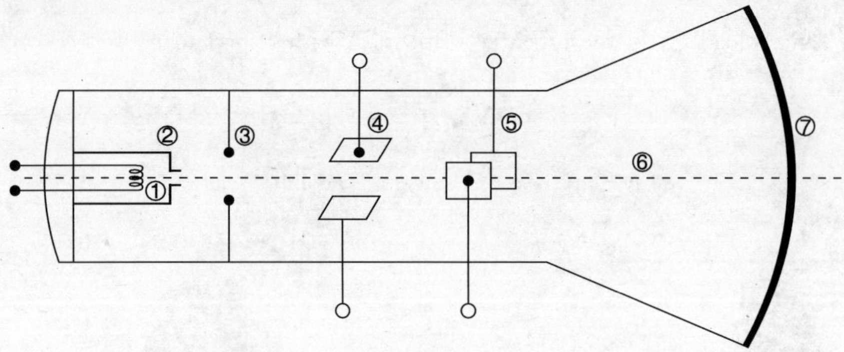

Figure 18 A Cathode Ray Tube. ① Cathode producing electrons due to heating of an element. ② Mod erator to focus the beam of electrons. ③ Anode to accelerate the electrons. ④ Set of horizontal (*Y*) plates. ⑤ Set of vertical (*X*) plates. ⑥ Beam of electrons. ⑦ Fluorescent screen.

The parts 1, 2 and 3 make up the electron gun.

The CRO works using a CRT. This is explained by a simplified diagram in Figure 19.

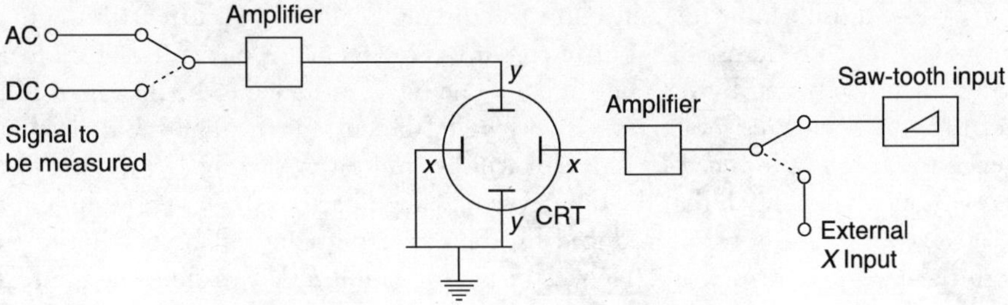

Figure 19 A Cathode Ray Oscilloscope.

The AC or DC signal to be measured is fed through the terminals shown on the left-hand side of Figure 19. In order to keep the signal within the measurement range, an amplifier is used through which the magnitude of signal is adjusted in steps. The signal measured is multiplied by the amplification factor to get the actual value.

The grounds plates of *X* and *Y* are grounded together. A saw-tooth signal is fed to the *X* plates. The signal to be measured rides over this saw-tooth wave which carries it forward with respect to time. This is explained in Figure 20.

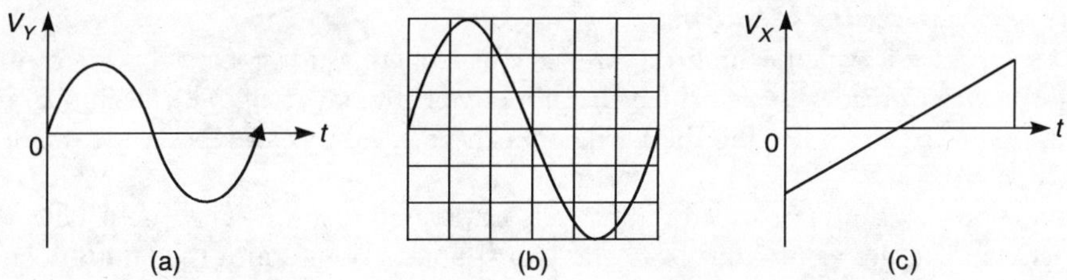

Figure 20 Superimposition of V_Y and V_X to show a waveform on screen: (a) Signal to be measured. (b) Screen of CRO. (c) Saw-tooth for horizontal deflection.

6.5 Measurement of Insulation Resistance

The most commonly used instrument for measurement of high (insulation) resistance is a Megger, which is shown in Figure 21.

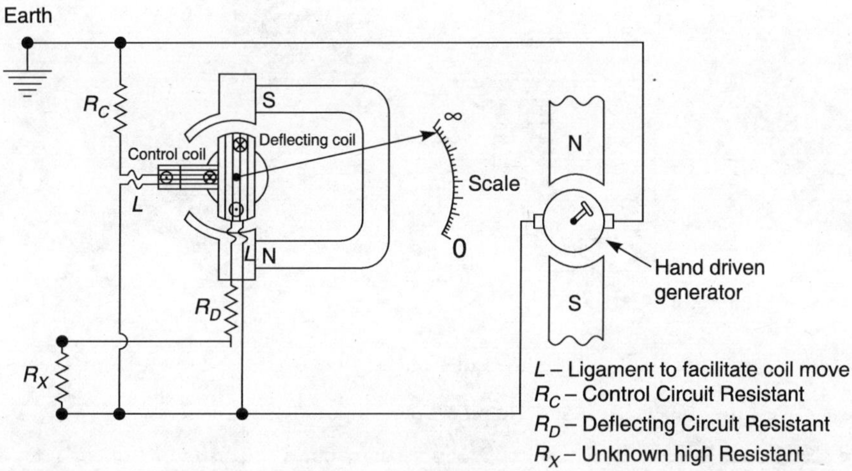

Figure 21 A Megger.

In a Megger, R_D is the resistance in the deflection coil circuit. It limits the current through the coil. The supply voltage appears directly across the control coil and is independent of the resistance to be measured. Hence a fixed torque is produced by the control coil. The high resistance to be measured is kept in series with the deflecting coil. If this resistance (R_X) is very high, the current through the deflecting coil will be very small; hence the torque produced by this coil will be minimum (nearly zero). Thus the cumulative effect of the two torques will be such that the pointer would align itself to show ∞ reading. On the contrary, if the resistance R_X is zero, the torque produced by the deflecting coil will be maximum (much higher than that produced by the control coil). Hence the pointer indicates 0 reading on the scale. This is a portable instrument free from the need of an external voltage supply. Hence, it is a popular device for the measurement of very high resistances like insulation resistance.

6.6 Magnetic Circuits and Transformers

The explanation of the operation of a transformer mainly requires an understanding of magnetic circuits. Some topics related to magnetic circuits are discussed in the following text.

6.6.1 Review of Laws of Magnetism

Magnets are ferrous metals which exhibit the property of attracting iron pieces. They are either available naturally or are artificially made. A magnet has the property of aligning itself (when held free) in the north–south direction with its "N" indicating the Earth's North pole. Like poles repel while the opposite poles attract each other.

A magnet develops lines of force called flux, which are elastic in nature. They are parallel to each other and never intersect. The flux comes out from the North pole and goes into the South pole as shown in Figure 22(a).

When a current of I amperes flows through a coil having N turns [wound over a magnetic core as shown in Figure 22(a)], the soft iron core gets magnetized. Such magnets are called Electromagnets. The polarity of such a magnet is decided by the sense of winding and the direction of current flowing through it. Figure 22(b) reveals that if we see the core from its end A, the current's direction appears to be anti-clockwise, making this end (end A) of North polarity. However, the core seen from the other end (B) indicates that the current's direction appears to be clockwise. Hence end B becomes the South pole as shown in Figure 22(b).

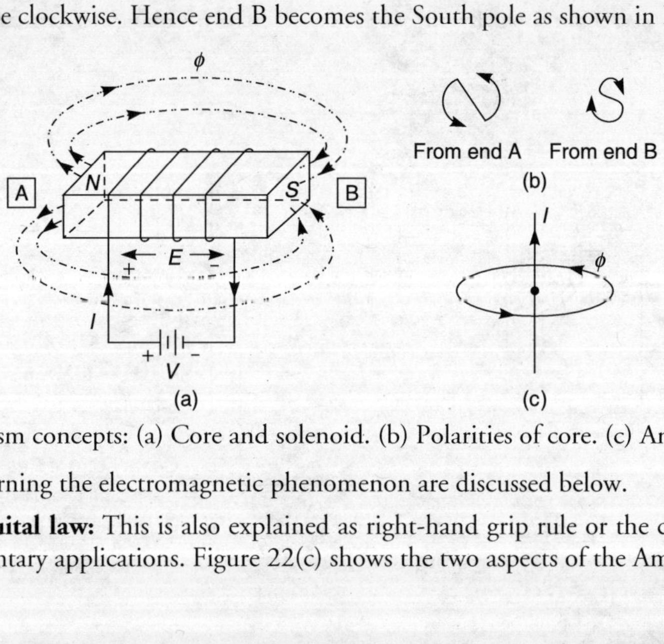

Figure 22 Magnetism concepts: (a) Core and solenoid. (b) Polarities of core. (c) Ampère's circuital law.

Some rules/laws governing the electromagnetic phenomenon are discussed below.

1. **Ampère's circuital law:** This is also explained as right-hand grip rule or the cork screw rule. It has two complementary applications. Figure 22(c) shows the two aspects of the Ampère's circuital law.

(a) If an electric current passes through a solenoid, resulting in a magnetic field, we can find the direction of flux. Wrap your right hand around the solenoid with your fingers pointing in the direction of current, your thumb points in the direction of flux.

(b) If an electric current passes through a straight wire, place your thumb pointing in the direction of the current. Then your fingers point in the direction of flux that would surround the conductor.

2. **Faraday's Law of Electromagnetic Induction:** This law states that if a current due to a voltage source causes flux inside a solenoid, which, in turn, causes development of an electromotive force (emf) in that coil. This emf is directly proportional to the number of turns of the coil and the rate of change of the flux.
3. **Lenz's Law:** This law states that the induced emf due to Faraday's law is such that it opposes the cause of flux, that is the current, and hence the applied voltage. The expression of induced emf is given by

$$E = -N\frac{d\phi}{dt}\,\text{V}$$

where N is the number of turns of the coil, ϕ is the flux developed. The minus sign in the above expression is due to Lenz's law to signify that this emf will oppose the voltage source that is causing flow of current in the circuit. In Figure 22(a), it is shown that E opposes V.

6.6.2 Energy Storage in a Magnetic Circuit

From Faraday's law, we know that in a solenoid the instantaneous value of induced voltage is given by

$$e = N\frac{d\phi}{dt} \tag{6.5}$$

where N is the number of turns and ϕ is the magnetic flux in webers. If the current through the coil is i A, then the instantaneous power p is given by

$$p = e \cdot i \tag{6.6}$$

The energy stored in the coil in time dt is given by

$$E = e \cdot i \cdot dt \tag{6.7}$$

Using Eq. (6.5), we get

$$E = N \cdot i \cdot d\phi \tag{6.8}$$

The total energy stored in the coil during a quarter cycle of sinusoidal flux will be

$$E_{\text{stored}} = \int_0^{\phi_{\max}} N \cdot i \cdot d\phi \tag{6.9}$$

This concept can be further understood by the following analysis.

Consider a magnetic circuit of length l and cross-sectional area a. The magnetic flux density is defined as

$$B = \frac{\phi}{a}\ \text{Wb/m} \tag{6.10}$$

or

$$\phi = B \cdot a \tag{6.11}$$

Differentiating, we get

$$d\phi = a \cdot dB \tag{6.12}$$

Substituting Eq. (6.12) into Eq. (6.9), we get

$$E_{\text{stored}} = \int_{0}^{B_{\max}} N \cdot i \cdot a \cdot dB \tag{6.13}$$

This can be further written as

$$E_{\text{stored}} = \int_{0}^{B_{\max}} \frac{N \cdot i}{l} \cdot a \cdot l \cdot dB \tag{6.14}$$

We know that if the magnetic field is uniform, then

$$H = \frac{N \cdot i}{l} \text{ AT/m} \tag{6.15}$$

and

$$v = a \cdot l \text{ m}^3$$

where v is the volume of the magnetic material. Substituting Eq. (6.15) into Eq. (6.14), we get

$$E_{\text{stored}} = v \int_{0}^{B_{\max}} HdB \tag{6.16}$$

If the magnetic circuit is linear, then

$$E_{\text{stored}} = v \cdot \frac{1}{2} HB$$

Now since $B = \mu_0 \mu_r H$,

$$E_{\text{stored}} = \frac{1}{2} v \frac{B^2}{\mu_0 \mu_r} \tag{6.17}$$

6.6.3 Analyses of Linear and Non-Linear Magnetic Circuits

With the previous knowledge of the properties of magnetic materials, we know that for solving problems based on magnetic circuits, it is often required to refer to the B–H relationship. Every magnetic material has its own B–H curve. The value of flux density B changes due to change in either the flux or the cross-sectional area of the magnetic path. Similarly, the value of H changes due to change in either the current, the number of turns or the length of the magnetic path. The B–H curve is mainly used to find the change in B corresponding to the change in H or vice versa.

The B–H curve for a linear magnetic circuit is a straight line. While solving the problems based on linear magnetic circuits, the value of B (or H) corresponding to the new H (or B) can be obtained using the given straight line relationship. However, for a non-linear magnetic circuit, it is necessary to have the B–H curve of the material used so as to get these values.

6.6.4 Per Unit Value

The **per unit value** of a quantity (also called **pu value**) is defined by the ratio of the given value to a base value of that quantity. For the case of a single-phase transformer, the per unit values of the various parameters (or impedances), that is, r_1, x_1, r_2, x_2, R_L, and X_m, are important.

The rating of the transformer is in volt-amperes (VA). Also given are the high-voltage and low-voltage values. These values give the base impedance (called Z_B) of the transformer.

Thus if, for example, a 1 KVA, 200V/100V transformer is considered, we have two base impedances: one for the high voltage (HV) side and one for the low voltage (LV) side.

$$Z_{B(\text{for HV side})} = \frac{V^2}{VA} = \frac{200^2}{1000} = 40\ \Omega$$

and

$$Z_{B(\text{LV side})} = \frac{V^2}{VA} = \frac{100^2}{1000} = 10\ \Omega$$

Now (if say) $r_{1e} = 0.5\ \Omega$, the per unit of the equivalent resistance will be (taking the HV side equivalent resistance)

$$\text{pu } r_{1e} = \frac{r_{1e}}{Z_{B(\text{HV side})}} = \frac{0.5}{40} = 0.0125$$

Now, referring the equivalent resistance to the LV side,

$$r_{2e} = r_{1e}\left(\frac{100}{200}\right)^2 = \frac{r_{1e}}{4} = \frac{0.5}{4}$$

The LV side equivalent resistance, in pu will be

$$\text{pu } r_{2e} = \frac{r_{2e}}{Z_{B(\text{LV side})}} = \frac{0.5}{4\times 10} = 0.0125$$

which is the same as the pu value obtained taking the HV side value. Thus we conclude that the pu value remains unchanged in moving from low voltage to high voltage side (and vice versa) in a transformer. Similarly, the pu value of reactance is given by

$$\frac{x_{1e(\text{HV side})}}{Z_{B(\text{HV side})}} = \frac{x_{2e(\text{LV side})}}{Z_{B(\text{LV side})}}$$

The pu values of R_L and X_m can also be found in a similar manner. If the pu value is multiplied by 100, the resulting quantity is called the percentage value. By using pu values, calculations become easier.

6.6.5 Phasor Diagram of a Single-Phase Transformer

The phasor diagram of a single-phase transformer shows the phasor relationship between various voltages and currents. The exact equivalent circuit is reproduced in Figure 23 whose phasor diagram would be drawn.

First of all it needs to be stated that $N_1 = N_2$ is taken (usually) for drawing the phasor diagram. Thus if the linking flux (which is common for both primary and secondary circuits) is taken as the reference, the induced voltage phasors E_1 and E_2 will be drawn at 90° lagging behind the flux, that is, along the negative Y-axis. The load voltage V_2 lags E_2 by a small angle. In the phasor diagram for a lagging power factor load, the current I_2 lags behind V_2 by an angle ϕ (say). The voltage drops I_2r_2 (parallel to I_2 along V_2) and I_2x_2 (perpendicular to I_2) are then added to V_2 to get E_2 as seen from the equivalent circuit shown in Figure 23.

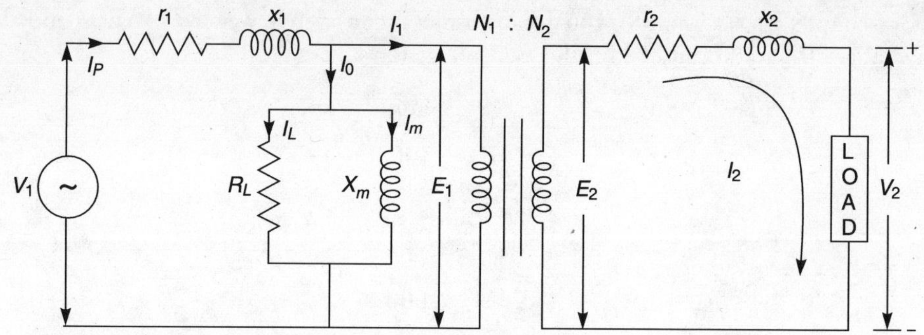

Figure 23 Exact equivalent circuit of a transformer for drawing the phasor diagram.

The no-load current I_0 is the vector sum of I_L and I_m. Addition of the current $I_1'(= I_2$ but 180° out of phase with it) gives the complete primary current I_p. Addition of the drops $I_p r_1$ and $I_p x_1$ to the magnitude E_1 gives the applied voltage V_1 as shown in Figure 24. For unity pf loads, I_2 is in line with V_2 and for leading pf loads I_2 is ahead of V_2 in the counter-clockwise direction. It can be noticed that for the leading pf case, the magnitude of voltage V_2 may be greater than that of E_2.

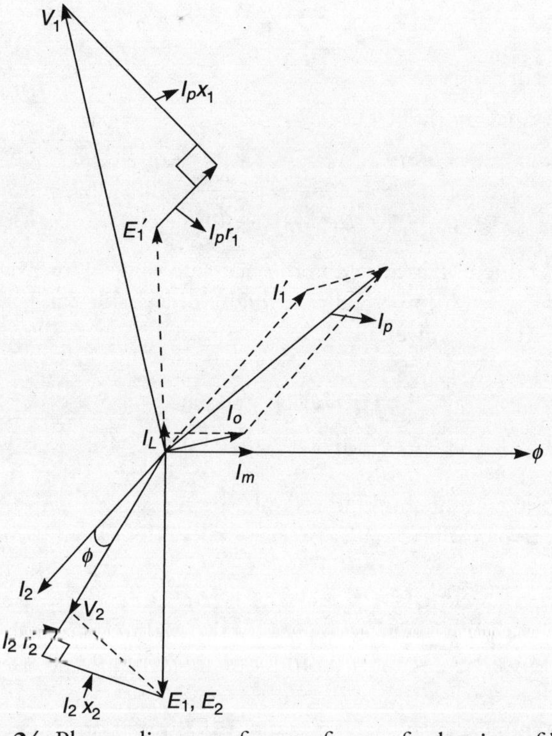

Figure 24 Phasor diagram of a transformer for lagging pf loads.

6.6.6 Open-Circuit and Short-Circuit Tests

The open-circuit and the short-circuit tests are conducted to find the various impedances featuring in the equivalent circuit and also to find the iron and copper losses in a single-phase transformer. Table 1 shows the salient features of the two tests.

Table 1 Comparison between the O/C and S/C Tests

Open-circuit test	*Short-circuit test*
1. Always done at rated voltage, reduced (no load) current.	1. Usually done at rated current, reduced voltage.
2. Done to find the iron losses, R_L and X_m.	2. Done to find the copper losses, r_{1e} and x_{1e}.
3. Usually conducted with the instruments in the LV side with the HV side kept open.	3. Usually done with the instruments in the HV side with the LV side kept shorted.

Open-Circuit Test

As mentioned in Table 1, this test is done to find the iron losses and shunt branch parameters R_L and X_m. The circuit arrangement for the text is shown in Figure 25. The test is done with the instruments connected in the LV side, with the HV side open-circuited. As can be seen, there is no output in the absence of a load. Also the only current flowing (in the primary winding) is the no load current I_0 which is 2–5% of the rated primary current, hence the copper loss in the primary winding is negligible ($I_0^2 r_1$ is small). Hence all the input power which will be registered by the wattmeter is the iron loss.

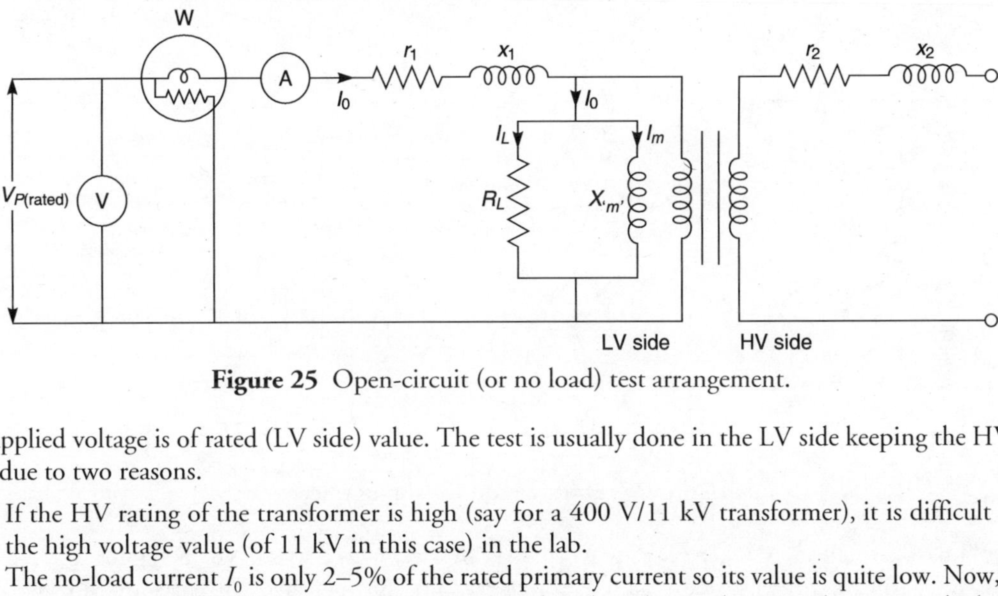

Figure 25 Open-circuit (or no load) test arrangement.

The applied voltage is of rated (LV side) value. The test is usually done in the LV side keeping the HV side open due to two reasons.

1. If the HV rating of the transformer is high (say for a 400 V/11 kV transformer), it is difficult to get the high voltage value (of 11 kV in this case) in the lab.
2. The no-load current I_0 is only 2–5% of the rated primary current so its value is quite low. Now, since the current in the LV side is (turns ratio times) higher than that in the HV side, we get a higher current value if the instruments are connected in the LV side. Hence a more accurate metering of the no-load current is observed this way.

Neglecting r_1 and x_1 (as $|I_0 r_1|$ and $|I_0 x_1|$ drops will be of very low value and $V_1 \sim E_1$), we get the following equivalent circuit under no-load (or open circuit) test (see Figure 26).

We have the readings of the three instruments W_i, V_r and I_0. We can say that $W_i = V_r I_0 \cos\phi_0$ (real power expression). This will give $\cos\phi_0$ and hence we get $\sin\phi_0$. Now $I_0 \cos\phi_0 = I_R$ and $I_0 \sin\phi_0 = I_m$ and $V_r / I_R = R_L$ and $V_r / I_m = X_m$.

Thus the values of the shunt branch parameters can be found.

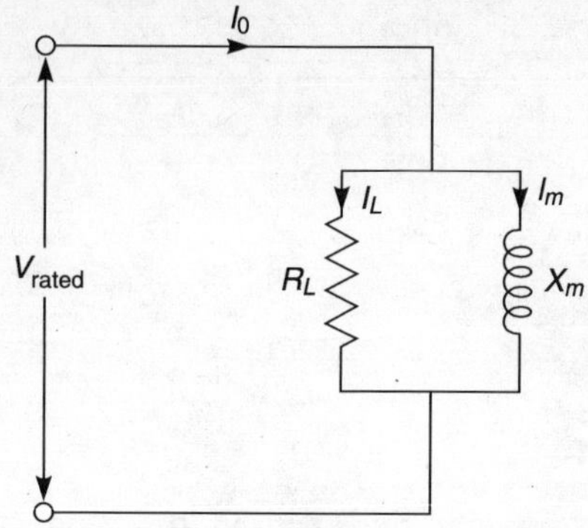

Figure 26 The equivalent circuit in an open-circuit test.

Short-Circuit Test

This test is done to find the copper loss and the values of series branch parameters r_{1e} and x_{1e}. The circuit arrangement is shown in Figure 27.

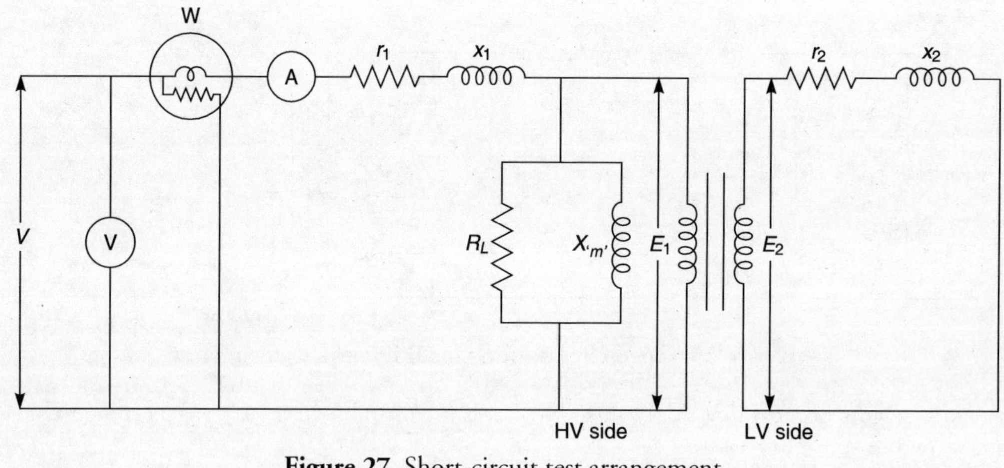

Figure 27 Short-circuit test arrangement.

The secondary side (usually taken as the LV side) is short circuited as shown in Figure 27. Then the applied voltage in the primary side is raised from zero value till the rated current starts flowing in the primary (indicated by the ammeter) and hence in the secondary side. It is observed that since the load impedance is zero (shorting means zero impedance), the voltage required to make the rated current flow in the primary (and hence also in the secondary) will be very less. This voltage is around 5% of the rated voltage of the primary side. Now normally the no-load current happens to be 2–5% of the rated primary current when the rated voltage is applied. In the short-circuit test where the applied voltage is only 5% of its rated value, the no-load current would be 5% of 2–5% of the primary rated current value. Obviously, the no-load current during the short-circuit test is almost negligible and is hence neglected. This also indicates that the branches X_m and R_L, where I_0 splits and flows, are of very high ohmic values (hence the current is less) so they can be neglected. Thus the equivalent circuit test referred to the primary side is as shown in Figure 28.

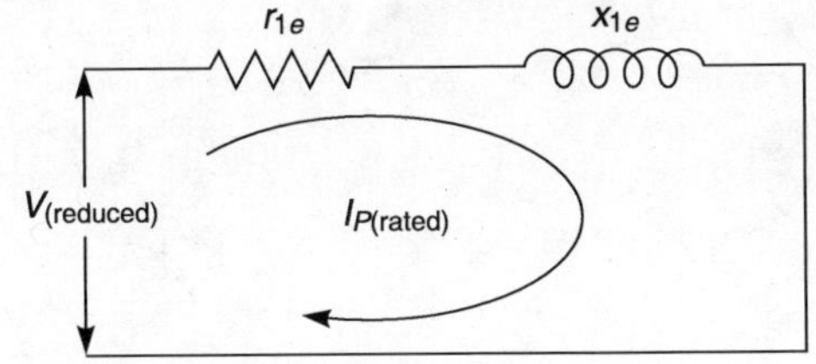

Figure 28 Transformer equivalent during short-circuit test.

The three meter readings (wattmeter, voltmeter, and ammeter) give the copper loss, rated current, and reduced voltage applied across the primary winding. Since I_0 is neglected, the iron loss is also neglected. Hence the wattmeter shows only the copper losses ($I_1^2 r_1$ and $I_2^2 r_2$) of the primary and secondary windings. So we have

$$W_{\text{Cu}} = V_{\text{reduced}} \times I_{\text{rated}} \times \cos\phi$$

where W_{Cu}, V_{reduced}, and I_{rated} are the wattmeter, voltmeter, and ammeter readings, respectively. The above equation gives the value of cos ϕ. Also

$$Z_{1e} = \frac{V_{\text{reduced}}}{I_{\text{rated}}}$$

$$r_{1e} = Z_{ie} \cos\phi$$

$$x_{1e} = Z_{ie} \sin\phi$$

Thus we get the rated copper loss along with the equivalent resistance and reactance referred to the primary winding. If r_{2e} and x_{2e} are desired, we can calculate them through $r_{2e} = r_{1e}(N_2/N_1)^2$ and $x_{2e} = x_{1e}(N_2/N_1)^2$.

Note that we cannot find individual resistance r_1 and r_2 separately. Similarly we cannot find x_1 and x_2.

6.6.7 Voltage Regulation

The voltage regulation in a single-phase transformer is defined as the voltage difference between no-load voltage and full load voltage taken as the percentage of no-load voltage (where the supply voltage and the power factor of the load are maintained constant). Thus,

$$\text{Voltage regulation} = \frac{V_{2(\text{NL})} - V_{2(\text{FL})}}{V_{2(\text{NL})}} \tag{6.18}$$

In some cases, the denominator contains $V_{2(\text{FL})}$ and

$$\text{Voltage regulation} = \frac{V_{2(\text{NL})} - V_{2(\text{FL})}}{V_{2(\text{FL})}} \tag{6.19}$$

Equation (6.18) is called "down-regulation" and Eq. (6.19) is called "up-regulation." Generally, Eq. (6.18) is used for a transformer [while Eq. (6.19) is used for alternators and transmission lines].

In order to derive the expression for voltage regulation, we represent the equivalent circuit of a single-phase transformer in the secondary side. Also since the shunt branch parameters R_L and X_m do not influence the voltage, they are not brought to the secondary side. The equivalent circuit thus looks as shown in Figure 29, where V_1' is the rated voltage referred to the secondary side, and r_{2e} and x_{2e} are the resistance and leakage reactance referred to the secondary side.

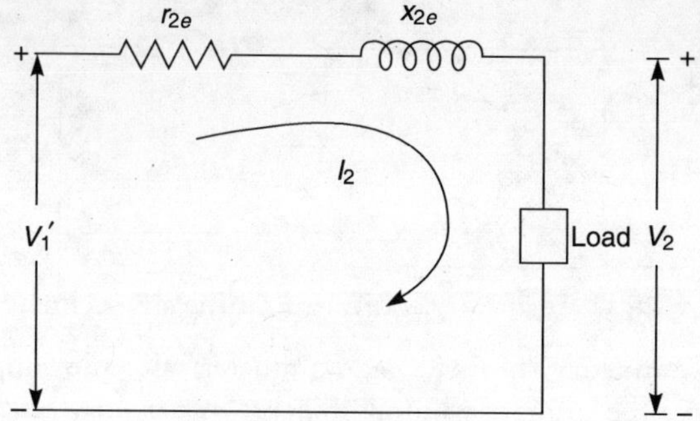

Figure 29 Approximate equivalent circuit of a transformer.

The phasor diagram for the circuit shown in Figure 29 looks as shown in Figure 30. Here the load considered is a lagging power factor load.

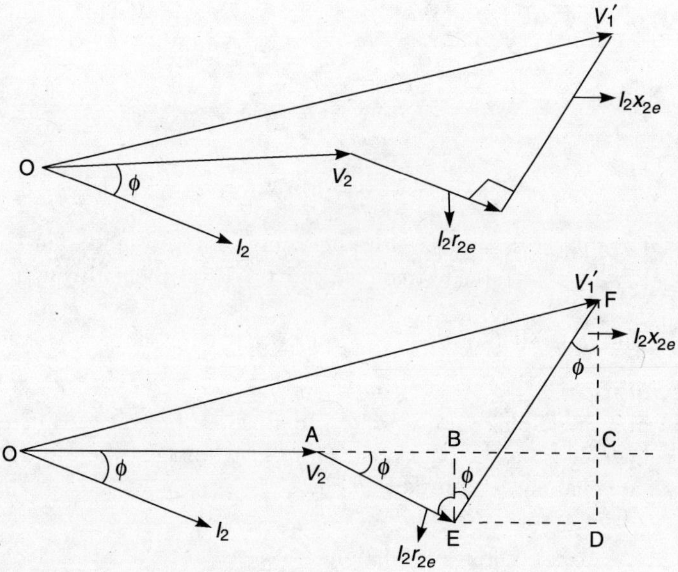

Figure 30 Phasor diagram to find voltage regulation (a) Phasor diagram for the circuit of Figure 29. (b) Details of Figure 29(a).

Applying the definition of (down) voltage regulation we get

$$\% \text{ Regulation} = \frac{V_1' - V_2}{V_1'}$$

From Figure 30(b), we can write:

$$V_1' = \text{OA} + \text{AE} + \text{EF}$$

Taking BE and FC to be very small, we can write AE ~ AB and OF ~ OC, which gives us the following relation:

$$V_1' = V_2 + \text{AB} + \text{ED}$$

Now AB = $I_2 r_{2e} \cos\phi$ and ED = $I_2 x_{2e} \sin\phi$ so

$$V_1' = V_2 + I_2 r_{2e} \cos\phi + I_2 x_{2e} \sin\phi$$

Thus,

$$\frac{V_1' - V_2}{V_1'} = \frac{I_2 r_{2e} \cos\phi + I_2 x_{2e} \sin\phi}{V_1'}$$

Hence, the right-hand side (RHS) can be written as

$$\frac{r_{2e} \cos\phi + x_{2e} \sin\phi}{V_1'/I_2}$$

Now V_1'/I_2 can be said to be equal to the base impedance Z_B (provided the load is rated and so is the current I_2). Also, the left-hand side (LHS) is the definition of regulation. Hence

$$\text{Regulation} = \frac{r_{2e} \cos\phi + x_{2e} \sin\phi}{Z_B} = \varepsilon_r \cos\phi + \varepsilon_X \sin\phi$$

where ε_r is the per unit resistance $= r_{2e}/Z_B$ and ε_X is the per unit reactance $= x_{2e}/Z_B$. Thus

$$\%\ \text{Regulation} = [\varepsilon_r \cos\phi + \varepsilon_X \sin\phi] \times 100 \qquad (6.20)$$

Cases of Zero and Maximum Voltage Regulation

1. **Zero Voltage Regulation:** The zero voltage regulation condition can be obtained by equating the RHS of Eq. (6.20) to 0, as follows:

$$\varepsilon_r \cos\phi + \varepsilon_X \sin\phi = 0$$

which gives

$$\tan\phi = -\frac{\varepsilon_r}{\varepsilon_X} \quad \text{or} \quad \phi = \tan^{-1}\left[-\frac{\varepsilon_r}{\varepsilon_X}\right]$$

We had initially taken a lagging load with the current lagging behind the voltage by ϕ degrees. Also we had taken the values of ϕ as positive in all our regulation calculations. Hence a negative ϕ at zero voltage regulation [i.e., $\phi = \tan^{-1}(-\varepsilon_r/\varepsilon_X)$] indicates a leading power factor load. So we conclude that "zero voltage regulation is possible only for capacitive loads."

2. **Maximum Voltage Regulation:** For finding the condition of maximum voltage regulation, we differentiate the expression of regulation with respect to ϕ and equate the answer to zero. Differentiating Eq. (6.20) w.r.t. ϕ, the RHS becomes

$$-\varepsilon_r \sin\phi + \varepsilon_X \cos\phi$$

For finding the maxima we equate this result to zero and solve. We get $\tan\phi = \varepsilon_X/\varepsilon_r$.

In a triangle if $\tan\phi = \varepsilon_X/\varepsilon_r$, then $\sin\phi = \varepsilon_X/\sqrt{\varepsilon_X^2 + \varepsilon_r^2}$ and $\cos\phi = \varepsilon_r/\sqrt{\varepsilon_r^2 + \varepsilon_X^2}$. Thus $\cos\phi = \dfrac{\varepsilon_r}{\varepsilon_Z}$ and $\sin\phi = \varepsilon_X/\varepsilon_Z$. Substituting these values of $\sin\phi$ and $\cos\phi$ in the original expression of regulation we get the value of the maximum voltage regulation as

$$\varepsilon_r \times \frac{\varepsilon_r}{\varepsilon_Z} + \varepsilon_X \times \frac{\varepsilon_X}{\varepsilon_Z} = \frac{\varepsilon_r^2 + \varepsilon_X^2}{\varepsilon_Z} = \frac{\varepsilon_Z^2}{\varepsilon_Z} = \varepsilon_Z$$

Thus we see that the maximum value of voltage regulation is equal to the per unit leakage impedance ε_Z

$$\varepsilon_Z = \frac{Z_{2e}}{Z_B} = \frac{\sqrt{r_{2e}^2 + x_{2e}^2}}{Z_B}$$

6.7 Performance of DC and AC Machines

The following section takes up certain issues which affect the performances of both DC and AC machines. The efficiency and starting and speed control of DC and the starting of the three-phase induction motors have been discussed.

6.7.1 Losses and Efficiency Calculation of DC Motors

The efficiency of the DC motor (or any motor for that matter) is always less than that of a transformer. This happens due to the presence of an additional loss, that is, the rotational loss in motors, which is totally absent in transformers. There are two parts to this rotational loss: frictional loss and windage loss.

Losses in DC Motors

There are three major losses occurring in a DC motor. They are:

1. Copper loss (electrical).
2. Iron loss (electrical).
3. Friction and windage losses (mechanical).

A DC motor carries DC field current in its stator winding and AC armature current in its rotor winding. In both these windings, corresponding I^2R losses (called **copper losses**) take place.

The current flowing in the armature winding is an AC current. The brush commutator arrangement converts it from DC to AC. The core of the rotor, on which the armature winding is placed, is made up of magnetic material. Hence, eddy current losses take place in DC machines. The rotation of the rotor is required to overcome the bearing and air friction losses known as friction and windage losses.

Efficiency of DC Motors

We know that in a DC motor the major losses taking place are the copper loss the iron loss, and the friction and windage loss.

The efficiency of DC motor is defined as

$$\eta = \frac{\text{Output power}}{\text{Input power}} = \frac{\text{Input power} - \text{Losses}}{\text{Input power}} = \frac{\text{Output power}}{\text{Output power} + \text{Losses}}.$$

Let the losses be represented as

$$\text{Copper loss} = (I_a^2 R_a + I_f^2 R_f) = P_c \text{ watts}$$

$$\text{Iron loss} = P_i \text{ watts}$$

$$\text{Friction and windage loss} = P_{\text{FW}} \text{ watts}$$

Therefore total losses $P_{\text{Loss}} = P_c + P_i + P_{\text{FW}}$. Thus,

$$\eta = \frac{P_{\text{Output}}}{P_{\text{Output}} + P_{\text{Loss}}}$$

and

$$\%\eta = \frac{P_{\text{Output}}}{P_{\text{Output}} + P_{\text{Loss}}} \times 100$$

6.7.2 DC Motor Starter

We know that initially, when the rotor is stationary, the back emf developed in the motor is zero. Hence all the applied voltage is fed to the armature winding. This results in a large amount of armature current to flow through it which may damage it. The expression of armature current is $I_a = (V_t - E_b)/R$, where V_t is the terminal (supply) voltage, E_b is the back emf (which is proportional to speed N), $R = R_a$ (for shunt motors) and $= R_a + R_f$ (for series motors).

It is obvious from above expression that I_a can be reduced during starting by inserting a high resistance in series with the armature winding. Once the motor picks up speed, E_b gets developed, which automatically reduces I_a. At such condition, the addition resistance can be gradually removed from the armature circuit to allow the motor to run normally.

A three-point DC motor starter for a DC shunt motor is explained here. As shown in Figure 31, it consists of a spring mounted handle (H) which makes contact with the external starting resistance (R_{ext}) in steps. When it is brought manually to its ON position, the "No-Volt Release" magnet gets magnetized as the circuit becomes complete. This magnet holds the handle to the ON position as long as the motor runs safely.

It is seen that at this position of the handle, the R_{ext} is completely removed from the armature current of the motor.

The three-point starter not only helps in starting a motor, but also prevents the motor from getting overloaded. In case a current greater than its rated value flows through the motor, another magnet, called "Overload Release," operates which short-circuits the "No-Volt Release" coil. Thus the latter gets de-energized thereby releasing the handle "H" to "OFF" position due to the action of the spring attached to it.

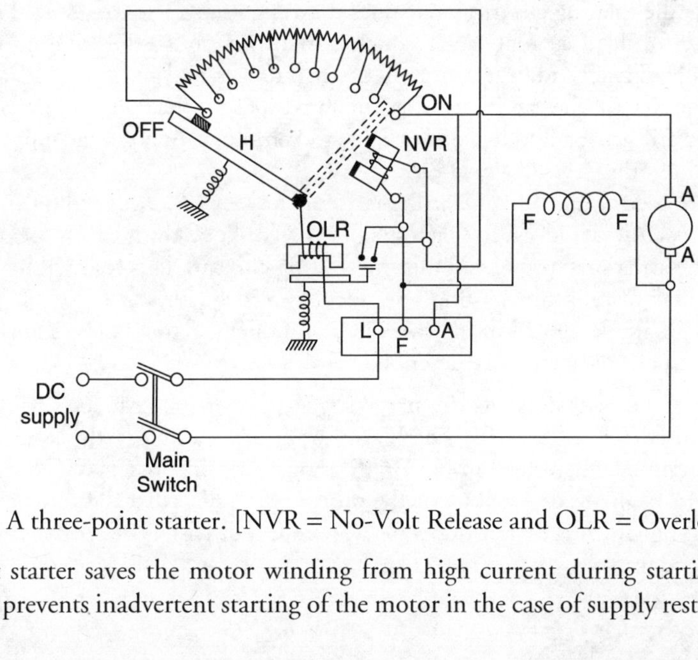

Figure 31 A three-point starter. [NVR = No-Volt Release and OLR = Overload Release]

Thus a three-point starter saves the motor winding from high current during starting as well as due to overloading. It also prevents inadvertent starting of the motor in the case of supply restoration after a break.

6.7.3 Speed Control of DC Motors

The expression governing the speed of DC motors is given by

$$N = \frac{V_t - I_a R_a}{k\phi} = \frac{E_b}{k\phi}$$

where N is the speed in rpm, V_t is the supply terminal voltage, $I_a R_a$ is the armature drop, ϕ is the flux due to field winding, and E_b is the back emf. It is clear from this expression that the speed can be controlled by field control (controlling ϕ) or armature control (controlling I_a). We describe them as follows:

1. **Field Control Method:** An external variable resistance is inserted in the field circuit of the motor. By varying its value, the field current can be controlled. This controls the value of ϕ, which in turn, controls speed. It can be appreciated that insertion of external resistance in the field circuit can eventually only reduce the field current (and hence ϕ). Hence the speed can only be increased by this method.
2. **Armature Control Method:** An external variable resistance is inserted in the armature circuit. This is controlled to control I_a, which in turn controls the $I_a R_a$ drop. It can be understood here that by this method the voltage available to the armature can get only reduced, hence resulting in lower speeds only.

   So we can see that the speed can be increased by using the field control method and can be reduced by using the armature control method. Usually the current in the armature circuit is much higher as compared to the field current. Hence the I^2R loss occurring in the external resistance is higher in armature control method.

   The speed of a DC motor can also be controlled by controlling the supply voltage V_t (which was kept constant in the field and armature control methods). The Ward–Leonard method of speed control is available for controlling the speed by controlling V_t. But this method is highly expensive because it requires an additional Motor–Generator (M–G) set for the same.

6.7.4 Methods of Starting Three-Phase Induction Motors

At the time of starting, the rotating electrical machines draw high starting current. This may be of the order of 5–7 times the rated full load current. Such a high current may burn the winding insulation resulting in damage to the motor. Hence, additional arrangements are required to limit the starting current from flowing through the motor, till the motor acquires the rated speed. Once the motor starts attaining speed, back emf accordingly starts getting developed. This, due to Lenz's law, opposes the supply voltage and hence the current stays within limits.

Several methods are used to limit high starting currents in motors. One method is to apply a reduced supply voltage to the armature winding, so that a smaller current flows through it. The other is to introduce high resistance into the armature current to limit the starting current. This resistance is gradually removed so that the I^2R loss taking place in this starting resistance is avoided.

Different methods are implemented in the two types of induction motors: the squirrel cage IM and the slip ring IM. These methods of starting are described below:

1. **Direct on line (DOL) starters:** As the name suggests, it operates direct on line and works like a switch with some additional features. The DOL starter provides protection against overload also. In Figure 32 it is seen that when the start button (S) is pushed, the contacts (C) of supply are made. The solenoid (A) gets magnetized which holds the plunger (P_A) such that the connection remains "ON" for the time during which the motor operates. There is a provision of breaking the circuit either by

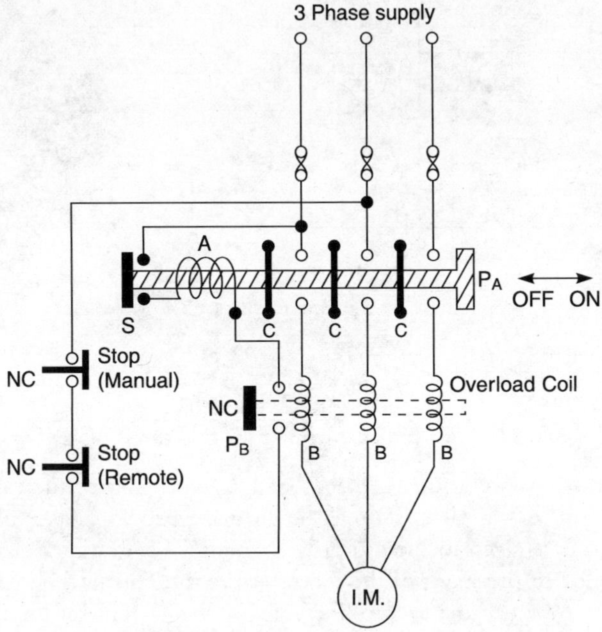

Figure 32 Direct online starter. NC implies normally closed switches.

a manual or by a remotely operated “OFF” switch. Similarly, another breaking switch is available in the circuit which operates to break the circuit and hence stop the motor under overload condition. This switch normally remains closed under normal operating conditions. In case the motor gets overloaded, the solenoids (B) in the three phases operate the plunger (P_B) and shift it such that it opens the circuit to stop the motor.

The main switch closes the starting circuit to start the motor. This circuit can be opened by any of the three circuit-opening mechanisms. If the circuit gets opened due to any of them the solenoid (A) gets de-energized and releases the main plunger (P_A) to the “OFF” position. This prevents the motor from starting inadvertently. In other words, if the motor stops due to overloading or failure of supply, it would not automatically start unless switched “ON” manually after checking the removal of the cause of stopping, hence ensuring the safe operation of the motor.

2. **Starting of induction motor by applying reduced supply voltage:** This method can be used for both the squirrel cage as well as for the slip ring induction motors.

   (a) **Auto-transformer starter:** The supply voltage is increased gradually at the time of starting through the three-phase auto-transformer as shown in Figure 33. The variable connections move simultaneously in the three-phase auto-transformer. Their position is kept at zero while starting the motor. They are shifted gradually to increase the supply voltage to the motor. Once the motor attains normal speed, the switch is thrown to a position where the normal rated voltage is applied to the motor. The starting current thus gets limited due to smaller voltage applied at the time of starting.

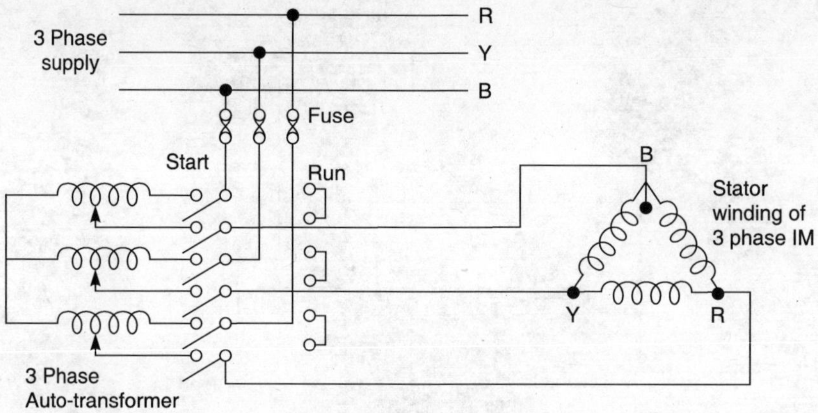

Figure 33 Auto-transformer starter of a three-phase IM.

(b) Star–delta starter: As we know, the phase voltage in a star connection is $1/\sqrt{3}$ times that in a delta connection. A star–delta starter, shown in Figure 34, makes use of this fact. The supply during starting is given to the motor by keeping its stator winding star connected. Voltage $V_{Line}/\sqrt{3}$ is applied to the stator of the motor at the time of starting. This allows less current to flow through it. For the same line-to-line voltage, the current for a star-connected load is one-third of that for a delta-connected load. Once the motor gains normal speed, the switch is operated to change the stator winding and the connection is switched to delta. Thus after starting, the normal supply voltage V_{Line} is fed to the motor.

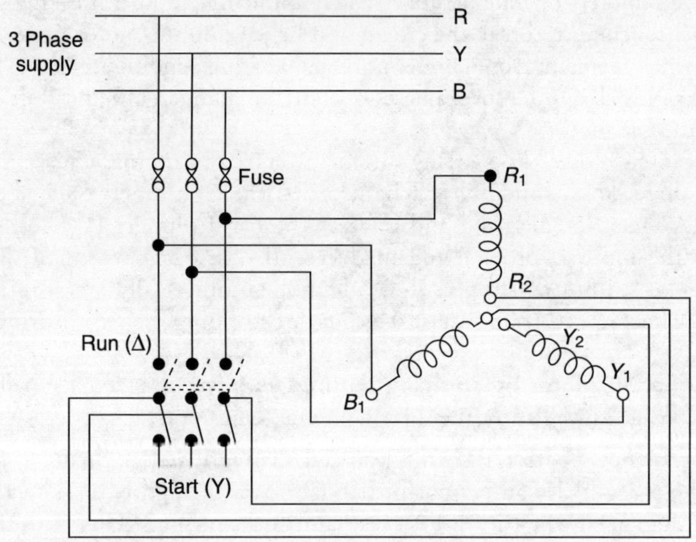

Figure 34 Star–delta starter of a three-phase IM.

3. Use of external rotor resistance in starting of slip ring (wound) induction motor: The rotor terminals are available through slip rings in wound-type three-phase induction motors. A set of three resistances connected in star, and having three variable jockey connections to it, is used for starting the three-phase induction motors as shown in Figure 35. The variable jokey connections are connected to the rotor winding through slip rings. Full resistances are placed into the rotor circuit during starting. They are then simultaneously reduced to zero in all the three phases after the motor picks up its normal speed. Thus the external resistance in the rotor circuit becomes zero and the motor starts drawing normal current. Therefore, using this arrangement, reduced starting current is allowed at the time of starting.

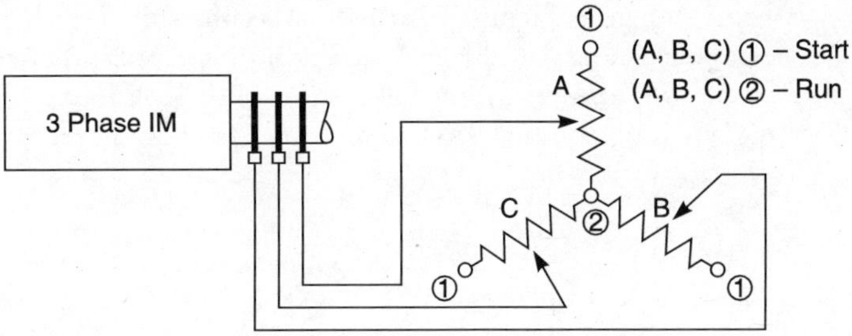

Figure 35 Rotor resistance in slip ring IM (jockeys A, B, and C move together).

4. **Starting of squirrel cage induction motor:** The rotor of the squirrel cage induction motor is constructed differently in order to introduce the current limiting feature in it. In place of a single layer of cage winding, a double cage construction is used in which two layers of cage winding are provided as shown in Figure 36. The outer winding (cage) "A" is made such that its resistance is lower than that of the inner winding (cage) 'B' while its inductive reactance is higher than that of "B".

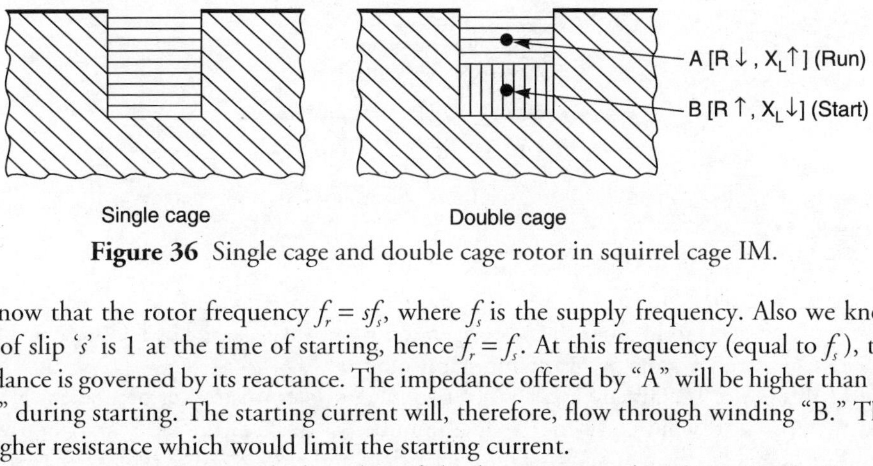

Figure 36 Single cage and double cage rotor in squirrel cage IM.

We know that the rotor frequency $f_r = sf_s$, where f_s is the supply frequency. Also we know that the value of slip 's' is 1 at the time of starting, hence $f_r = f_s$. At this frequency (equal to f_s), the winding impedance is governed by its reactance. The impedance offered by "A" will be higher than that offered by "B" during starting. The starting current will, therefore, flow through winding "B." This winding has higher resistance which would limit the starting current.

Once the motor gains speed, the value of slip becomes very low (0.01–0.05). Hence the rotor frequency becomes very small. Now the impedances of windings are mainly governed by their resistances. Thus, under running condition, the current gets shifted from the inner winding "B" and the outer winding "A," which is the path having lower resistance. Thus, the double cage rotor offers higher resistance at the time of starting and lower resistance during normal operation.

6.7.5 Stepper Motor

Stepper motors operate on getting digital signals (from PLCs or computers) and give discrete mechanical movement as output. They do not require digital-to-analog converters for their operation. The output of a stepper motor can be in the form of either velocity or position.

The construction of motor is described in Figure 37. The stator and rotor of this motor have teeth cut on their inner and outer peripheries, respectively. The arrangement is such that the alternate teeth of the stator become North and South poles, respectively. The input is given to the stator teeth in sequence so that the successive teeth become North and South poles one after the other. These stator poles attract the rotor poles. Thus the rotor moves in the direction in which the stator poles keep getting magnetized. For example, let there be 8 poles as shown in Figure 37. If pole 1 becomes the North pole then pole 2 will become the South pole. Next, poles 3 and 4 will become the North and South poles, respectively, and so on. The process continues in succession and the rotor follows the stator poles in steps. These motors can be designed to rotate in discrete steps at precise angles. The motor may fail to respond to a fast-changing input at the stator teeth.

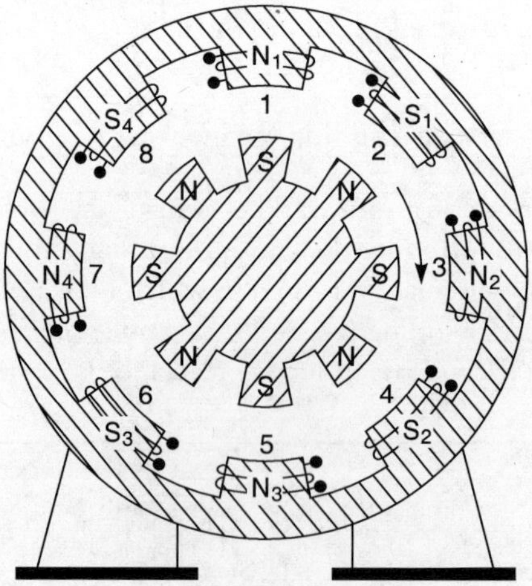

Figure 37 Stepper motor.

Stepper motors find application in printers, scanners, photocopiers, tape drivers, numerically controlled machines, printers, etc.

6.7.6 Universal Motors

Universal motors are named so because they can operate for both AC as well as DC inputs. These motors are also known as AC series motors or AC commutator motors.

The stator of this motor is laminated. It has got two salient poles. Instead of permanent magnets, these are wound with series field winding having a small number of turns of thicker wire. The rotor contains armature winding similar to that of a DC motor. The field and the armature windings are connected in series through the commutator placed on the rotor. Its construction is complex.

The universal motor can be controlled either as a phase-angle drive or as a chopper drive. The speed obtained for DC input is higher than that for AC input. The efficiency and the reliability of these motors are low but its power-to-weight ratio is good. As compared to AC or DC motors, universal motors have higher starting torque, smaller size, and the ability to operate at very high speeds. The universal motors are generally used to operate electrically powered tools and domestic appliances.

Solved Examples

Example 1

In the circuit of Figure 38, calculate the current I, voltage V, and power.

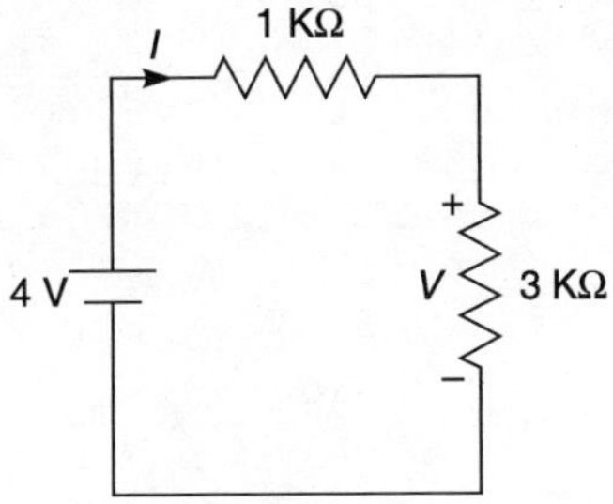

Figure 38

Solution

$$I = \frac{V}{R_T} \Rightarrow I = \frac{4}{4} \qquad [\because R_T = (3+1)\,\text{K}\Omega]$$

Therefore

$$I = 1 \text{ mA}$$

Now

$$V = 3 \times 1 = 3 \text{ V}$$

Also

$$\text{Power} = V \times I = 4 \times 1 = 4 \text{ mW}$$

Example 2

In the circuit of Figure 39, show that $V_2 = V_1 \dfrac{1}{1 + \dfrac{R_1}{R_2}\left(\dfrac{1}{1+\alpha}\right)}$.

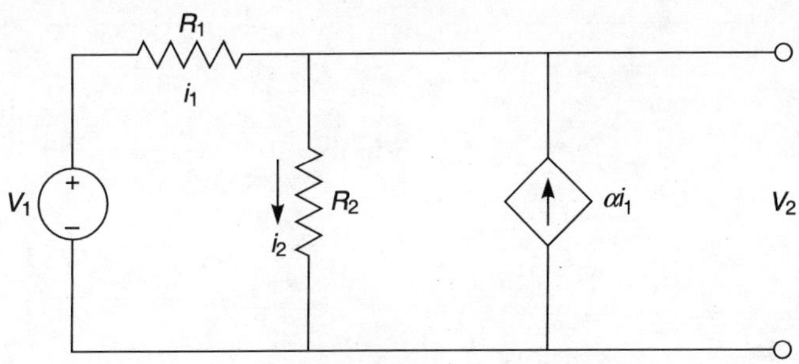

Figure 39

Solution

We have

$$V_1 = i_1R_1 + i_2R_2$$

$$i_1 + \alpha i_1 = i_2$$

and

$$V_2 = i_2R_2$$

Now

$$\frac{V_2}{V_1} = \frac{i_2R_2}{i_1R_1 + i_2R_2} = \frac{(1+\alpha)i_1R_2}{i_1R_1 + (1+\alpha)i_1R_2}$$

$$\frac{V_2}{V_1} = \frac{(1+\alpha)R_2}{R_1 + (1+\alpha)R_2} = \frac{1}{\dfrac{R_1}{R_2(1+\alpha)} + 1}$$

$$V_2 = V_1 \frac{1}{1 + \dfrac{R_1}{R_2}\left(\dfrac{1}{1+\alpha}\right)}$$

Hence proved.

Example 3

In the network of Figures 40(a) and (b), find I separately when the battery position is changed. Use reciprocity theorem.

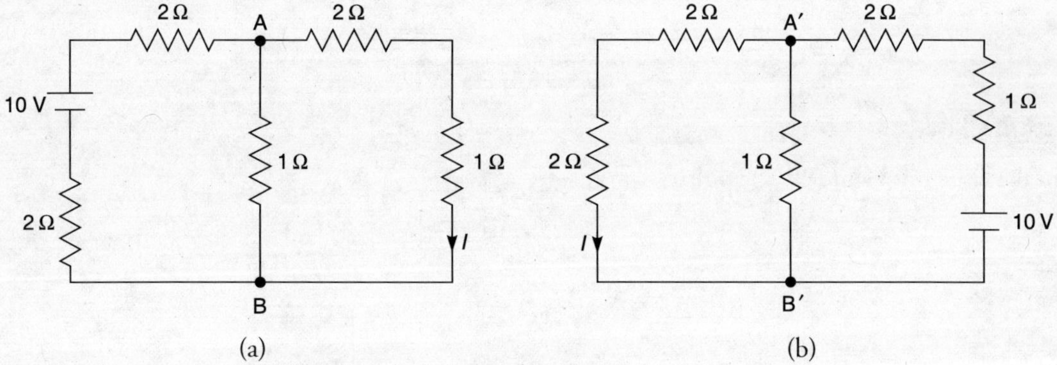

Figure 40

Solution

Referring to Figure 40(a), the equivalent resistance between A and B is $\dfrac{3\times 1}{3+1} = \dfrac{3}{4}\ \Omega$.

Hence, total circuit resistance $= 2 + 2 + \dfrac{3}{4} = \dfrac{19}{4}\ \Omega$.

Current drawn from battery $= \dfrac{10}{(19/4)} = \dfrac{40}{19}$ A and current $I = \dfrac{40}{19}\cdot\dfrac{1}{1+2+1} = \dfrac{10}{19}$ A.

Again, in Figure 40(b) the net resistance between A′ and B′ is $\dfrac{4\times 1}{4+1}=\dfrac{4}{5}\ \Omega$.

Total resistance $=\dfrac{4}{5}+2+1=\dfrac{19}{5}\ \Omega$.

Current drawn from battery $=\dfrac{10}{19/5}=\dfrac{50}{19}$ A.

and $I=\dfrac{50}{19}\dfrac{1}{1+2+2}=\dfrac{10}{19}$ A.

Thus, it has been shown that current in each case in Figures 40(a) and (b) is the same, irrespective of placement of battery in either branch. Thus reciprocity theorem is verified.

Example 4

Verify the reciprocity theorem for the network shown in Figure 41.

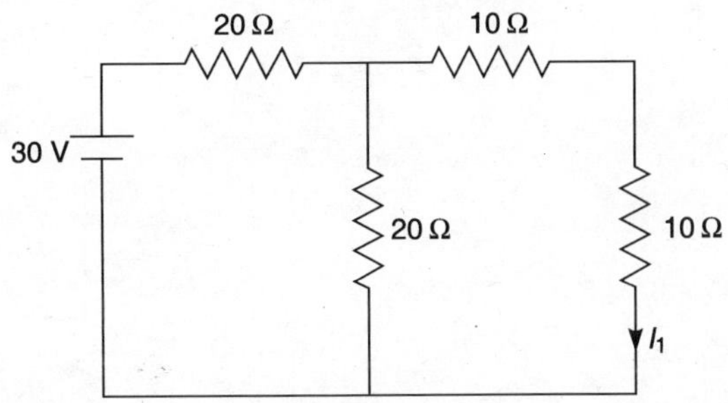

Figure 41

Solution

We have

$$R_T=[(10+10)\|20]+20=30\ \Omega$$

Current delivered by the battery $I=\dfrac{30}{30}=1$ A.

Now calculate I_1:

$$I_1=1\times\frac{20}{20+10+10}=0.5\ \text{A}$$

So

$$R_T=(20\|20)+10+10=30\ \Omega$$

The current drawn by battery is $=\dfrac{30}{30}=1$ A.

Now the source is interchanged as shown in Figure 42.

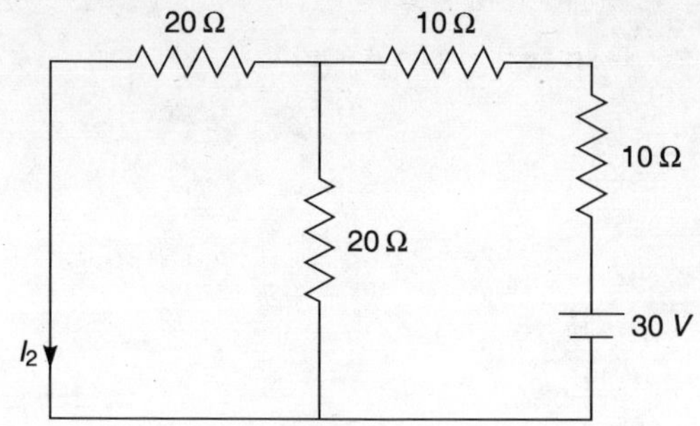

Figure 42

Calculating current, we get

$$I_2 = 1 \times \frac{20}{20+20} = 0.5\,\text{A}$$

Therefore $I_1 = I_2$. Hence reciprocity theorem is verified.

Example 5

Using Millman's theorem, find the current through the 10 Ω resistance in the network shown in Figure 43 and find the voltage drop.

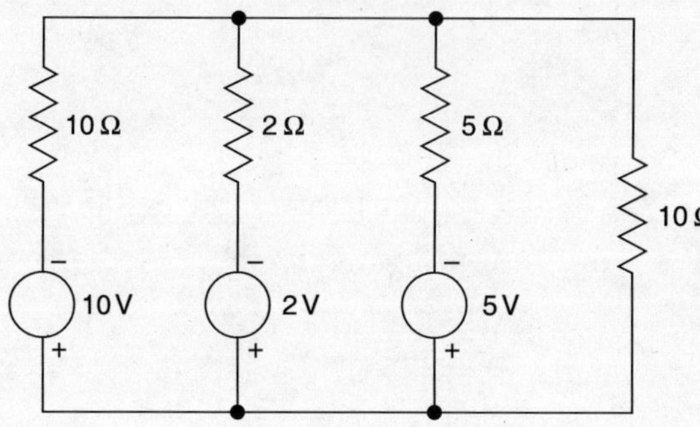

Figure 43

Solution

$$V = \frac{-V_1G_1 - V_2G_2 - V_3G_3}{G_1 + G_2 + G_3}$$

Now

$$R = \frac{1}{G_1} = \frac{1}{G_1 + G_2 + G_3}$$

$$= \frac{1}{\dfrac{1}{R_1} + \dfrac{1}{R_2} + \dfrac{1}{R_3}}$$

$$= \frac{1}{\frac{1}{10}+\frac{1}{2}+\frac{1}{5}}$$

$$= \frac{1}{\frac{1+5+2}{10}} = \frac{10}{8}\ \Omega = \frac{5}{4}\ \Omega$$

So

$$V = \frac{-10\times 1/10 - 2\times 1/2 - 5\times 1/5}{4/5} = \frac{-3\times 5}{4} = \frac{-15}{4}\ \text{V}$$

The load $R_L = 10\ \Omega$ resistance is now connected across Millman's equivalent circuit shown in Figure 44.

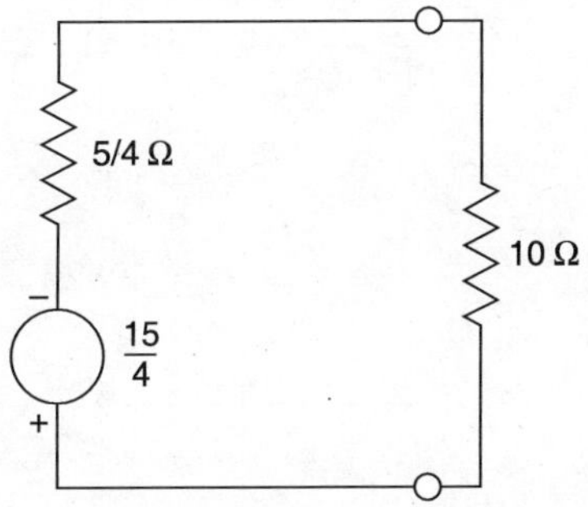

Figure 44

$$I = \frac{-15/4}{5/4+10} = \frac{-1}{3}\ \text{A}$$

Therefore the current through 10 Ω resistance is 0.33 A, while drop across it is $10 \times 0.33 = 3.3$ V.

Example 6

Using Millman's theorem, find the current through 5 Ω resistance in the network shown in Figure 45.

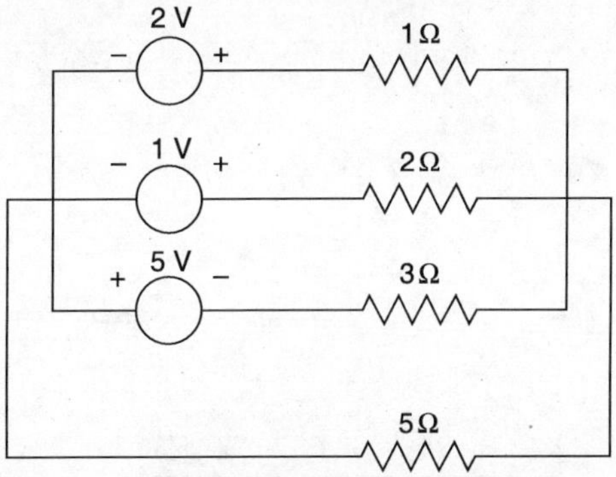

Figure 45

Solution

Applying Millman theorem, we can write

$$V = \frac{G_1V_1 + G_2V_2 - G_3V_3}{G_1 + G_2 + G_3}$$

where $G_1 = \frac{1}{R_1}$ and so on.

$$= \frac{1\times 2 + 1/2\times 1 - 1/3\times 5}{1/1 + 1/2 + 1/3} = \frac{5}{11}\ \text{V}$$

and

$$R = \frac{1}{1/1 + 1/2 + 1/3} = \frac{6}{11}\ \Omega$$

Now for Millman's network shown in Figure 46, we obtain the current through the 5 Ω resistance as

$$I_{5\Omega} = \frac{V}{R} = \frac{5/11}{(5+6/11)} = 0.082\ \text{A}$$

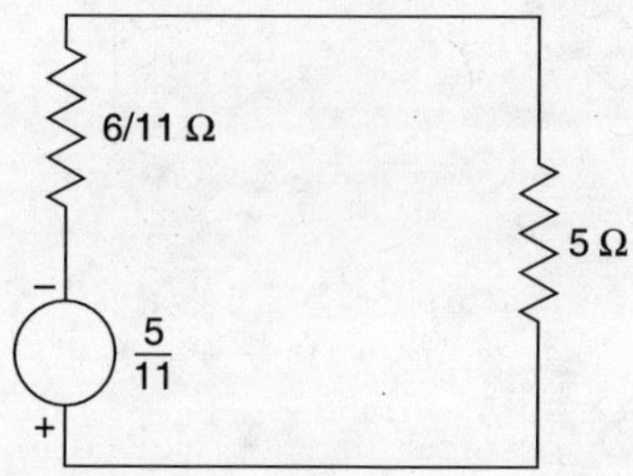

Figure 46

Thus the current through the 5 Ω resistance is 0.082 A.

Example 7

Find the current through the capacitor by superposition theorem for the network shown in Figure 47.

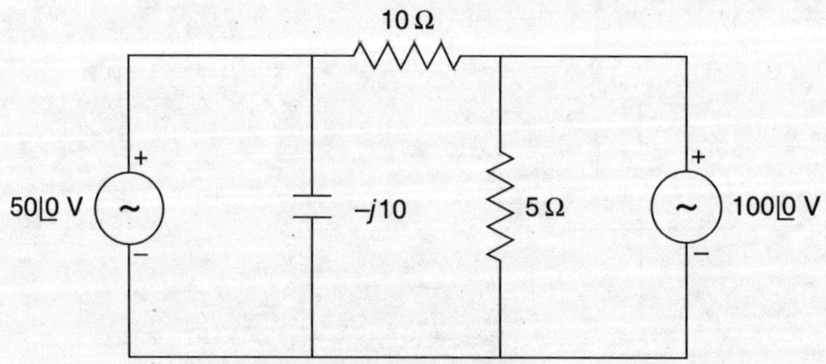

Figure 47

Solution

First we keep the source $50\underline{|0^\circ}$ active and deactivate the other source [Figure 48(a)].

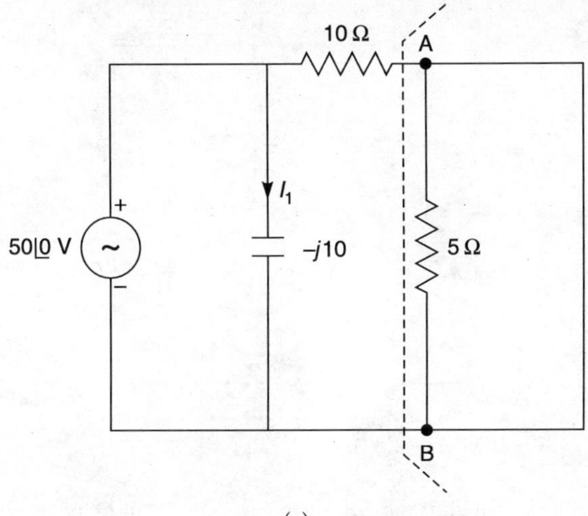

(a)

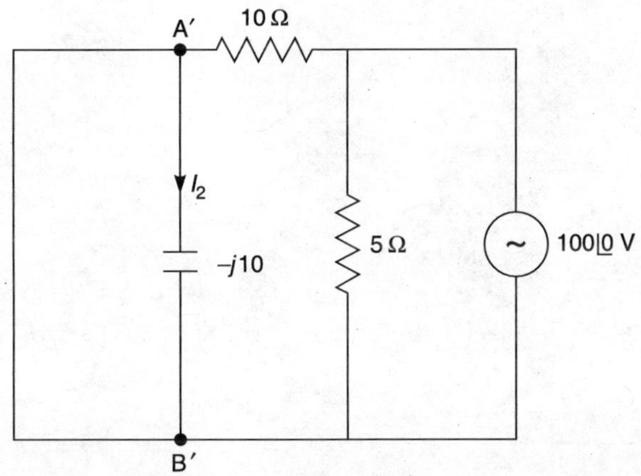

(b)

Figure 48

Here nodes A and B have the same potential, so they are shorted. The whole current passes to the shorted branch and no current flows in the 5 Ω resistance. Now, current through capacitor is

$$I_1 = \frac{50\angle 0^\circ}{-j10^\circ} = \frac{50\angle 0^\circ}{10\angle -90^\circ} = 5\angle 90^\circ$$

Next we make the $100\angle 0^\circ$ V source active and make the other source inactive [see Figure 48(b)].

In the present network, nodes A′ and B′ are at the same potential, so no current flows through A′B′.

So $I_2 = 0$ and net current $= I_1 + I_2 = 5\angle 90^\circ + 0 = 5\angle 90^\circ$ A.

Example 8

Using superposition theorem, find the voltages across $(5 + j10)\ \Omega$ in the network shown in Figure 49.

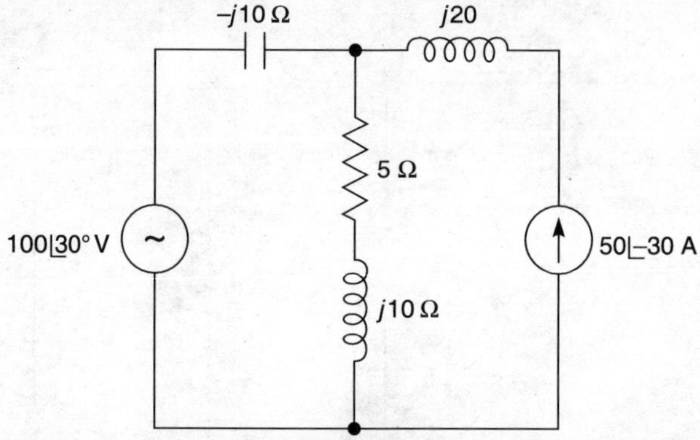

Figure 49

Solution

Considering only $100\angle 30^\circ$ first, the network of given Figure 49 can be represented by the network shown in Figure 50(a).

$$I_1 = \frac{100\angle 30^\circ}{5 + j10 - j10} = \frac{100\angle 30^\circ}{5} = 20\angle 30^\circ \text{ A}$$

$$\begin{aligned} V_{\text{PQ}} &= (20\angle 30^\circ)(5 + j10) \\ &= -13.40 + j223.21 \\ &= 223.61\angle 93.43^\circ \end{aligned}$$

Now considering $50\angle -30^\circ$ A, the given network can be represented as shown in Figure 50(b).

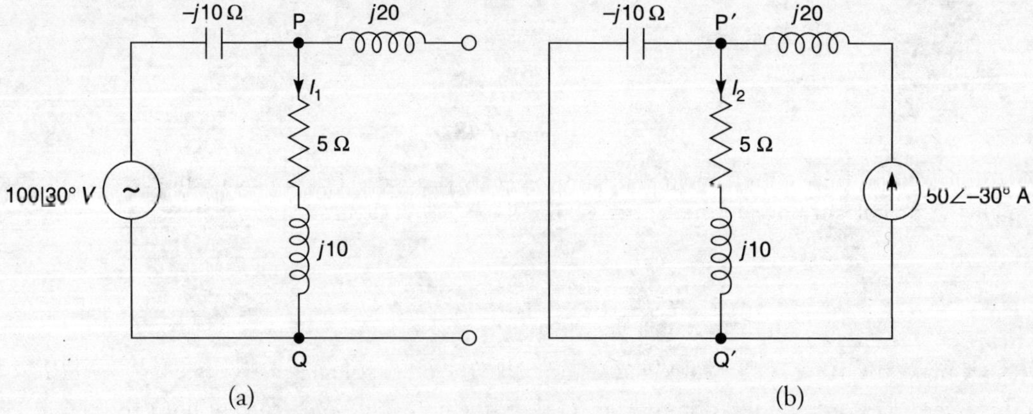

Figure 50

Here

$$I_2 = 50\angle{-30^\circ}\frac{-j10}{5+j10-j10} = -50 - j86.60 = 100\angle{-120^\circ}$$

$$\begin{aligned} V_{P'Q'} &= I_2 \times (5+j10) \\ &= (-50-j86.60)\times(5+j10) \\ &= 616 - j933 \\ &= 111'8.001\angle{-56.566^\circ} \end{aligned}$$

By using superposition principle

$$\begin{aligned} V &= V_{PQ} + V_{P'Q'} \\ &= -13.40 + j223.21 + 616 - j933 \\ &= 602.6 - j709.79 \\ &= 931.09\angle{-49.67^\circ} \end{aligned}$$

So voltage across $(5+j10)$ is $931.09\angle{-49.67^\circ}$, with positive terminal at P.

Example 9

Find Thevenin's equivalent network for Figure 51.

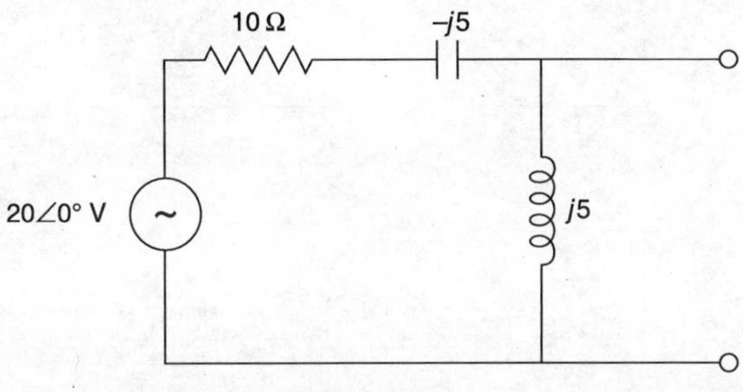

Figure 51

Solution

Calculating the V_{Th} of the given network

$$V_{Th} = 20\angle{0^\circ}\frac{j5}{10-j5+j5} = 10\angle{90^\circ}\text{ V}$$

Calculation of Z_{Th}:

$$\begin{aligned} Z_{Th} &= (10-j5)\,\|\,j5 \\ &= 2.5 + j5 \\ &= 5.60\angle{63.43} \end{aligned}$$

Thevenin's equivalent network is shown in Figure 52.

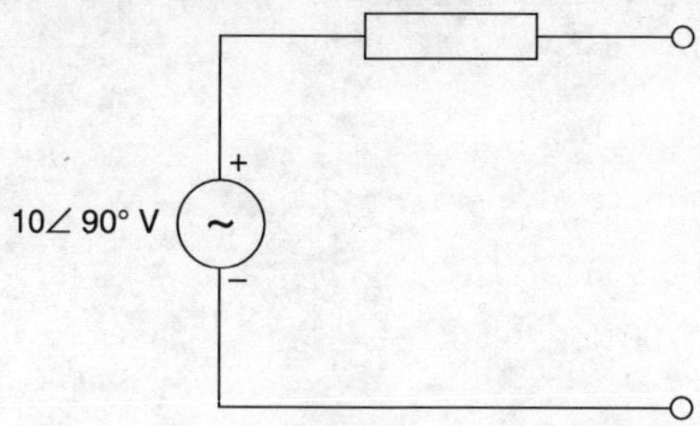

Figure 52

Example 10

Obtain Thevenin's equivalent of the network shown in Figure 53 across AB.

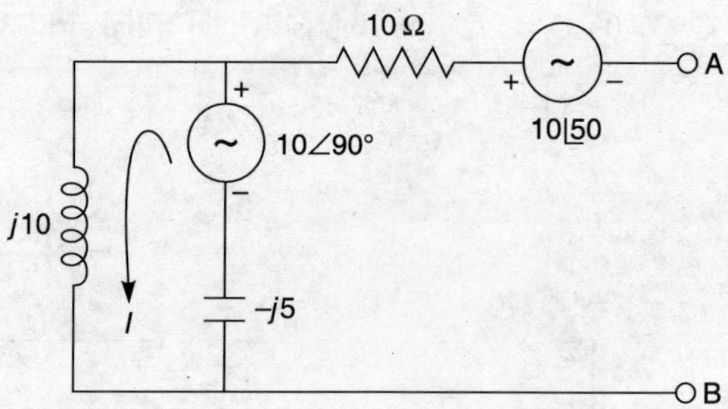

Figure 53

Solution

Calculation of V_{Th}:

$$I = \frac{10\angle 90^\circ}{j10 - j5} = \frac{10\angle 90^\circ}{j5} = 2 \text{ A}$$

$$\begin{aligned} V_{\text{Th}} &= 5\angle 90^\circ I + 10\angle 90^\circ - 10\angle 50^\circ \\ &= 5\angle 90^\circ \times 2 + 10\angle 90^\circ - 10\angle 50^\circ \\ &= -6.43 + j12.34 \\ &= 13.91\angle 117.52^\circ \text{ V} \end{aligned}$$

Calculation of Z_{Th}:

$$\begin{aligned} Z_{\text{Th}} &= (j10 \| -j5) + 10 \\ &= \frac{50}{5\angle 90^\circ} + 10 = 10\angle -90^\circ + 10 \\ &= 10 - j10 \\ &= 14.14\angle -45^\circ \end{aligned}$$

Thevenin's equivalent network is shown in Figure 54.

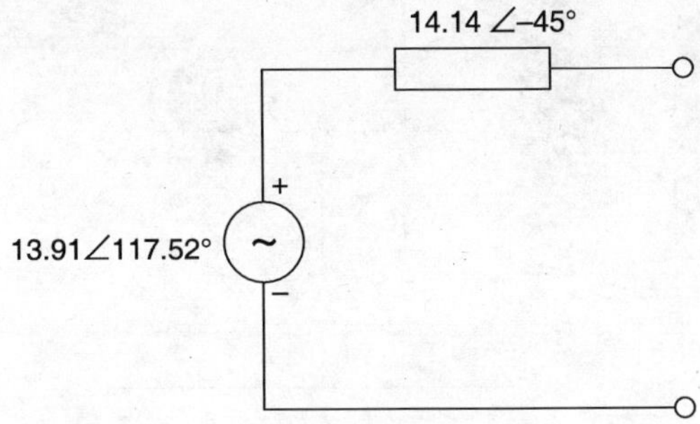

Figure 54

Example 11

Find Norton's equivalent circuit across AB for the network shown in Figure 55.

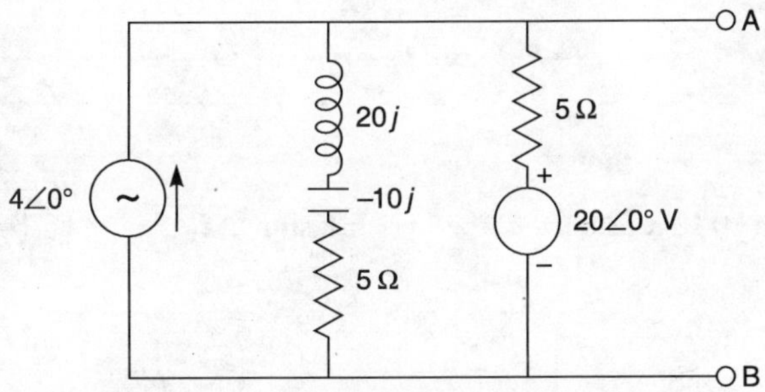

Figure 55

Solution

Let us apply short circuit across AB. Obviously no current would pass through the $(5 + j20 - j10)\ \Omega$ impedance branch and the total current through the short circuit becomes additive [Figure 56(a)]. Now

$$
\begin{aligned}
I_{SC} &= 4\angle 0^\circ + \frac{20\angle 0^\circ}{5} \\
&= 4\angle 0^\circ + 4\angle 0^\circ \\
&= 8\angle 0^\circ \text{ A}
\end{aligned}
$$

Calculation of Z_{in}:

Now the network is modified as shown in Figure 57(b).

$$Z_{in} = 5 \| (5 + 10j) = 3.75 + j1.25 = 3.95\angle 18.43\ \Omega$$

So Norton's equivalent circuit, shown in Figure 57(c).

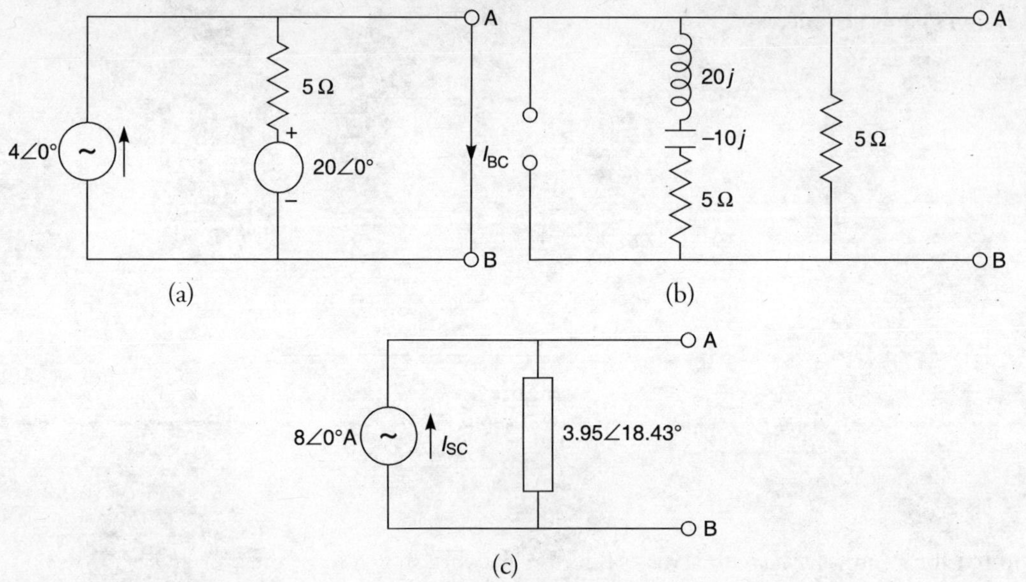

Figure 56

Example 12

Find current through the 10 Ω resistance of the network shown in Figure 57 by using Norton's theorem.

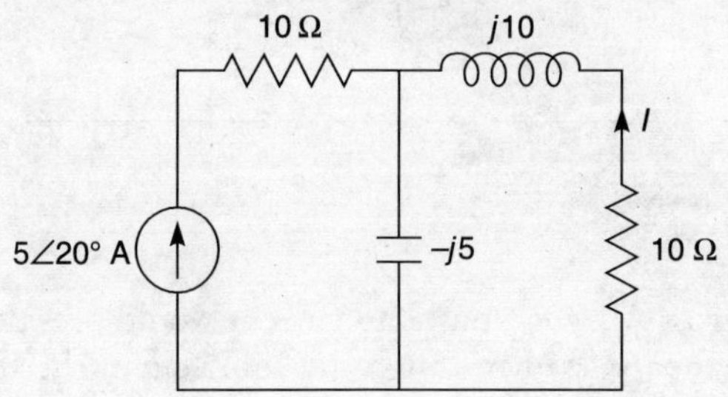

Figure 57

Solution

Calculation of Z_{in}: For calculation of Z_{in} the network becomes as shown in Figure 58(a).

$$Z_{in} - j10 - j5 = j5 = 5\angle -90^\circ\ \Omega$$

Calculation of I_{SC}:

Taking Figure 58(b) for calculation of I_{SC}, we short the load resistance branch. Now we calculate I_{SC} by current divider rule

$$I_{SC} = 5\angle 20^\circ \frac{-j5}{j10 - j5} = -5\angle 20^\circ\ \text{A}$$

The corresponding Norton's equivalent circuit is shown in Figure 58(c).

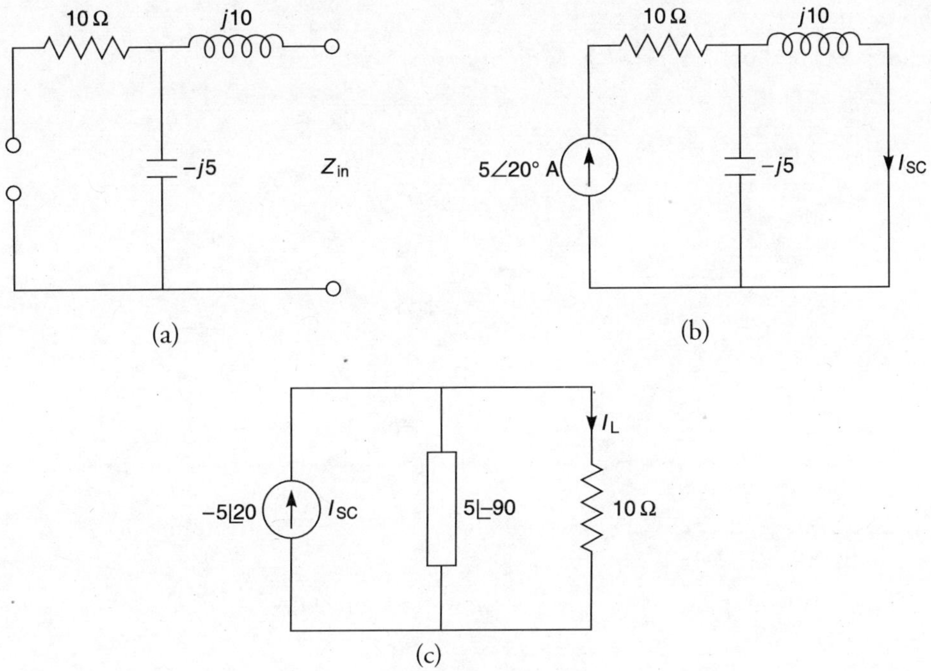

Figure 58

We now calculate load current I_L by using the current divider rule:

$$I_L = -5\angle 20^\circ \frac{5\angle -90^\circ}{10 + 5\angle -90^\circ}$$
$$= -(1.624 - j1.54)$$
$$= -1.624 - j1.54$$
$$= 2.236\angle 136.56^\circ \text{ A}$$

Example 13

Find the value of C which gives maximum power across the load of the network shown in Figure 59.

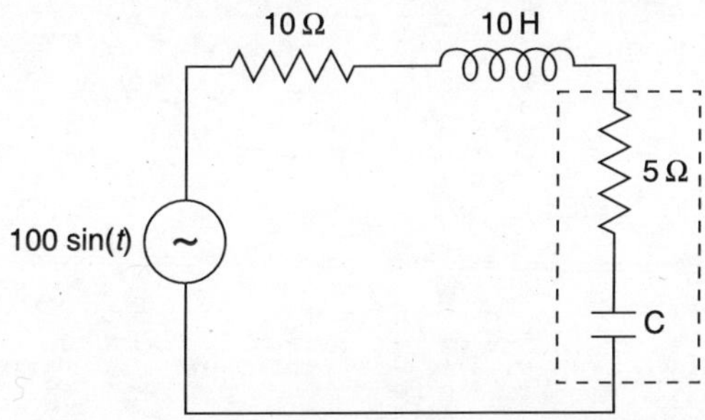

Figure 59

Solution

Here we have for 100 sin(t), $\omega = 1$. Now

$$X_L = j\omega L = j10$$

$$X_C = \frac{-j}{\omega C} = \frac{-j}{C}$$

For maximum power

$$|X_L| = |X_C|$$

$$\Rightarrow j10 = \frac{j}{C}$$

$$\Rightarrow C = 0.1 \text{ F}$$

Example 14

Find R_L for maximum power of the network shown in Figure 60.

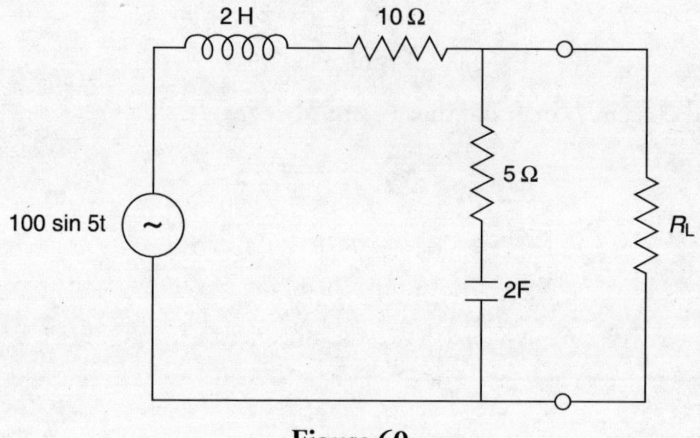

Figure 60

Solution

We know that

$$\omega = 5$$

$$X_L = j\omega L = 5 \times 2j = j10\ \Omega$$

$$X_C = \frac{-j}{\omega C} = \frac{-j}{5 \times 2} = -j0.1\ \Omega$$

Redrawing the network as shown in Figure 61, we get the equation for maximum power transfer as

$$R_{L_{\max}} = Z_{\text{th}}$$

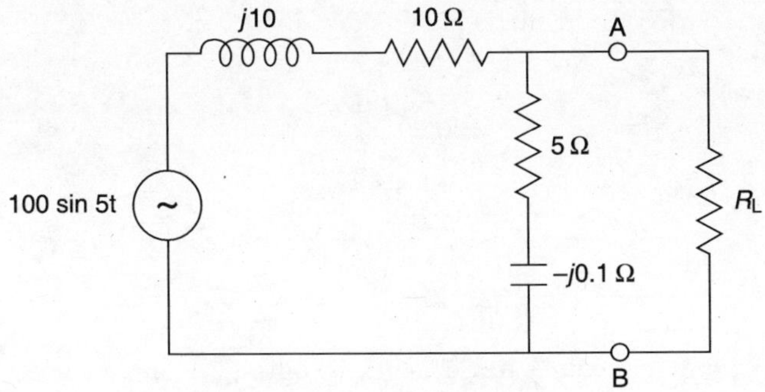

Figure 61

Calculation of Z_{th} across AB

$$Z_{Th} = (10 + j10)\|(5 - j0.1) = 3.87 + j0.71$$
$$R_{L_{max}} = Z_{th} = 3.93\angle 10.43°\ \Omega = 3.93\ \Omega$$

Example 15

A single-phase transformer rated 2000/200 V, 10 kVA, has an equivalent resistance and leakage reactance referred to the low voltage side as 0.02 Ω and 0.08 Ω, respectively. Find the per unit value of the resistance and reactance referred to the high voltage and low voltage sides.

Solution

The base impedance in the low voltage side is

$$Z_{b(\text{for LV side})} = \frac{200^2}{10 \times 10^3} = 4\ \Omega$$

Thus the per unit resistance will be

$$\varepsilon_r = \frac{0.02}{4} \times 100\% = 0.5\%$$

and the per unit reactance will be

$$\varepsilon_x = \frac{0.08}{4} \times 100\% = 2\%$$

The equivalent resistance and reactance referred to the high voltage side will be

$$r_{1e} = \left[\frac{2000}{200}\right]^2 \times 0.02 = 2\ \Omega \text{ and } x_{1e} = \left[\frac{2000}{200}\right]^2 \times 0.08 = 8\ \Omega.$$

where as the $Z_{b(\text{for HV side})} = \dfrac{2000^2}{10 \times 10^3} = 400\ \Omega$

Thus the per unit value referred to the high voltage side will be

$$\varepsilon_r = \frac{2}{400} \times 100\% = 0.5\%$$

and the per unit reactance will be

$$\varepsilon_x = \frac{8}{400} \times 100\% = 2\%$$

Thus we see that the per unit value **does not** change on shifting the impedance to either the low voltage or the high voltage side.

Example 16

Find the power factor of the load if the voltage regulation has to become zero in Example 1.

Solution

We know that the expression for the voltage regulation is

$$R = \varepsilon_r \cos(\phi) + \varepsilon_x \sin(\phi)$$

Where R is the voltage regulation, ε_r is the per unit resistance, ε_x is the per unit reactance and ϕ is the power factor of the load.
Now for zero R, we have

$$0 = \varepsilon_r \cos(\phi) + \varepsilon_x \sin(\phi)$$

or $\tan\phi = -\varepsilon_r / \varepsilon_x$ from which $\cos\phi$ can be found.
Here in our case, $\varepsilon_r = 0.5\%$ or 0.005 and $\varepsilon_x = 2\%$ or 0.02 so substituting the values we get $\tan\phi = -0.005/0.02$ from which $\phi = -14.04°$.
This gives $\cos\phi = \cos(-14.04°) = 0.97$ leading.

Example 17

What is the value of maximum possible voltage regulation in Example 1.

Solution

As already derived in the chapter, we have the value of maximum regulation equal to that of the per unit impedance, that is,

$$\varepsilon_z = \sqrt{\varepsilon_r^2 + \varepsilon_x^2}$$

Here in our case, $\varepsilon_r = 0.5\%$ or 0.005 and $\varepsilon_x = 2\%$ or 0.02.

Substituting the values we get $\sqrt{0.005^2 + 0.02^2} = 0.0206$ i.e., 2.06%

Summary

This chapter discusses some additional topics which were not the part of the first five chapters of the book. The concept of dependant sources has been introduced. Some important theorems have been explained like reciprocity theorem and Millman's theorem. These theorems are useful in solving the circuits by simplifying them. Also the various theorems discussed earlier have been re-explained for AC circuits. These include the Thevenin theorem, Norton theorem, superposition theorem and the maximum power transfer theorem.

A basic introduction to the transient analysis has been made through the definitions of time constants for the inductive and capacitive circuits. Also the concept of the energy storage has been dealt with. This is followed by the explanation of the method for solving the unbalanced three-phase circuits.

Two significant instruments, namely, the Cathode Ray Oscilloscope and the Meggar (for measuring the insulation resistance) have next been explained.

The phasor diagram, open and short circuit tests and the voltage regulation of a single phase transformer have been taken up next with relevant solved examples.

The chapter concludes with the description of the starting of DC and three-phase induction motors along with the speed control of the former. The basics of the stepper motor and the universal motor have been clarified.

Short Answer Questions

1. Which values of current and voltage are used in the expressions for stored energy in inductive and capacitive circuits, respectively?
2. If the load is purely resistive and variable and the source impedance is complex, for which value of load resistance the power transferred to it will be maximum?
3. What will be the value of voltage regulation for an ideal transformer?
4. When will the value of voltage regulation be zero for a practical transformer?
5. A motor is reported to be not working. Which instrument will be made use of after ascertaining that the power supply is proper?
6. What will be the exact value of the copper loss occurring in an ideal transformer under open-circuit conditions?
7. The phasor diagram of a single-phase transformer reveals that the secondary output voltage is greater than the secondary induced voltage. What may be concluded from this information?
8. If the impedance of the source is a 3 ohms resistor in parallel with an inductive reactance of 4 ohms, what will be the value of the load resistance which will ensure that maximum real power is transferred to the load? (Given that the load is complex with both components variable.)
9. If the frequency of the input supply is 1 rad/sec, find the nature and value of the reactive component of the load in Q8?
10. What is the reason for the use of starter in a DC motor in spite of the fact that it is self-starting?

Review Questions

1. Explain the principle of operation of a three point starter used in the DC motor.
2. Why is starter required for a three-phase induction motor? Explain the various types of starters used to start it.
3. Explain the open circuit and short tests for a single phase transformer with the relevant experimental set ups.
4. Explain the difference between the analysis of AC and DC circuits. What are the values of the inductance and capacitance in DC circuits?

Problems

1. Calculate the Power absorbed by each element in Figure P1.

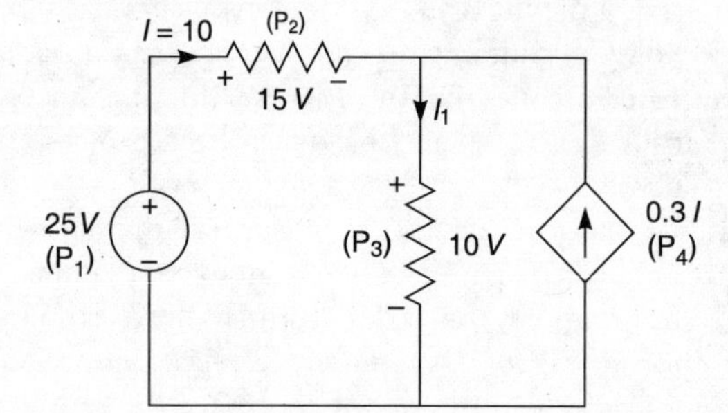

Figure P1

2. Verify reciprocity theorem for the circuit of Figure P2.

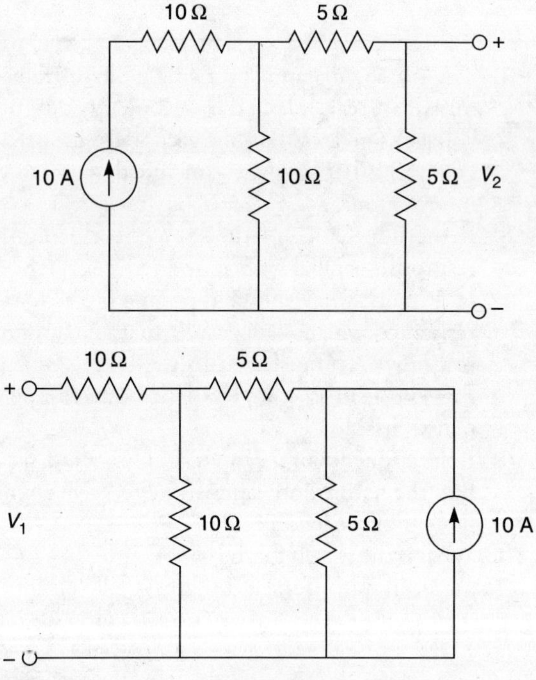

Figure P2

3. Using Millman's theorem, find the current through R_L in the network shown in Figure P3 and find the voltage drop, given R_L is 10/3 Ω.

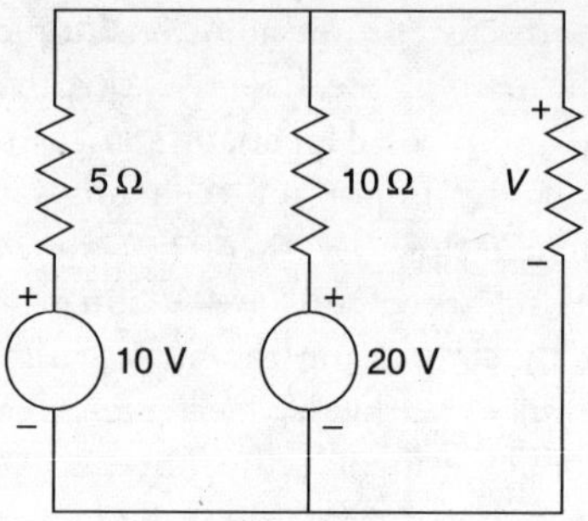

Figure P3

4. Determine current I in Figure P4 using superposition theorem.

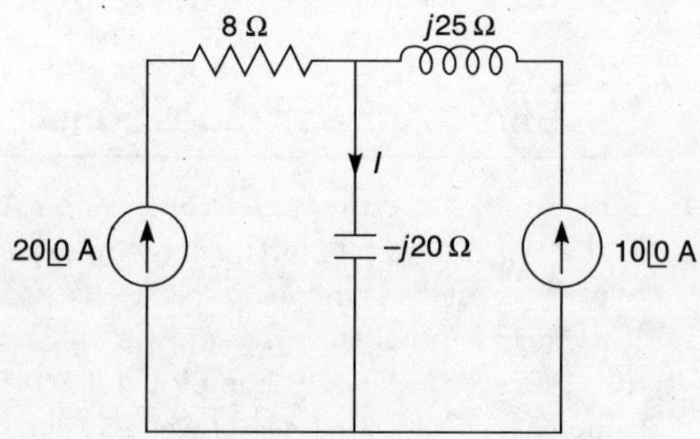

Figure P4

5. In the circuit shown in Figure P5, find current through inductor by using Thevenin's theorem.

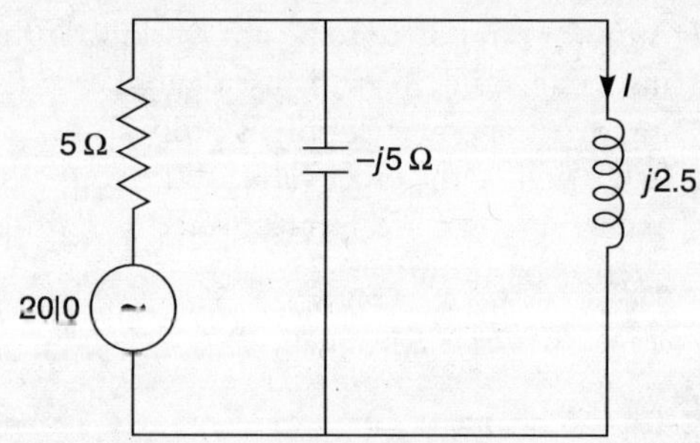

Figure P5

6. Given the network of Figure P6, find Norton's equivalent circuit to the left side of terminals *a-b*.

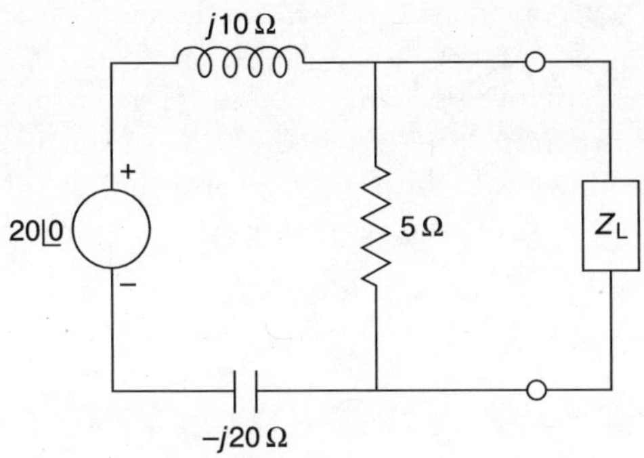

Figure P6

7. Find R_L for maximum power of the network shown in Figure P7.

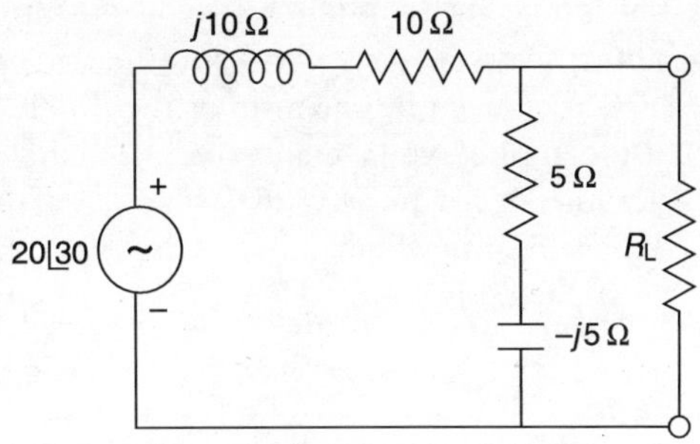

Figure P7

8. Prove that for zero voltage regulation in a single phase transformer, the power factor of the load should be leading.

Answers

Short Answer Questions

1. The peak value is always used in the expressions of energy stored.
2. In such a case, the initial differentiation will be done with respect to the load resistance itself and finally the value of load resistance at which maximum power will be received by it will be equal to the *magnitude of the source impedance.*
3. Zero (as the leakage reactance and the resistance of the windings will be zero here).
4. This will happen when the inductive effect of the practical transformer will be neutralized by a capacitive load, that is, when the load is capacitive.
5. In such a case, the insulation of the windings is checked with respect to the ground to ensure that no part of the winding is touching the body of the motor. This is done by an instrument called Megger.
6. For an ideal transformer the iron loss will be zero and also the magnetizing current will be zero. Hence there will be zero no load current and so the exact value of the copper loss under open-circuit conditions will be zero.
7. If the secondary output voltage is greater than the secondary induced voltage it indicates negative voltage regulation meaning whereby that the load is capacitive in nature.
8. We have

$$z = \frac{3 \times j4}{3 + j4} = \frac{j12(3 - j4)}{(3 + j4)(3 - j4)}$$

$$= \frac{48 \quad j36}{25} = \left(\frac{48}{25} + j\frac{36}{25}\right)\Omega$$

Hence the resistance value will be $\frac{48}{25}\Omega$.

9. The reactive component of the source is $\frac{36}{25}\Omega$ as can be seen from Q8. Hence according to complex conjugate rule, the load should be capacitive of ohmic value equal to $\frac{48}{25}\Omega$. Hence

$$\frac{1}{\omega C} = \frac{48}{25}\Omega \Rightarrow \omega C = \frac{25}{48}$$

Since $\omega = 1$ so $C = \frac{25}{48}$ farads.

10. The armature winding resistance in a DC motor is very less (<1 ohm) and the armature current is given by the ratio of the difference of the applied voltage and the back emf and armature resistance. At the time of starting, the back emf (which directly depends upon the speed) is zero. Hence the armature current may become too large. This may damage the armature winding. Hence, a resistance is connected in series with the armature winding at the time of starting to limit the high starting armature current. This resistance arrangement is called a starter in a DC motor.

Problems

1. $P_1 = -250$ W, $P_2 = 150$ W, $P_3 = 130$ W, $P_4 = -30$ W where positive sign, supplies power and negative sign, absorbs power.
2. $V_1 = V_2 = 25$ V
3. $I = 2$A and Voltage drop is 20/3 V.
4. 30 $\angle 0°$A.
5. $V_{Th} = 14.144\angle{-45°}$ V and $Z_{th} = 3.53$ $\angle{-45°}$, current through inductor is 5.65 $\angle{-45°}$A.
6. $Z_{in} = 4.47\angle{-26.56°}$, $I_{SC} = 2\angle 90°$.
7. 5.995 Ω.

Model Question Paper 1

Electrical Engineering

Time: 3 hours **Maximum marks:** 100

Note: All sections are compulsory.

Section A

1. Attempt the following: (10 × 2 = 20)

(i) Is 'L' a linear element? Explain the answer.

(ii) If resistance of each branch is 3 ohms in a delta connected load, what would be the resistance of each branch in its star equivalent?

(iii) Draw suitable waveforms to show the following:

$$v = V_m \sin \omega t; \quad i = I_m \sin(\omega t + \theta)$$

Label the quantities suitably.

(iv) For a series *R-L-C* circuit draw the phasor diagram and show the power factor angle in it. Assume the circuit draws inductive current.

(v) Give the relationship between quality factor, resonant frequency and bandwidth for a series *R-L-C* circuit.

(vi) What is the major difference between PMMC type and dynamometer type of instruments?

(vii) Show the residual magnetism in a **B–H** curve.

(viii) Name the constant losses taking place in a transformer.

(ix) Name two motors used for constant speed operation.

(x) Define the term "slip".

Section B

Attempt any THREE of the following: (10 × 3 = 30)

2. Applying mesh analysis, obtain the current through 5 ohm resistance in the following circuit:

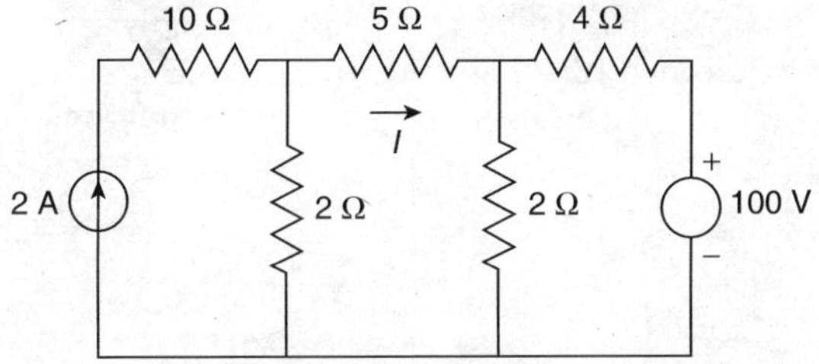

3. Voltages across R, L, C connected in series are 5, 8 and 10 volts respectively. Calculate the value of supply voltage at 50 Hz. Also find the frequency at which this circuit would resonate.
4. Find the current in phase R in the following circuit:

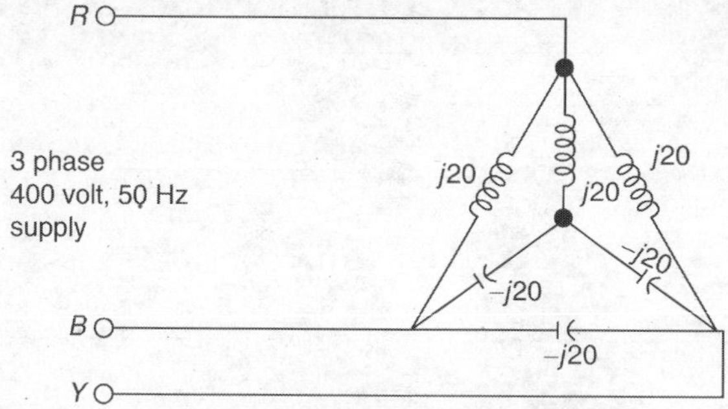

5. The maximum efficiency of a 100 kVA, 1100/440 volt, 50 Hz transformer is 96%. This occurs at 75% of full load at 0.8 pf lagging. Find the efficiency of transformer at ¾ FL at 0.6 pf leading.
6. The armature resistance of a 200 V DC shunt motor is 0.12 ohm. It runs at 600 rpm at constant torque load and draws a current of 21 ampere. Calculate its new speed if the field current of the motor is reduced by 10%.

Section C

Attempt ALL questions. $(10 \times 5 = 50)$

7. Attempt any TWO of the following:

   (a) Find the equivalent resistance for the following circuit and hence calculate the current supplied by the source.

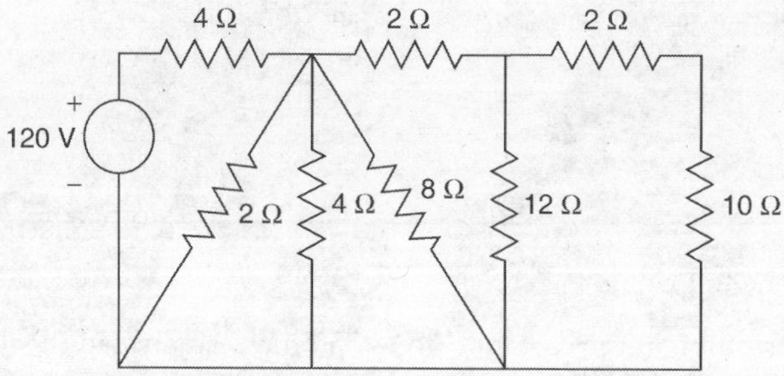

   (b) Explain the duality between a Thevenin's and a Norton's equivalent circuits.

   (c) Find the current flowing through 10 Ω resistance in the following circuit. Use Superposition theorem.

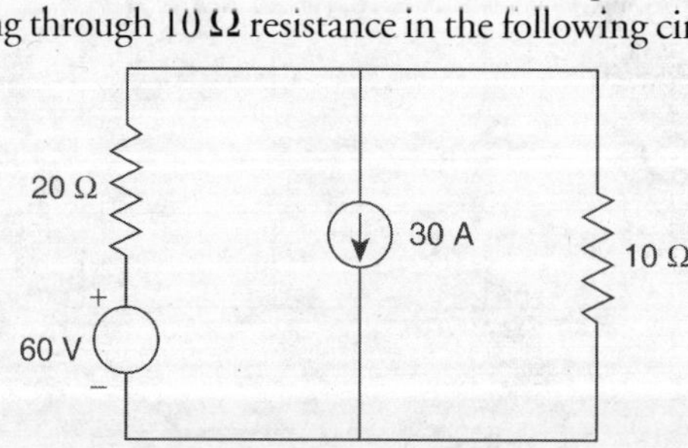

8. Attempt any TWO of the following:

(a) If load draws a current of 10 A at 0.8 pf lagging; when connected to 100 V supply, calculate the values of real, reactive and apparent powers. Also find out the resistance of the load.

(b) Deduce the formula for half power frequencies for a series *R-L-C* circuit under resonance. Why are they called 'half power frequencies'?

(c) The two branches of a parallel circuit draw currents I_1 and I_2 such that $I_1 = 10\sqrt{2}\sin\omega t$ and $I_2 = 5\sqrt{2}\sin(\omega t - 60°)$. What is the total current drawn by them?

9. Attempt any TWO of the following:

(a) Discuss about induction type instrument in detail.

(b) With the help of connection diagram and the related phasor diagram show that $\omega_1 + \omega_2$ gives the total 3ϕ power in a two wattmeter method of power measurement.

(c) An ammeter has internal resistance of 2 ohms and it is designed to measure 1 A current. What would you do to make this meter read 15 A current?

10. Attempt any TWO of the following:

(a) Draw and explain **B–H** curve. What is meant by saturation, coercive force, residual magnetism? Show them in the diagram.

(b) Write down the expression of efficiency for a transformer. How is it affected by change in power factor?

(c) Discuss in brief about autotransformer.

11. Attempt any TWO of the following:

(a) Define 'slip'. Discuss the principle of operation of 3ϕ induction motor.

(b) What do you know about splitting of phase in a single-phase induction motor? How does it help in starting the capacitor start single-phase induction motor?

(c) Describe briefly the construction and working of a synchronous generator.

Model Question Paper 2

Electrical Engineering

Time: 3 hours **Maximum marks:** 100

Note: All sections are compulsory.

Section A

1. ALL parts are compulsory. All questions carry equal marks. $(10 \times 2 = 20)$

(i) Match the following:

(a) Linear	(A) Transistor
(b) Unilateral	(B) Inductor
(c) Energy storage	(D) Resistor
(d) Active	(C) Diode

(ii) If all the three branches of a delta-connected network have resistance of 3 Ω, what will be the value of resistance in each branch in the equivalent star connection?

(a) 1 Ω (b) 3 Ω
(c) 9 Ω (d) 1/3 Ω

(iii) The following is caused in series resonant circuit:

(a) Current magnification (b) Voltage magnification
(c) Filtering (d) Rectification

(iv) Which of the following is incorrect in a series *R-L-C* circuit?

(a) $Q = \dfrac{\omega_0 L}{R}$ (b) $BW = \dfrac{Q}{\omega_0}$

(c) $\omega_0^2 = \dfrac{1}{LC}$ (d) $\omega_2 - \omega_1 = \dfrac{R}{L}$

(v) For the same real power requirement, the current drawn is maximum, if the power factor is

(a) Unity (b) 0.6 leading
(c) 0.65 lagging (d) 0.8 leading

(vi) The scale of the following type of instrument is cramped:

(a) PMMC (b) Attraction-type MI
(c) Induction (d) Dynamometer type

(vii) Which of the following is not a power system component:

(a) Transformer (b) Generator
(c) Fan (d) Transmission line

(viii) Eddy-current loss is proportional to

(a) Frequency (b) Square of frequency
(c) Frequency to power 1.6 (d) None of these

(ix) Which of the following is incorrect?

(a) In IM, field winding is on stator, armature winding is on rotor
(b) In IM, armature winding is on stator, field winding is on rotor
(c) In synchronous machine, field winding is on rotor, armature winding is on stator
(d) In DC machine, armature winding is on rotor, field winding is on stator

(x) Which of the following is not a self-starting motor?

(a) DC series motor (b) DC shunt motor
(c) Synchronous motor (d) Three-phase induction motor

Section B

2. Attempt any THREE of the following: $(10 \times 3 = 30)$

(a) Solve the following circuit (Figure 1) by using Thevnin's theorem to find the current I in the load $R_L = 4\ \Omega$.

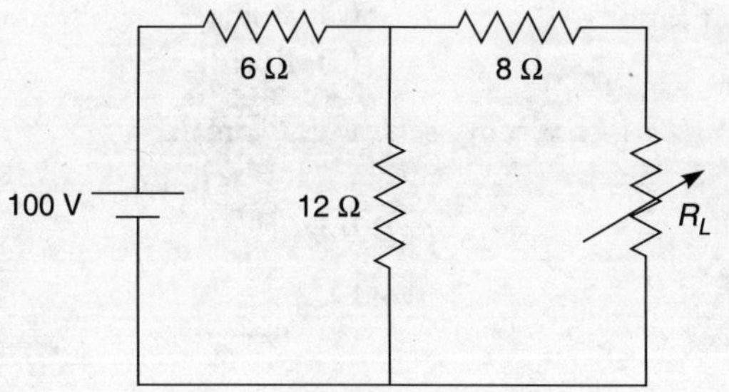

Figure 1

(b) A coil having a resistance of 10 Ω and an inductance of 0.05 H is connected across a 230 V, 50 Hz AC supply. Calculate: (i) current, (ii) power factor, (iii) active power and (iv) apparent power. It is desired to improve power factor to 0.8. What value of capacitance should be connected in series?

(c) Three voltages represented by the equations

$$e_1 = 15 \sin \omega t, \quad e_2 = 5 \sin\left(\omega t + \frac{\pi}{6}\right), \quad e_3 = 10 \cos \omega t$$

are series drops in an AC circuit. Represent these voltages by phasors, and calculate an expression for the resultant voltage drop. Check the result so obtained graphically.

(d) A 25 kVA, 2000 V/200 V, single-phase 50 Hz transformer has a primary resistance of 1.25 Ω and reactance of 2.5 Ω. The secondary resistance and reactance are 0.0225 Ω and 0.05 Ω, respectively. The number of load current of transformer is 1.5 A at 0.3 power factor. Determine

(i) Equivalent resistance, reactance and impedance referred to secondary
(ii) Total copper loss
(iii) Efficiency at full load 0.9 pf lagging

(e) Double revolving field theory relates to which type of rotating machine? Explain the working of that machine on the basis of this theory.

Section C

Note: ALL questions are compulsory. All questions carry equal marks. $(10 \times 5 = 50)$

3. Attempt either (a) or (b).

(a) Explain Norton's theorem in detail, and show its relationship with Thevnin's theorem.
(b) Solve the circuit shown in Figure 2 to get the value of the current in the 20 Ω resistor by using superposition theorem, and verify the same by mesh analysis.

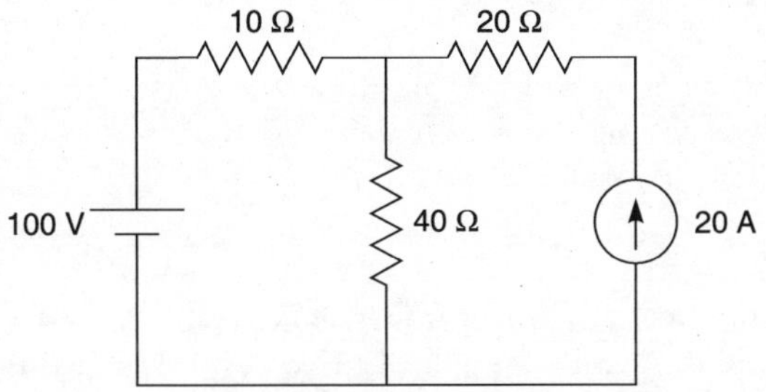

Figure 2

4. Attempt (a) and either (b or c).

(a) Two impedances $Z_1 = (10 + j15)$ ohms and $Z_2 = (6 - j8)$ ohms are connected in parallel. The total current supplied is 15 A. What is the power taken by each of the impedances? Also, calculate the reactive powers and the apparent powers of each impedance. Why does the sum of each apparent power not equal to the total apparent power?
(b) Refer to the circuit shown in Figure 3. Find: (i) rms line current, (ii) power dissipated, (iii) power factor.

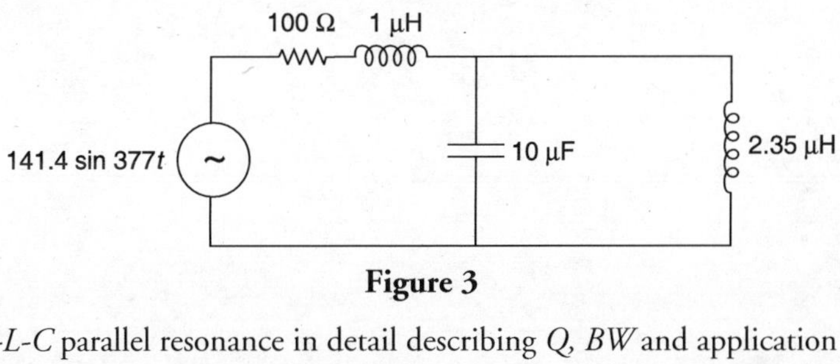

Figure 3

(c) Describe *R-L-C* parallel resonance in detail describing *Q, BW* and application of the same.

5. Attempt (a or c) and (b or d).

(a) A 3-phase voltage source has a phase voltage of 120 V and supplies star-connected load having impedance $(36 + j48)$ in per phase. Calculate

(i) The line voltage
(ii) The line current
(iii) The power factor
(iv) The total 3-phase power supplied to the load

(b) A balanced-star connected inductive load is connected to a 400 V, 50 Hz AC supply. Two wattmeters used to measure supply power indicate 8000 W and 4000 W, respectively. Determine: (i) line current, (ii) impedance of each phase and (iii) resistance and inductance of each phase.

(c) Give the construction and working of a dynamometer-type wattmeter.

(d) A single-phase energy meter has a constant of 1200 revolutions per kWh. When a load of 200 W is connected, the disk rotates at 4.2 revolutions/minute. If the load is on for 10 hours, how many units are recorded as an error? Also, find percentage error.

6. Attempt either (a) or (b).

(a) Core of a magnetic circuit is of mean length 30 cm and uniform cross-sectional area 3 cm^2. The relative permeability of the core material is 800. An air gap of 1.5 mm is cut in the core, and 1500 turns are wound on the core. Determine the inductance of the coil, if the fringing effect is neglected.

(b) The maximum efficiency of a 100 kVA, 50 Hz, single-phase transformer is 98%, and it occurs at 80% full load at 0.9 pf lagging. Find the copper loss and iron loss at full load and half full load. Also, calculate the efficiency at full load 0.65 pf.

7. Attempt (a or b) and (c).

(a) A 200 V DC shunt motor has an armature resistance of 0.1 Ω and a field resistance 140 Ω. The full load line current is 40 A and the full load speed is 1500 rpm. If the brush contact drop is 2 V, find the speed of the motor at half load.

(b) A four-pole, three-phase induction motor is energized from a 50 Hz supply, and is running at a load condition for which the slip is 0.01. Determine: (i) rotor speed (in rpm), (ii) rotor current frequency (in Hz), (iii) speed of the rotor's rotating magnetic field (in rpm) with respect to the stator frame, (iv) speed of the rotor's rotating magnetic field (in rpm) with respect to the rotor and (v) speed of the rotor's rotating magnetic field (in rpm) with respect to the stator's magnetic field.

(c) Describe the principle of operation of a synchronous motor.

Model Question Paper 3

Electrical Engineering

Time: 3 hours **Maximum marks:** 100

Note: All sections are compulsory.

Section A

1. ALL parts are compulsory. All questions carry equal marks. $(10 \times 2 = 20)$

(i) A current of 25 A flows through a parallel combination of R_1 (= 2 Ω) and R_2 (= 3 Ω). The current through R_1 will be

(a) 15 A (b) 10 A
(c) 5 A (d) 5 A

(ii) A pure inductor is

(a) Linear and unilateral element (b) Active and bilateral element
(c) Passive and nonlinear element (d) Bilateral and linear element

(iii) Form factor is the ratio of

(a) Average value to rms value (b) Peak value to average value
(c) rms value to average value (d) Peak value to rms value

(iv) The relationship between bandwidth and quality factor in a series resonant circuit is

(a) $Q = \dfrac{R}{BW}$ (b) $Q = \dfrac{\omega}{BW}$

(c) $Q = \dfrac{BW}{\omega}$ (d) $Q = \dfrac{LC}{BW}$

(v) In two-wattmeter method of measurement of power, if one of them reads negative value, then

(a) CC and PC connections are interchanged
(b) CC or PC connections are interchanged
(c) Load terminals are interchanged
(d) No change is made

(vi) The range of an ammeter can be extended by

(a) Connecting a high resistance in series with it
(b) Connecting a low resistance in series with it

(c) Connecting a high resistance in parallel with it
(d) Connecting a low resistance in parallel with it

(vii) The unit of magnetizing force is

(a) Ampere per meter (b) Weber per square meter
(c) Ampere-turn per meter (d) Weber per meter

(viii) The maximum efficiency of a transformer occurs when

(a) Copper loss is one-fourth of iron loss (b) Copper loss is half of iron loss
(c) Copper loss is equal to iron loss (d) Copper loss is double the iron loss

(ix) Which of the following machines are nonself-starting type?

(a) Three-phase induction motor and single-phase induction motor
(b) DC series motor and single-phase induction motor
(c) Synchronous motor and single-phase induction motor
(d) Three-phase induction motor and synchronous motor

(x) Which motor is used in electric traction system?

(a) DC series motor (b) DC shunt motor
(c) Synchronous motor (d) Single-phase induction motor

Section B

2. Attempt any THREE of the following: $(10 \times 3 = 30)$

(a) Determine the power loss in the 10 Ω resistor by using Thevenin's theorem, as shown in Figure 1.

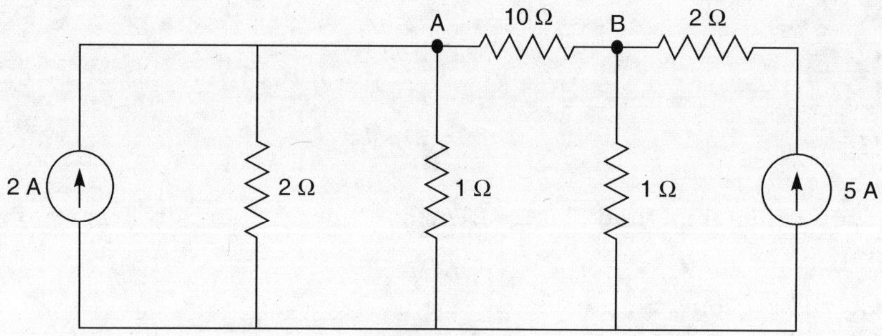

Figure 1

(b) A coil of power factor 0.9 is in series with a 120 μF capacitor. When connected to a 50 Hz supply, the potential difference across the coil is equal to the potential difference across the capacitor. Find the resistance and inductance of the coil.

(c) A balanced star-connected inductive load is connected to a 400 V, 50 Hz AC supply. Two wattmeters used to measure the supply power indicate 8000 W and 4000 W, respectively. Determine: (i) line current, (ii) impedance of each phase and (iii) resistance and inductance of each phase.

(d) A 120 kVA, 2000/200 V 1-phase transformer takes a current of 50 A and 2400 W at 100 V, when the low-voltage winding is short-circuited. Determine the load voltage when delivering full-load current at 0.8 pf lagging, the supply voltage being 2000 V.

(e) The armature of a six-pole DC shunt motor takes 300 A at the speed of 400 revolutions per minute. The flux per pole is 75 mWb. The number of armature turns is 500. The torque lost in windage, friction and iron losses can be assumed as 2.5%. Calculate

(i) Torque developed by the armature
(ii) Shaft torque
(iii) Shaft power in kW

Section C

Note: ALL questions are compulsory. All questions carry equal marks. (10 × 5 = 50)

3. Attempt either (a) or (b).

(a) For the circuit shown in Figure 2, determine the value of R_L to get the maximum power. Also, find the maximum power transfer to load.

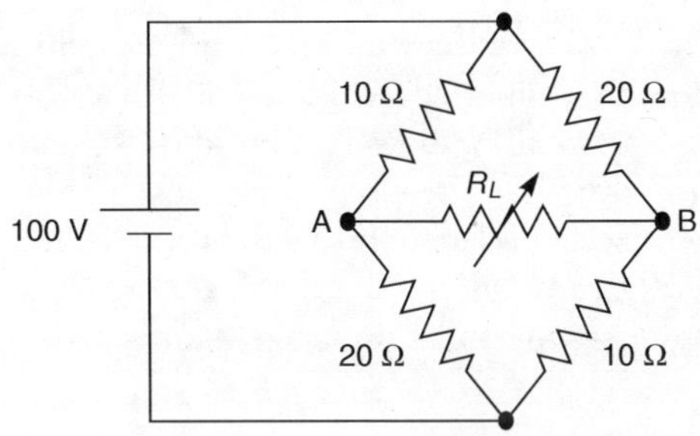

Figure 2

(b) Find the maximum power that can be transferred to the load resistance R_L in the circuit shown as in Figure 3.

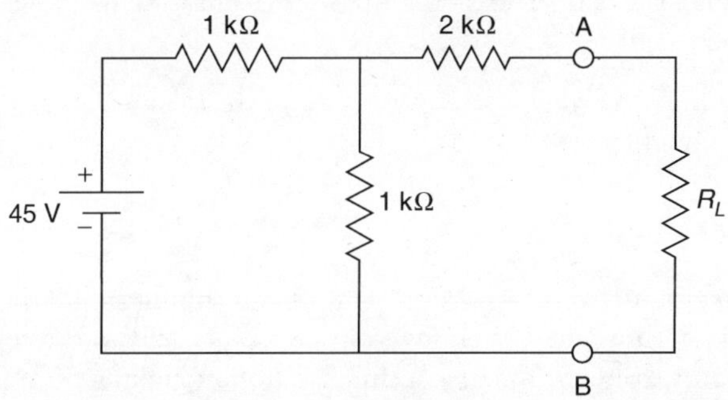

Figure 3

4. Attempt (a) and (b or c).

(a) Three impedances $Z_1 = 10 + j31.4\ \Omega$, $Z_2 = 6 + j8\ \Omega$, and $Z_3 = 3 - j4\ \Omega$ are connected in parallel across a 200 V AC supply. Calculate:

(i) Total admittance, conductance and susceptance of the whole circuit.
(ii) Total current, power factor and power

(b) A coil of power factor 0.9 is in series with a 120 μF capacitor. When connected to a 50 Hz supply, the potential difference across the coil is equal to the potential difference across the capacitor. Find the resistance and inductance of the coil.

(c) For an *R-L-C* series circuit of resistance 1000 ohms, inductance of 100 mH and capacitance of 10 μF, determine: (i) the resonant frequency, (ii) *Q* factor of the circuit at resonance and (iii) the half-power frequencies.

5. Attempt (a or b) and (c or d).

(a) A symmetrical star connected three phase load is made of three identical coils having an internal resistance of 10 ohm and inductance of 138 mH. The power taken by this load when connected to a balanced three phase three wire supply of 400 volt, 50 Hz is measured by two identical wattmeters. Determine the following.

(i) Expected readings of both the wattmeters
(ii) The total three-phase power.

(b) Each phase of delta-connected load has a resistance 25 Ω. and an inductance of 0.25 H and a capacitance of 120 μF. The load is connected across a 400 V, 50 Hz three-phase supply. Determine the line current, volt-amperes, active power and reactive volt-amperes.

(c) What do you understand by attraction-type and repulsion-type moving iron instruments? Explain.

(d) Two ammeters, one with a current scale of 10 A and resistance 0.01 Ω and the other with a current scale of 15 A and resistance 0.005 Ω, are connected in parallel. What can be the maximum current carried by this parallel combination so that no meter reading goes out of the scale?

6. Attempt either (a) or (b)

(a) An iron ring has a mean diameter of 25 cm and a cross-sectional area of 4 sq. cm. It is wound with a coil of 1200 turns. An air gap of 1.5 mm width is cut in the ring. Determine the current required in the coil to produce a flux of 0.48 mWb in the air gap if the relative permeability of the iron under these conditions is 800. Neglect magnetic leakage and fringing.

(b) The iron and full load copper loss in a 40 kVA 1-phase transformer are 450 W and 850 W respectively. Find

(i) Efficiency at full load when the power factor of the load is 0.8 lagging.
(ii) The maximum efficiency.
(iii) The load at which the maximum efficiency occurs.

7. Attempt (a or b) and (c)

(a) A DC shunt generator delivers 50 kW at 250 V when running at 500 rpm. The armature and field resistances are 0.05 ohm and 125 ohms, respectively. Calculate the speed of the same machine and developed torque when running as a shunt motor and taking 50 kW at 250 V. Allow 1 V per brush for contact drop.

(b) A four-pole, 50 Hz, three-phase induction motor has rotor resistance and reactance of 0.025 Ω and 0.02 Ω per phase, respectively. Determine

(i) The value of speed at which the maximum torque occurs.
(ii) The value of the external rotor resistance per phase to be inserted to obtain 60% of the maximum torque at the starting.

(c) What is the reason of nonself-starting feature of synchronous motor? How is the motor started?

Model Question Paper 4

Electrical Engineering

Time: 3 hours **Maximum marks:** 100

Note: All sections are compulsory.

Section A

1. ALL parts are compulsory. All questions carry equal marks. $(10 \times 2 = 20)$

(i) The form factor of the square wave will be

(a) 1.57 (b) 1.41
(c) 1.11 (d) 1.0

(ii) If P is the true power, Q is the reactive power and S is the apparent power, the power factor is represented by

(a) P/S (b) P/Q
(c) Q/S (d) $S/(PQ)$

(iii) The phase angle between V and I for parallel resonance will be

(a) 0° (b) 45°
(c) 90° (d) 180°

(iv) The superposition is based on the concept of

(a) Duality (b) Reciprocity
(c) Nonlinearity (d) Linearity

(v) Which of the following can give a four-wire supply system?

(a) Star (b) Delta
(c) Star and delta both (d) None of these

(vi) Dynamometer-type instruments are used to measure

(a) DC quantities (b) AC quantities
(c) DC and AC quantities (d) None of these

(vii) In moving iron instrument, the deflection is proportional to

(a) $d\theta/dL$ (b) $d^2\theta/dL^2$
(c) $dL/d\theta$ (d) $d^2L/d\theta^2$

(viii) An ideal transformer does not change

(a) Power (b) Voltage
(c) Current (d) All of these

(ix) In DC motor if the speed is high, then the back-emf will be

(a) Low (b) High
(c) May be low or high (d) None of these

(x) The synchronous speed is the speed of

(a) Stator (b) Rotor
(c) Magnetic flux (d) All of these

Section B

2. Attempt any THREE parts of the following: All questions carry equal marks (10 × 3 = 30)

(a) State and derive the Thevenin's theorem. Also compare it with Norton theorem. Determine the equivalent resistance appearing at the battery terminal of Figure 1.

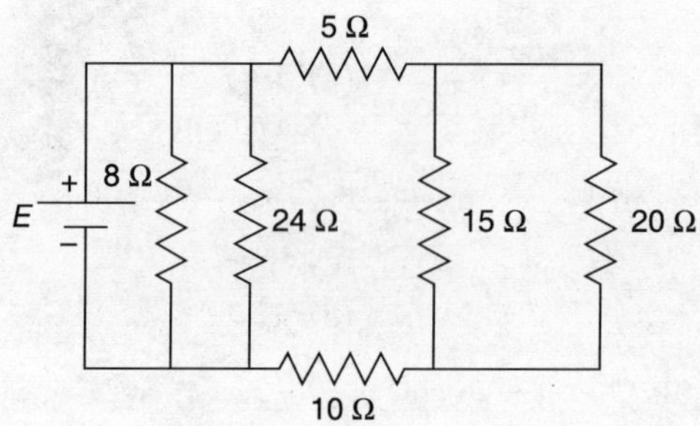

Figure 1

(b) Describe the terms: resonant frequency, voltage resonance and current resonance.
A capacitor is connected in parallel with a coil having L = 5.52 mH and R = 10 Ω to a 100 V, 50 Hz supply. Calculate the value of the capacitance for which the current taken from the supply is in the phase with voltage.

(c) Draw the equivalent circuit of the transformer. What are the various losses in a transformer? Where do they occur and how do they vary with the load?

(d) What are the factors that affect the speed of DC motor?
A 20 kW, 200 V shunt generator has an armature resistance of 0.05 Ω and a shunt field resistance of 200 Ω. Calculate the power developed in the armature when it delivers rated output.

Section C

Note: ALL questions are compulsory. All questions carry equal marks. (10 × 5 = 50)

3. Attempt any TWO parts of the following:

(a) Define the ideal voltage source and an ideal current source. Also, state the maximum power transfer theorem.

(b) For the circuit shown in Figure 2, find: (i) total impedance, (ii) total current and (iii) total power absorbed and power factor.

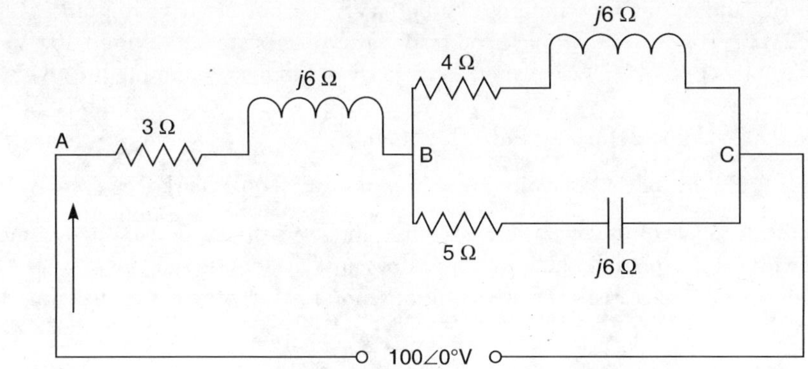

Figure 2

(c) In the circuit of Figure 3: (i) transform delta ABC to star and (ii) find the current I' in 20 Ω resistance by using the superposition theorem.

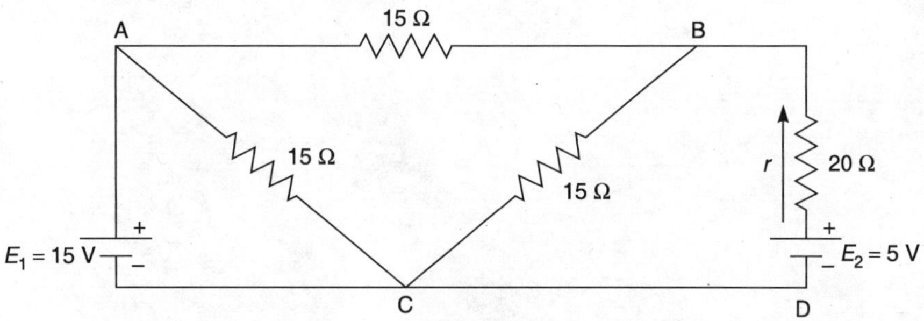

Figure 3

4. Attempt any TWO parts of the following. All questions carry equal marks.

(a) What is the meaning of a low power factor in the circuit? Also derive the expression and phasor diagram for *R-C* parallel network.

(b) Derive the expression for resonant frequency, bandwidth and quality factor for the *R-L-C* series circuit.

(c) State the reason why moving-iron instruments can be used for AC and DC measurements both.

5. Attempt any TWO parts of the following. All questions carry equal marks.

(a) Three identical impedances are connected to a 200 V, three phase supply and each takes 500 W and power factor 0.8 lagging. If these impedance are now connected in delta across the same supply and power is measured by two-wattmeter method, find the readings of the wattmeter.

(b) Differentiate between controlling torque and damping torque. Is it possible to convert an ammeter into a voltmeter and vice-versa? If yes, then how?

(c) Describe the construction and working of dynamometer-type wattmeter.

6. Attempt any TWO parts of the following. All questions carry equal marks.

   (a) Differentiate between feeder, distributor and service main in context of electrical supply system.
   (b) The primary and secondary windings of a 500 kVA transformer have resistances of 0.42 Ω and 0.0011 Ω, respectively. The primary and secondary voltages are 6600 V and 400 V, respectively, and the iron loss is 2.9 kW. Calculate the efficiency of full load, assuming the power factor of the load to be 0.8.
   (c) Describe the **B–H** curves for DC and AC excitation.

7. Attempt any TWO parts of the following. All questions carry equal marks.

   (a) Explain the principle of the operation and application of a three-phase induction motor.
   (b) Explain why a single-phase induction motor does not develop the starting torque.
   (c) Discuss the main characteristics of a synchronous motor. Also, write the applications of alternators.

Index